Principles and Techniques of Electron Microscopy

Biological Applications

Volume 3

Edited by

M. A. HAYAT

Associate Professor of Biology
Newark State College
Union, New Jersey

VNR

VAN NOSTRAND REINHOLD COMPANY

New York/Cincinnati/Toronto/London/Melbourne

Van Nostrand Reinhold Company Regional Offices:
New York Cincinnati Chicago Millbrae Dallas

Van Nostrand Reinhold Company International Offices:
London Toronto Melbourne

Library of Congress Catalog Card Number: 70-129544
ISBN: 0-442-25674-4

Manufactured in the United States of America

Published by Van Nostrand Reinhold Company
450 West 33rd Street, New York, N.Y. 10001

Published simultaneously in Canada by Van Nostrand Reinhold Ltd.

15 14 13 12 11 10 9 8 7 6 5 4 3 2 1

Library of Congress Cataloging in Publication Data

Hayat, M A, date-
Principles and techniques of electron microscopy.

Includes bibliographies.
1. Electron microscope—Collected works. I. Title.
QH212.E4H38 578'.4 70-129544
ISBN: 0-442-25674-4

It is a pleasure to dedicate this volume to:

Thomas F. Anderson, Gunther F. Bahr, Pierre Baudhuin,
V. E. Cosslett, Robert T. De Hoff, G. Dupouy, Christian de Duve,
Robert D. Heidenreich, August Hennig, J. Hillier,
Edward Kellenberger, A. K. Kleinschmidt, Franz Perrier,
E. Ruska, F. Lenz, and Friedrich Thon.

Preface

This volume presents six techniques and approaches currently being employed for the study of the structure, composition, size, and location of cellular components. Discussions of relatively new techniques in terms of their application (e.g., dark-field and in-focus phase contrast) are also included. These techniques with interference contrast and electron manipulation of elastic and inelastic scattering components should prove useful in elucidating the structure at the molecular and atomic levels.

The theory, principle, and application of each technique are presented in depth. The various procedures are self-explanatory and sufficiently detailed so that they can be carried out to completion without outside help. Areas of disagreement and potential research problems have been pointed out. It is within the scope of this volume to provide the reader with operating instructions for an electron microscope including possible problems encountered and their remedies.

In preparing this volume, we enlisted the help of eleven scientists from several countries, each of whom is an authority in his field. As a result of this approach, an authoritative description of the most modern and reliable techniques has been compiled in this volume.

This is the third in a series of five volumes. The first volume presents a detailed discussion on fixation, embedding, sectioning, staining and support films. The second volume contains a thorough explanation of negative staining, freeze-drying and freeze-substitution, freeze-etching, shadow casting and replication, high-resolution shadowing, and autoradiography.

I owe special thanks to Drs. Gunther F. Bahr, Arthur L. Cohen, Albert V. Crewe, and Harry Johnson for their help in the preparation of this volume. With pleasure I acknowledge the cooperation and friendship shown by George Narita and Alberta Gordon of Van Nostrand Reinhold Company. For the patience and accuracy with which the manuscript was typed, I am grateful to Miss Patricia Guempel.

M. A. Hayat

Contents

2 ELECTRON MICROSCOPY OF SELECTIVELY STAINED MOLECULES

T. Koller, M. Beer, M. Müller, and K. Mühlethaler

3 HIGH RESOLUTION DARK-FIELD ELECTRON MICROSCOPY

Jacques Dubochet

5 ELECTRON MICROSCOPIC EVALUATION OF SUBCELLULAR FRACTIONS OBTAINED BY ULTRACENTRIFUGATION

Russell L. Deter

6 STEREOLOGICAL TECHNIQUES FOR ELECTRON MICROSCOPIC MORPHOMETRY

Ewald R. Weibel with the collaboration of Robert P. Bolender

Contents of

1 The Electron* Microscope

Saul Wischnitzer

Electron Microscope Laboratory, Department of Biology
Yeshiva University, New York City

*This chapter contains in part material originally published in the author's book, "*Introduction to Electron Microscopy*, Pergamon Press, Elmsford, N.Y. 1970, and is reprinted with the publisher's permission.

INTRODUCTION

In 1878 Ernst Abbe, in opposition to many contempory microscope designers and optical manufacturers, expressed the view that optical technology had reached a fundamental limit and that no further improvement in resolution was possible. He declared that light was a wave motion whose half wavelength marked an impenetrable limit of resolution for any optical magnifier at about the quarter-micron level. Abbe realized that to investigate the molecular realm a new form of short-wavelength radiation was needed, but he noted that new instruments "which in the future will perhaps aid our senses in the investigation of the ultimate elements of the universe more powerfully than present-day microscopes, will have little more in common with them than the name." Time has proven Abbe's resolution limit postulate correct, but his prediction regarding the nature of an ultramicroscope has in large measure turned out to be inaccurate.

The development of the electron microscope was delayed by lack of insight into the physical nature of the radiation used to form the image, as well as by technological considerations. The possibility of developing an electron microscope came about indirectly as a result of an attempt to explain the action of the cathode ray tube in terms of geometric optics. Once the basic idea of magnifying an electron optical image was established, electron microscopes were developed in the decade between 1931 and 1941 in several countries (viz., Germany, Holland, U.S., and Japan) where a high level of technology existed (Fig. 1.1). By 1947 the electron microscope had reached the stage where it could be said that it really resembled the original concept that had eluded Abbe's vision, namely a microscope that behaved like an optical microscope, but which employed radiation of a much shorter wavelength.

LIMIT OF RESOLUTION

The ideal lens system is one that provides an exact image of an object. However, this goal is unattainable because of a phenomenon called *diffraction,* which is a consequence of the interaction of wavelike illumination beams and objects. Diffraction is the bending or spreading of light into the region behind an obstruction that the waves have passed.

Diffraction is very significant so far as the formation of images by lenses is concerned. Thus, if a strong beam of light illuminates a pinhole

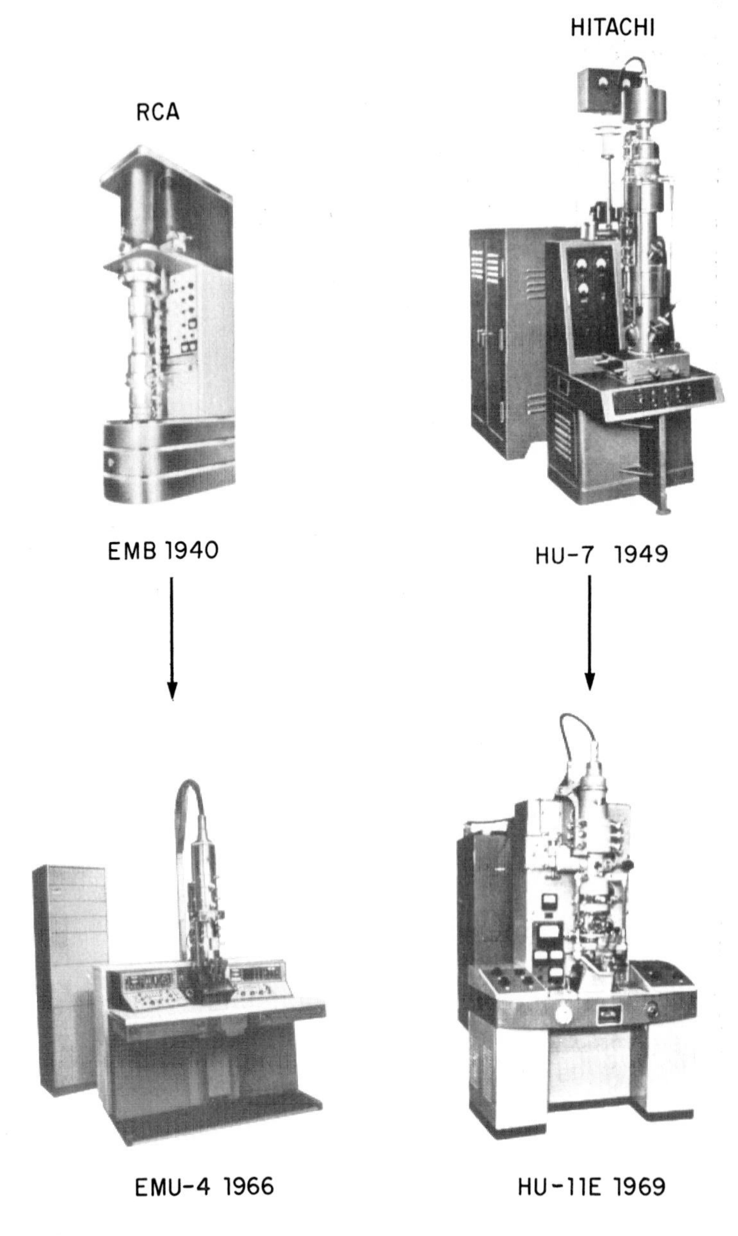

Fig. 1.1 Progress in the development of the electron microscope.

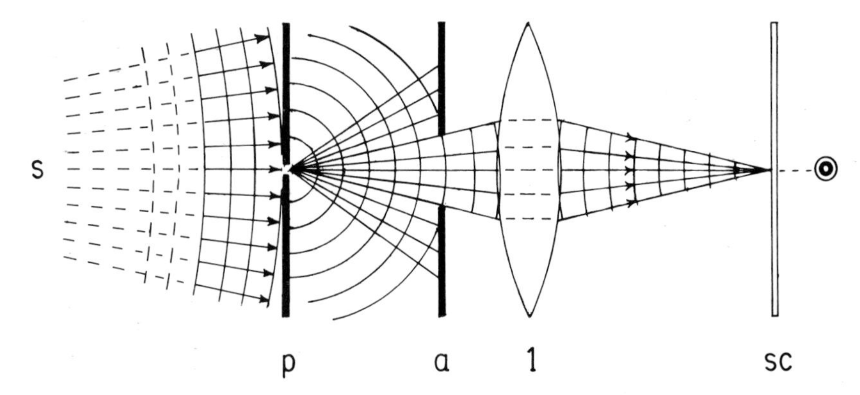

Fig. 1.2 Diffraction of light waves. S: light from distant source; l: biconcave lens; p: pinhole aperture; SC: viewing screen; a: circular limiting aperture.

in a screen the image obtained is a bright central disk surrounded by diffused rings of light (Fig. 1.2). This type of image is known as an *Airy disk.*

The study of the diffraction of light was put on a quantitative basis by the work of Sir George Airy, Lord Rayleigh, and Ernst Abbe during the nineteenth century. The latter mathematically defined the magnitude of the diffraction effect as

$$d = \frac{0.612\ \lambda}{n \sin \alpha}$$

where

d = radius of the first dark ring of an Airy disk measured at the minimum
λ = wavelength of image-forming radiation
n = index of refraction of the medium between the point source and the lens, relative to free space
α = half the angle of the cone of light from the specimen plane accepted by the front lens of the objective

The term d provides a numerical value for the limit of resolution of any diffraction-free optical system. Thus, considering the case of two point sources located a short distance apart in the object plane, the object points are resolvable if the disks they produce do not overlap to any appreciable extent and if the points are separated by a distance greater than d (Fig. 1.3).

From Abbe's equation it is apparent that to attain the maximum resolution with a light microscope, as indicated by a minimum value for

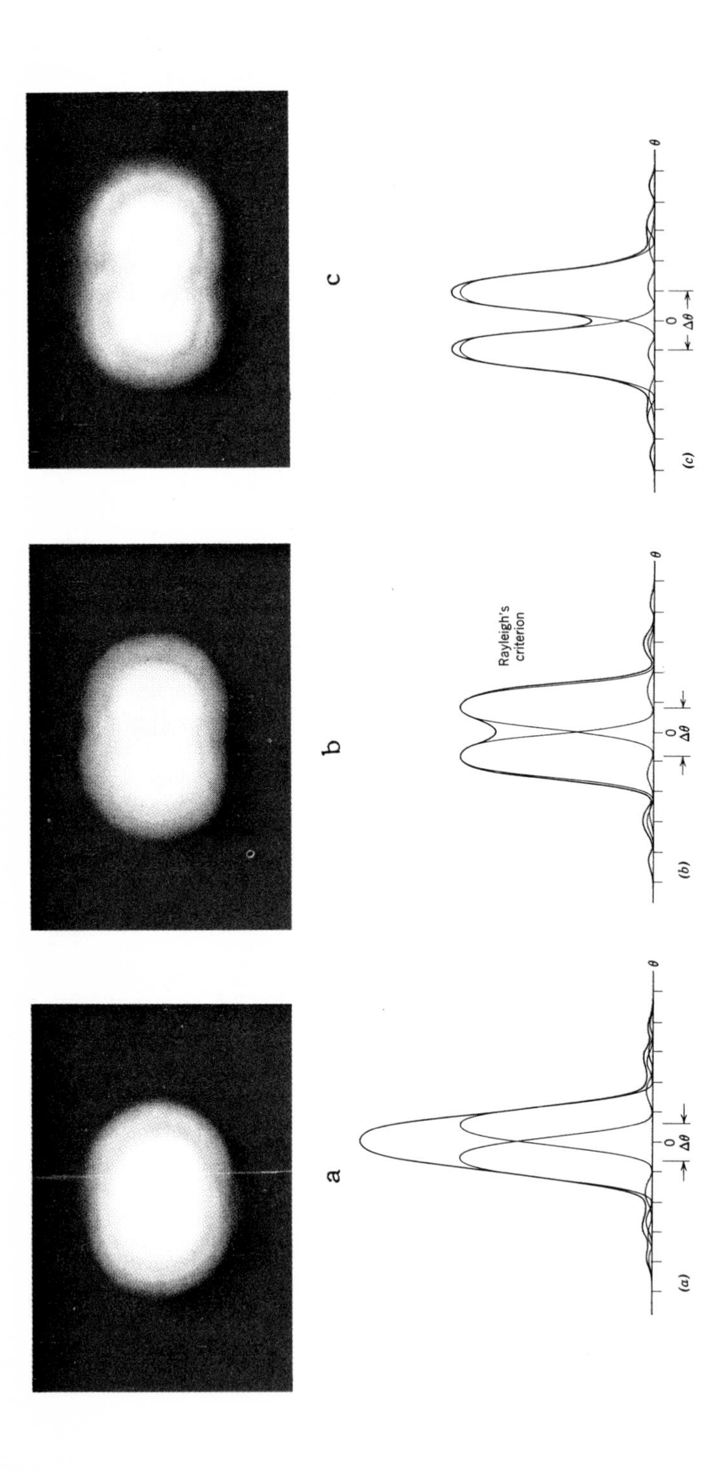

Fig. 1.3 (top) The images of two distant point objects formed by a converging lens as they appear in the focal plane of the lens. a: The angular separation is so small that the images are not resolved. b: The objects are farther apart and the images are just resolvable. c: The objects are still farther apart and the images are well resolved.

Fig. 1.3 (bottom) The images of two distant point objects are formed by a converging lens whose diameter (=10 cm) is 200,000 times the effective wavelength (= 5000 A). Sketches of the images as they appear in the focal plane of the lens are shown with the corresponding intensity plots below them. (*a*) The angular separation of the objects (see vertical ticks) is so small that the images are not resolved. (*b*) The objects are farther apart and the images meet Rayleigh's criterion for resolution. (*c*) The objects are still farther apart and the images are well resolved.

d, the optical condition should be such that d has a minimal value, while n and sin α should be maximal. Since the value of n and α cannot be significantly increased beyond about 1.5 and near 90° respectively, the only factor that can be readily altered to attain a decrease in d and thereby improve resolving power is the value of λ. Electromagnetic radiation of shorter wavelength than ultraviolet light cannot be focused. However, a beam of electrons can be focused, has a characteristic wavelength capable of significantly improved resolution, and interacts 10^6 times more strongly with matter than does X radiation. Thus, Abbe's equation can be rewritten, after substituting for λ with the de Broglie equation (see below), as

$$d = \frac{(0.61)(12.3)}{n \sin \alpha \sqrt{V}}$$

Since electron microscope aperture angles are always very small, sin $\alpha \simeq \alpha$, and since both the object and the image are usually in field-free space in an electron microscope, the refractive index is $n = 1$. Then, the above equation can be written as

$$d \approx \frac{7.5}{\alpha \sqrt{V}} \text{ Å}$$

Thus, in the electron microscope the level of resolution is basically determined by the accelerating potential and angular aperture of the objective lens.

Substituting specific realistic values in the above equation ($\simeq = 10^{-2}$ rad, $V = 10^5$ V), we get

$$d \approx 2.4 \text{ Å}$$

NATURE OF ELECTRON BEAMS

The fact that moving electrons might be used as a kind of "illumination" was implicit in the theory advanced by de Broglie who stated that moving particles have wavelike properties. By combining some of the principles of classical physics with the quantum theory, he concluded that a wavelength can be assigned to moving particles that can be calculated from the equation

$$\lambda = \frac{h}{mv}$$

where

λ = wavelength of the particle
h = Planck's constant
m = mass of the particle
v = velocity of the particle

which, when the known values are substituted, becomes

$$\lambda = \frac{12.3}{\sqrt{V}} \text{ Å}$$

This formula tells us that the wavelength of a beam of electrons is dependent upon the potential, V, through which it has been accelerated. Thus, for example, if the latter were 60,000 V (60 kV), the wavelength of the electron beam would be about 0.05 Å, which is 1/100,000 as long as the wavelength of ordinary green light.

In 1927, Thomson and Reid in England, and Davison and Germer in the United States, experimentally measured the wavelength of an electron beam and found it to be in agreement with that calculated from de Broglie's formula. Thus it became apparent that high-velocity electrons, as a result of their small wavelength, might provide a suitable form of radiation to fulfill the requirements for high-resolution microscopy (Fig. 1.4).

COMPONENTS OF THE ELECTRON MICROSCOPE

Both the light and electron microscopes are analogous so far as the arrangement and function of their components are concerned. Thus both

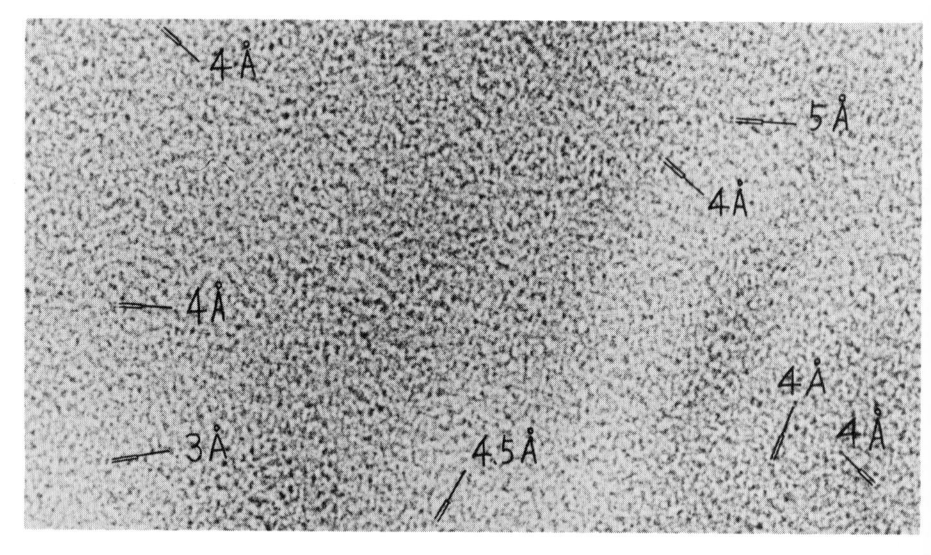

Fig. 1.4 High resolution. Resolution in the 3 to 5 Å range using a K_2PtCl_4 crystal specimen examined with a Hitachi HU-11A microscope. (*Courtesy of Perkin-Elmer Corp., Norwalk, Conn.*)

microscopes when used for photographic purposes can be conveniently divided into three systems, each of which contains the same basic components.

Illuminating System. This system serves to produce the required radiation and to direct it onto the specimen.

(a) Source: Emits radiation that is used to form an image.
(b) Condenser lens: Regulates the convergence (and thus the intensity of the illuminating beam on the specimen).

(Interposed between the illuminating and imaging systems is the specimen stage.)

Imaging System. This system comprises the lenses that together produce the final magnified image of the specimen.

(c) Objective lens: Focuses the beam that has passed through the specimen so as to form a magnified intermediate image.
(d) Projection lens (or ocular): Magnifies a portion of the intermediate image to form the final image.

Image Recording System. This system converts the radiation into a permanent image that can be directly viewed.

(e) Photographic emulsions: Usually carry out the recording of the image.

THE ELECTRON MICROSCOPE (FIG. 1.5)

Illumination System

This system contains two units: the electron gun that is the source of the electrons, and the condenser that regulates the intensity of the beam and directs it onto the specimen.

Electron Gun

The parts of the electron gun have the properties of and act like an electrostatic lens. The gun consists of three components: the filament, shield, and anode.

FILAMENT

The prerequisites for the electron source are that it be small, symmetrical, and intense. This is usually met by the use of a V-shaped piece

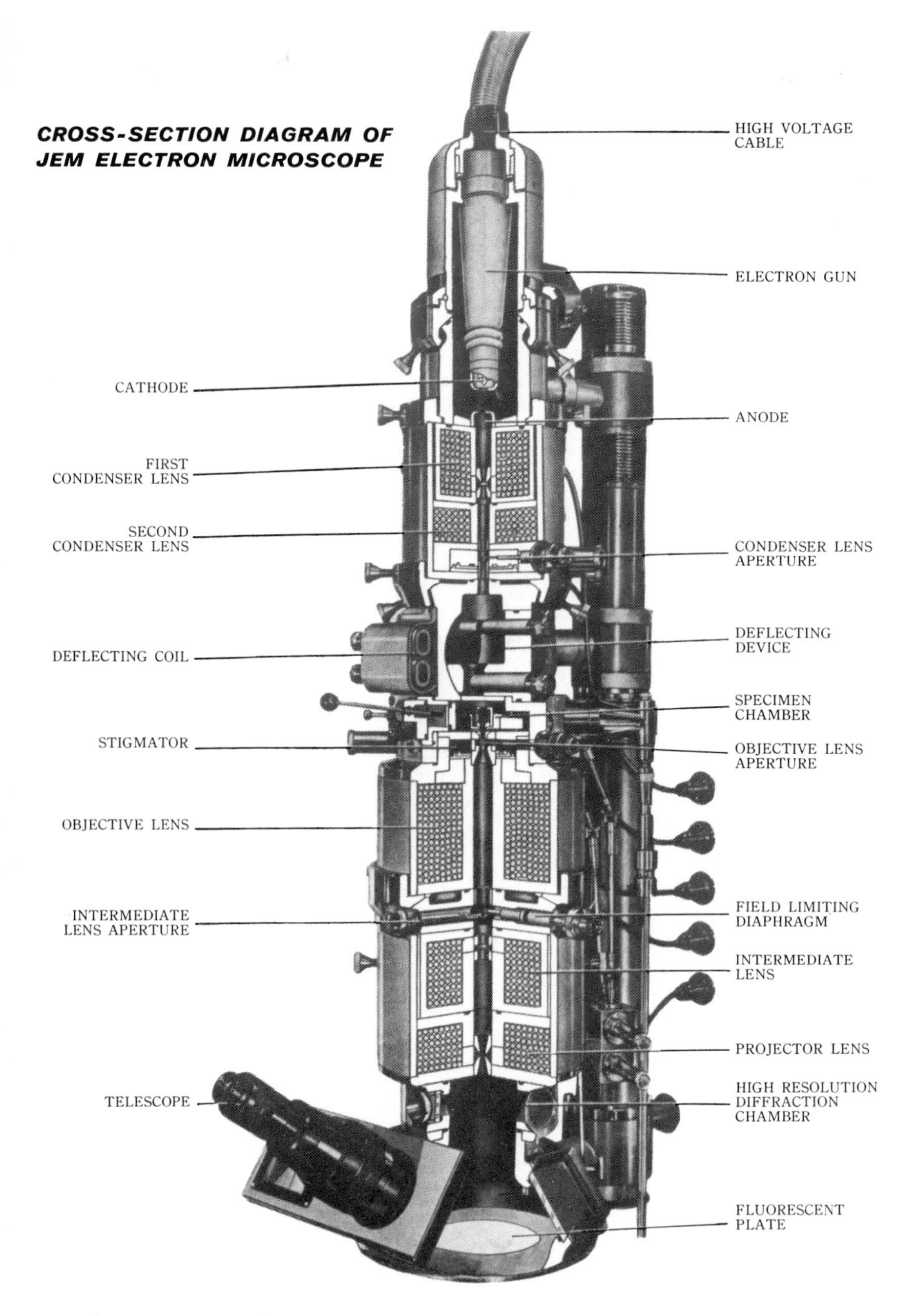

Fig. 1.5 Schematic cross section of the column of a (JEM) electron microscope.

of pure tungsten wire, about 0.1 mm in diameter, which can be electrically heated to incandescence and thus serve as a thermionic cathode. The filament, which is also known as the cathode or emitter, is attached to the negative terminal of the high-voltage supply (usually through a resistance network).

SHIELD

The shield, the equivalent of the grid of a triode, is an apertured cylinder. It lies immediately in front of the filament and its aperture, which is usually 1–3 mm in diameter, is centered over the filament tip. A relatively small potential difference is applied between the filament and the shield. The shield is also known as the focusing electrode or grid cap.

ANODE

The anode or positive electrode is an apertured disc, which like the remainder of the microscope is grounded and is attached to the positive terminal of the high-voltage supply. The cathode is, therefore, held at a large negative potential with respect to the anode. The anode is coaxially aligned with the filament. The emitted electrons are accelerated in the space between the filament and anode by the potential difference across this gap. After leaving the electron gun the electrons pass down the field-free space in the column at a constant velocity.

SELF-BIASED GUN

In practice all microscopes currently manufactured have incorporated the biased system in their electron source and the type usually used is called the self-biased gun. In such an arrangement, the shield aperture acts as a strongly converging lens.

OPERATION OF THE SELF-BIASED GUN

When a microscope is operated with a self-biased gun at constant accelerating potential and fixed condenser current (with the lenses set to focus the image of the filament on the viewing screen), and the filament current is gradually increased to saturation, the following changes in the image of the source can be observed at low magnifications on the final screen (Fig. 1.6).

With a rise in filament current (i.e., filament temperature), a small

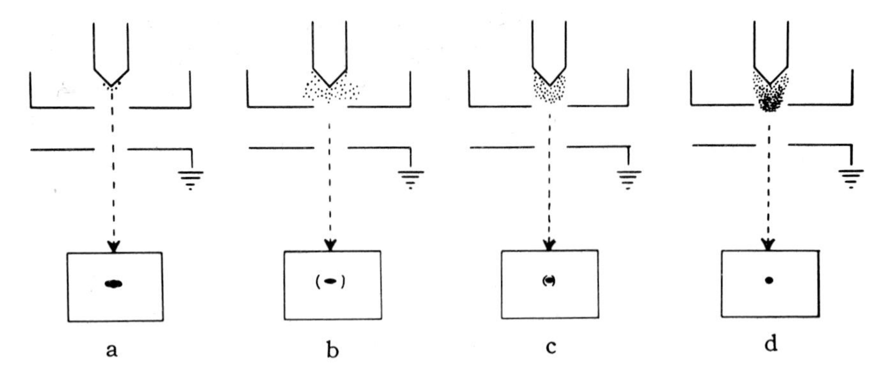

Fig. 1.6 Emission patterns with increasing filament current. a: The initial spot represents emission directly from the filament tip. b: In the second stage a ring of satellites is seen, produced by electron emission from parts of the filament away from the tip. c: A further increase in electron emission results in an accumulation of electrons (i.e., formation of a space charge from which electrons are extracted). d: At saturation the space charge has become intense and only electrons from its axial portion are extracted to produce the intense spot.

faint spot becomes visible. A slight increase in filament current will intensify and expand the spot, and gradually also produce around it a bright ring broken at two opposite points. With a small additional increase in filament current, the central spot emerges with the other components, the breaks close, and the ring contracts toward the center. Finally, at saturation the image is a single bright spot that is larger, brighter, and more uniform than the initial central one. Any further increase in the filament heater current produces no appreciable change in the intensity or appearance of this spot.

These changes in the emission pattern are interpreted as being due to the rapid build-up of space-charge-limited conditions as the emission is increased. Thus the initial spot seen (Fig. 1.6a) is due to emission from the tip of the filament and represents an image of the filament itself, whose surface structure is sometimes visible. The incomplete ring seen in the second stage (Fig. 1.6b) is produced by the electron emission from parts of the filament away from the tip (which are observed out-of-focus and astigmatically distorted). Gradually during this stage the space charge is formed around the axis. At this time, the anode field is still powerful enough to extract electrons directly from the filament through a "canal" in the negative field formed by the electron cloud. With further slight increase in electron emission, the accumulation of electrons in the region and on the shield is sufficient to cancel the force of the extracting anode field, and thus further removal of electrons directly from the filament itself is prevented (Fig. 1.6c).

Under these conditions, electrons appear to be removed from the space charge rather than from the filament. When "saturation" is reached, the space charge becomes very intense and the negative shield field, acting with the positive anode field, becomes sufficient, so that the only electrons that are extracted are those from the axial portion of the cloud. These produce the final intense spot seen (Fig. 1.6d). With any further increase in the filament current, there is no additional increase in electron extraction, i.e., the beam current remains constant.

From what has just been said, it follows that when an electron microscope with a self-biased gun is put into operation, the filament current should be increased up to the point where the beam-current meter reading levels off. Since this condition is reached quite rapidly, little further control of illumination through manipulation of the filament temperature can be attained and in most microscopes, therefore, the main function of the condenser is to provide such a control.

A self-biased gun needs to have only positioning controls for adjustment in a plane perpendicular to the axis of the instrument and/or a tilting arrangement to achieve centration.

BRIGHTNESS OF ELECTRON SOURCE

An electron microscope can use only a very small beam angle due to the spherical and chromatic aberrations of the objective lens; therefore, only equally small illuminating beam angles are useful. As a result very high beam brightness is required.

Brightness* may be improved by the use of *pointed filaments.* Pointed filaments (Fig. 1.7) are short pieces of tungsten wire etched to a fine

*"Brightness," as usually used here, refers to source or source image brightness. The latter corresponds to the image of the source on the specimen at cross-over.

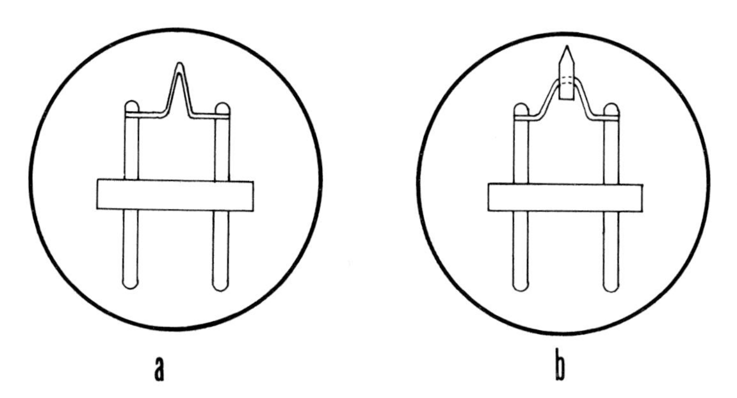

Fig. 1.7 Filaments for the electron gun. a: Standard. b: Pointed.

point (about 0.1μ in diameter) or a single tungsten crystal attached to a supporting, electrically heated tungsten filament. The tip is away from the asymmetrical fields that surround the supporting piece and, as a result, a very small symmetrical source is obtained. More importantly, the accelerating potential gradients are very high near the tip, resulting in field emission (electrons "tunnel" through the potential barrier of the work function) that gives higher brightness even at low beam current. Very sharp points at high voltage gradients are required to achieve useful currents with cold pointed cathodes. With a filament-heated point, thermally assisted field emissions (thermal-field emission) permit useful brightness gains with easily fabricated comparatively "dull" points. Pointed filaments are not yet in wide use.

Condenser Lens

The basic function of the condenser lens is to focus the electron beam emerging from the gun onto the specimen to permit optimal illuminating conditions for visualizing and recording the image. This depends upon the condenser's focal length (i.e., the distance from the center of the lens or lens aperture to the point of convergence of the rays at cross-over, for a source at infinity).

The condenser lens of the electron microscope is a relatively weak lens having a focal length of the order of a few centimeters which can be adjusted over a considerable range of values. As is the case in vertical movement of the substage condenser (with fixed focal length) of the light microscope, changing the focal length of the electron condenser lens shifts the position of the image of the electron source along the optical axis of the instrument. Consequently, two interrelated properties are simultaneously controlled:

(a) Aperture angle (convergence) of the illuminating pencils* accepted by the objective
(b) Intensity of the illumination at the specimen level

CONDENSER LENS OPERATION

Until a few years ago, it was common practice for the condenser setting for viewing and focusing to be set differently from that for photography. However, it was found that by changing the "in-focus" position of the condenser just prior to photography a number of possible errors

*A pencil of rays is made up of the cone of rays that participates in the formation of the image of a single point of the specimen.

could be introduced. These include specimen drift and changes in focus due to changes in the weak lens effects of charged apertures. To avoid these errors the same condenser setting is now usually maintained throughout the visual and photographic procedures for high resolution. The ideal operating conditions are obtained by

1. Providing the condenser with a physical aperture of such a dimension that the illuminating aperture angle at cross-over is equal to the optimal objective aperture
2. Adjusting the bias potential of the gun or the axial position of the shield to produce an illumination intensity that is just adequate for visualization and photography at a given magnification

For electron microscopes where the flexibility for making the above adjustments is not present, an attempt is made to approximate the ideal conditions.

Since the focal length of the condenser can be relatively long, the magnetic field for this lens can be relatively weak. For this reason, some condenser lenses are not equipped with pole pieces and usually they have a smaller coil than the other lenses. The condenser is located between the illuminating source and the specimen so that it forms an image of the source on the specimen. Provision is usually made to allow a physical aperture to be inserted into this lens. The condenser is also usually provided with means for centering in a plane perpendicular to the axis of the instrument.

DOUBLE CONDENSER

The shorter the focal length of the condenser, the smaller is the illuminated area at cross-over for a given electron gun. However, as pointed out previously, the focal points in short focal length lenses usually are located between the pole pieces of a lens. Since both a short focal length condenser and an objective have focal planes located within their pole pieces, it is necessary to add a second long focal length lens to the condenser to permit the imaging of such a small cross-over on the specimen plane within the objective. Such a two-lens condenser is known as a double condenser.

Through the use of a double condenser, controlled small-area illumination can be obtained. Each of the lenses can be individually excited. The first lens is strong and produces a (variable) demagnification of the image of the source (as does the ordinary condenser). The second condenser lens projects this reduced image onto the specimen. The advantage of the double condenser is that it provides a means of reducing the beam diameter at the plane of the specimen. However, such a system does not itself increase the "brightness" of the beam since this limit is set by the aperture angle of the condenser, which cannot usefully exceed that of the objective. The double condenser eliminates unnecessary irradiation of specimen areas outside of the field of view and, therefore, also cuts down background scattering from such areas.

Imaging System

The imaging system consists of an objective lens and one or more projector lenses.

Objective Lens

The objective lens forms the initial enlarged image of the illuminated portion of the specimen in a plane that is suitable for further enlargement by the projector lens. The objective magnetic lens is the most critical component of the microscope.

OBJECTIVE LENS OPERATION

In the electron microscope, focusing is accomplished by varying the focal length of the objective lens. As already discussed, this usually involves changing the current that passes through the coil. Since the adjustments in focal length may vary over a relatively wide range, while at the same time extreme precision in focusing is also required, several controls of progressively increasing sensitivity are usually provided.

As the components of the electron microscope that already have been discussed, so too must the objective lens be aligned or centered. Adjustments to move the lens (or its pole pieces) in a plane perpendicular to the optical axis are provided for in most instruments.

The scattered electrons that strike the lens elements may result in the formation of deposits on their surfaces. These deposits tend to accumulate an electrostatic charge during operation, which can modify the lens field significantly, producing astigmatism, distortion, and/or image shifts. To minimize this problem, lenses must be periodically cleaned quite meticulously.

Projector Lens

The projector lens, as its name implies, serves to project the final magnified image on the screen or photographic emulsion. This lens has a construction similar to the objective lens, except that it has no provision to accommodate the specimen holder.

The properties of the projector lens contribute to the final formation of the electron image in the following important respects.

a. a broad range of focal lengths
b. the great depth of focus of the resulting high magnification system
c. a difference in the manifestation and a less serious effect of its aberrations on resolution

The projector lens of a two-stage unit (i.e., an imaging system made up of one objective and one projector lens) is usually provided with a focal length with a greater range than the objective. The fact that the pole pieces used with this lens can be made symmetrical, since the specimen does not have to be accommodated, permits use of a shorter focal length. The full axial length of the magnetic field, rather than only its upper portion, takes part in the imaging process, as is the case for the objective lens.

Image Recording System

Fluorescent Observation Screen

In the electron microscope, the instantaneous and continuous transformation of an electron image into a visible one is accomplished by means of a "fluorescent"* material that has the property, when bombarded by radiation of one wavelength, of emitting light of a longer wavelength. For electron microscopy, sulfides containing a trace of some metal are used; when irradiated with electrons they act by emitting light in the visible range. A yellow mixture of zinc and cadmium sulfides (plus a trace of an activator like silver or copper) has been found to be one of the most suitable materials for use in electron microscopy. This mixture, in the form of finely divided particles, is coated on a plate with the aid of a dilute binder such as gum arabic or collodion. The visible image can be viewed directly or observed through a low-power optical magnifier.

The screen provides the opportunity for:

*Fluorescence and phosphorescence are distinguished by the time between the excitation (collision with an electron) and the emission of light. The so-called fluorescence of the screen material continues for periods longer than 10^{-8} sec after excitation (usually for periods up to 10^{-1} sec) and, therefore, is more properly phosphorescence than fluorescence. Zinc and cadmium sulfide are typical electron phosphors.

a. Orientation and location of the desired field of view
b. Study of the general characteristics of the image of the specimen
c. Accurate focusing of the lenses
d. Alignment of the entire instrument

The resolution obtainable in the image seen on the screen depends primarily on the grain size of the fluorescent material and secondly on the spreading of the beam as it penetrates into it. It has been determined that a coating of 50-100μm in thickness, which is desirable for work in the 50 $\simeq$ 100kV range, can, with sufficient image contrast, provide a resolution of 35–50μm. Since this is better than the limit of the resolving power of the unaided eye, a 3×–10× viewing telescope or magnifier can be profitably used. In any case, the contrast of the image observable on the screen is limited and thus the finer details of the image may not be visible. The photographic method of image translation overcomes this latter limitation.

Photographic Recording

A photographic recording material consists of a plastic or glass base coated with an emulsion that is made up of a layer of gelatin in which is embedded an electron-sensitive silver halide. The electron beam acts by the liberation of free silver halide grains and produces, after conventional development, a photographic negative of the final electron image. A print of the image on the film is known as an *electron micrograph* and serves as a permanent record. The fine-grained negative contains a more detailed and higher contrast image than that produced on the fluorescent screen. By photographically enlarging such a negative, a micrograph can be obtained that makes the detailed information of the final image visible to the eye. Since such a record is superior, both qualitatively and quantitatively, to the image visible on the viewing screen, photographic image translation has become the routine means used in the critical study of material that is being investigated with the electron microscope.

Operational Requirements

In order that one may be able to obtain electron micrographs of high resolution, a number of requirements have to be fulfilled. These include alignment of the lenses, detection and elimination of lens asymmetry (i.e., minimizing residual astigmatism), and elimination or control of potentially disturbing conditions, such as specimen, mechancal, and thermal draft, stray magnetic fields, instrument vibrations, etc.

Alignment

Alignment of the components of the column is necessary to minimize the effects of circuit instability that especially degrade off-axis resolution (through field aberrations), as an increasing function of distance from the optic axis. As already discussed, the extent of image magnification and rotation is dependent upon the accelerating voltage and lens current. Thus slight electrical disturbances will induce image movement, the magnitude of which will increase directly with the distance from the optical axis (Fig. 1.8). This movement is relatively insignificant near an optical axis but can be the major source of loss of resolution at large distances from the axis.

The usual method of alignment is based on voltage centration. This involves determining the point in a focused image that remains stationary

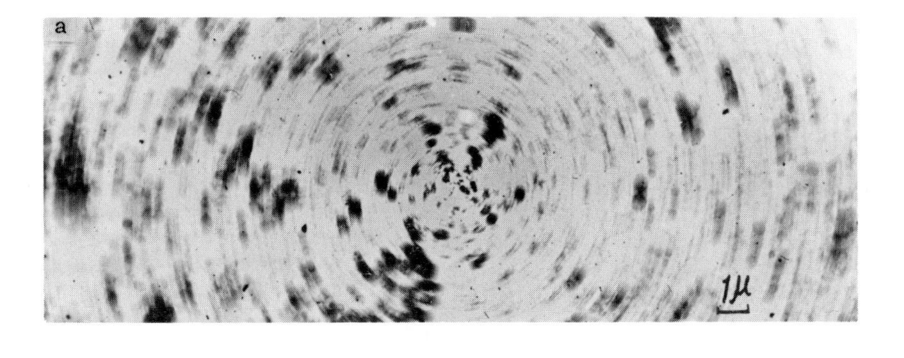

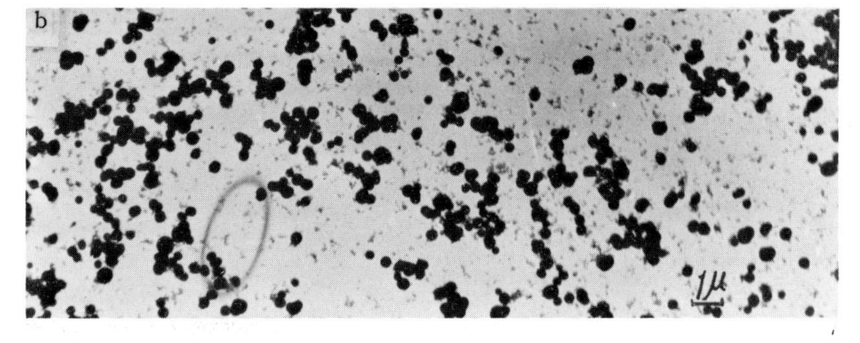

Fig. 1.8 Effect of voltage fluctuation on off-axis resolution. a: An image in the presence of large, artificially superimposed fluctuations of 5 percent of the acceleration voltage. Note how rapidly resolution is degraded as one proceeds peripherally from the voltage center. Near the optic axis chromatic field aberration is smaller than axial chromatic aberration, whereas off-axis it is many times larger. b: A similar field in the absence of such fluctuations.

(namely, around which rotation occurs) when the accelerating potential is varied with fixed lens currents, i.e., the point at which the image is least sensitive to field aberrations as a result of variation in the high tension. This voltage center defines the position of one of the optical axes of the magnetic lens system. The magnetic center, which defines the other optical axis, is determined by locating the stationary point in a focused image when the lens current is varied, the accelerating potential being fixed. The two axes are not necessarily coincident.

Voltage centration is the usual method of choice because the accelerating potential is both less well regulated and more susceptible to erratic fluctuation and, therefore, the image movements due to voltage fluctuations will be minimal when voltage centration is employed. The microscopist is most fortunate when the mechanical axis of the magnetic lenses of his microscope closely coincides with both optical axes. Under such circumstances, the rotational effects of fluctuations in both voltage and current on resolution are minimized. In practice, however, the optical axes usually do not coincide with the mechanical axis. The voltage and magnetic centers represent the points of the image in the field of view for which the vector sums of all displacements resulting from fluctuations of the accelerating potential lens current are zero. That the two centers do not coincide is due to the fact that machine tolerances to attain preset alignment are not easily achieved and/or inhomogeneities in the iron or stray magnetic fields induce curvature of the optical axes. Thus, commercial electron microscopes come equipped with controls that make alignment possible.

Early electron microscopes permitted wide translation of the components. Unfortunately, the extent of initial misalignment could be considerable and correction could be a very tedious process. Current microscopes are machined so that only that amount of translation of the components can be made as to compensate for the maximum divergence between the mechanical and voltage centers that may result during manufacture. Total "loss" of the beam at low magnifications due to misalignment of objective, intermediate, or projector lenses is, therefore, now much less likely.

The following seven alignment conditions must be met (usually in the order given) for an electron microscope having a single condenser lens.*

*For most electron microscopes the double condenser system, where present, is made as an integral unit. The manufacturing process is usually such that internal alignment tolerances can be met and thus no special additional alignment controls are needed. Thus alignment is straightforward and parallels that for a single condenser. This is true when the first condenser lens is regarded as a preset adjustment of spot size, and that variations of lens current are confined to the second condenser during the alignment procedure.

a. The beam emerging from the electron gun must be centered relative to the condenser lens aperture.
b. The gun and condenser as a unit must be centered relative to the objective axis and specimen.
c. The tilt of the illumination system must be adjusted to be made parallel with the objective lens axis.
d. The projector lens must be aligned relative to the objective lens axis.
e. The intermediate lens must be aligned relative to the objective lens axis.
f. The objective lens must be adjusted relative to the center of the screen (to provide voltage alignment). If substantial movement of the objective lens is required for this step, steps (a) to (e) may have to be repeated, then followed by a very slight readjustment of the objective lens alignment.
g. The objective aperture must be aligned relative to the objective lens axis.

ELECTRON GUN-CONDENSER ALIGNMENT

The electron source must be positioned on the condenser lens axis; otherwise uniform illumination will be obtained only at one condenser setting (i.e., when the beam is a narrow pencil). The effect of a misaligned electron gun is illustrated in Fig. 1.9. It can be seen that with an increase in condenser lens strength, the illumination, as indicated by its region of maximal intensity (cross-over), will move laterally. Moreover, as the condenser current is varied over a wide range on either side of cross-over, the illumination sweeps across and off the field rather than spreading uniformly about the center. Alignment is attained by manipulating screws of the electron gun and condenser that translate these in planes at right angles to the optic axis until symmetrical expansion of the illumination spot on either side of cross-over is attained.

Gun adjustment will normally need only infrequent checking, since it will remain fairly steady during the life of a filament (with the possible exception of its last hour). When filament replacement takes place, it must be adjusted by mechanical centration of the tip of the filament (e.g., using a center jig), and by setting correct height of the filament tip relative to the grid cap (e.g., using a depth guage), since this adjustment fixes the magnitude of the beam current and brightness at saturation in fixed-bias guns. When these adjustments are made and the gun is reassembled, gun-condenser alignment may not be necessary if the previous filament had been carefully centered.

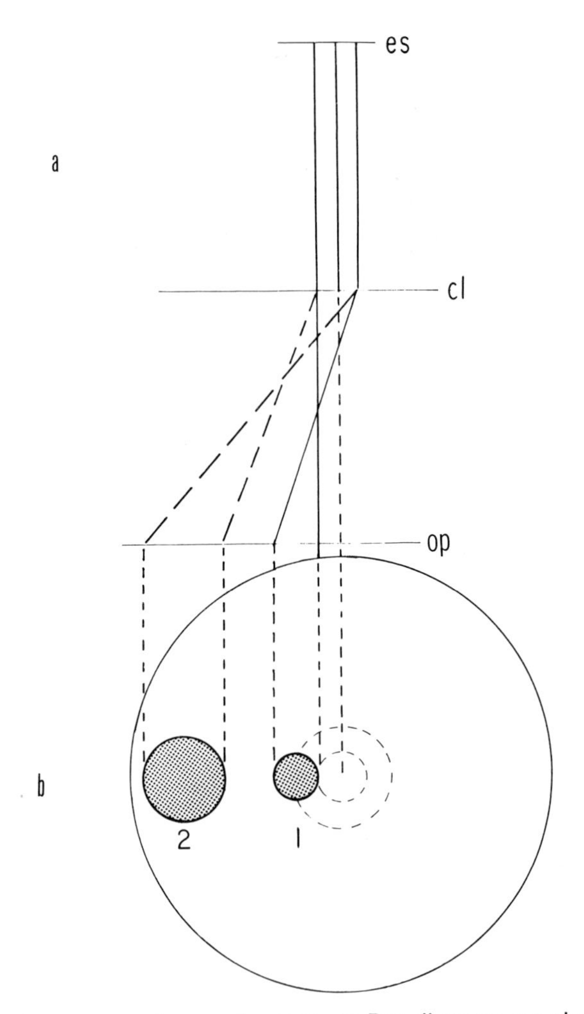

Fig. 1.9 Electron gun-condenser alignment. a: Ray diagram; es: electron source; cl: condenser lens; op: object plane. b: View on final image screen.

If a condenser aperture is introduced it may cause illumination sweep unless properly centered by lateral translation.

ILLUMINATION-OBJECTIVE LENS ALIGNMENT

A specimen is brought to focus at low magnification and the condenser is adjusted to cross-over (the brightest image of the source in the plane of the specimen). The condenser-gun assembly is translated to bring the brightest spot to the center of the field (Fig. 1.10).

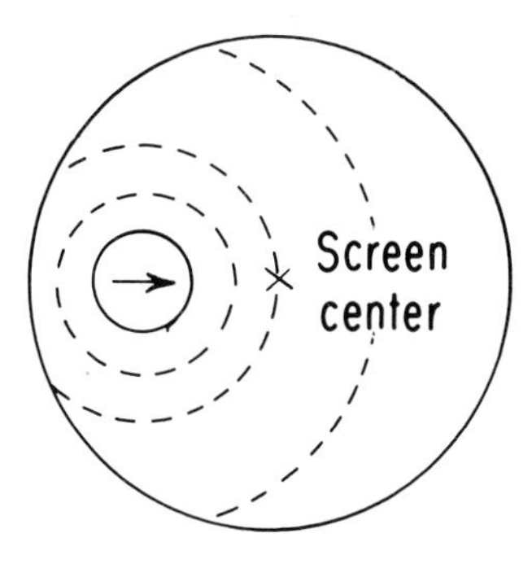

Fig. 1.10 Illumination—objective lens centration showing progressively focused illumination on final viewing screen.

ADJUSTMENT OF ILLUMINATION TILT

If the axis of illumination is inclined at an angle to the axis of the objective lens, movement of the image will take place with a change in the objective lens current. This can be seen in simplified form in Fig. 1.11, where, when a large tilt error is present, the image points appear

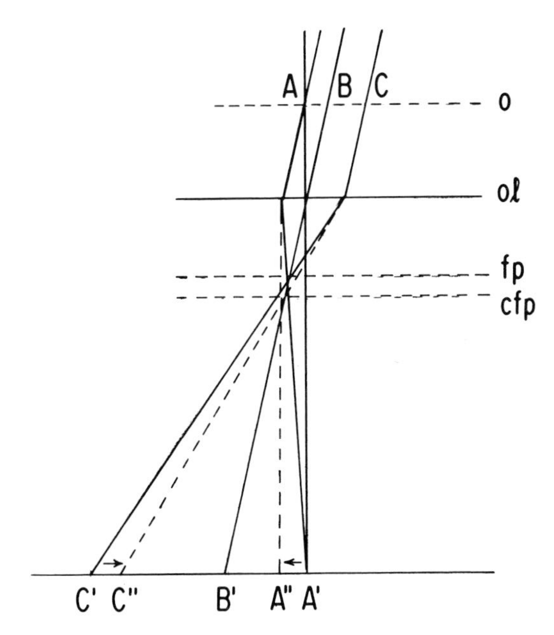

Fig. 1.11 Adjustment of illumination tilt. A ray diagram showing the effect of tilt in causing sweep of the image. o: Object; fp: focal plane; ol: objective lens; cfp: changed focal plane.

to move from both sides towards central magnification point. This slight movement will be magnified by the other lenses so that the final image of the central point may come to lie so far from the screen center as to be entirely off the screen. As a result, a sweep of the image across the screen will take place as the objective lens current is altered. In addition to sweep, a rotation of the image will be seen at the final screen.

Adjustment is made by placing some focused feature of the image in the center of the screen and then focusing the objective lens. Any image movement away from the screen center is adjusted with the tilt controls, which swing the illumination axis, so that the image is restored to its initial position. When properly adjusted, it should be found that variation of the objective lens current will only result in rotation of the image around the screen center.

Tilt adjustment should again be corrected after (d) projector and (e) intermediate lens adjustment are made. This is due to the fact that since image sweep due to tilt is observed on the final fluorescent screen (rather than objective image plane), any misalignment of the above lenses will also cause A' in Figure 1.11 to be displaced from the screen center, and the final resultant image motion on the screen will not readily be related to the misalignment of the particular contributing elements.

OBJECTIVE LENS-PROJECTOR LENS ALIGNMENT

The tolerances for alignment of these two lenses are such that normally they are adequately satisfied by original factory alignment.

OBJECTIVE LENS-INTERMEDIATE LENS ALIGNMENT

This can be achieved by use of translation controls on the intermediate lens or on the entire upper half of the column with respect to the objective lenses.

OBJECTIVE LENS-SCREEN ALIGNMENT

By varying the voltage or by varying the current of the objective lens the point that does not move, e.g., around which rotation occurs, can be found. This point is then translated to the center of the field.

After this step is completed, it is necessary to return and "touch up" the alignment of the gun-condenser assembly to the objective axis (step b).

OBJECTIVE APERTURE ALIGNMENT

The majority of the instruments are equipped with a removable objective aperture holder that permits cleaning or change of the aperture. When the aperture must be removed, realignment of the aperture with respect to the optical axis will be necessary. The general principles in achieving this are as follows.

1. The specimen or thin support film is brought into focus so as to provide necessary scattered electrons.
2. The beam is brought to cross-over.
3. The current going to the filament transformer is reduced to the point where emission of electrons takes place only from the center of the filament.
4. The lens system is then set so that the plane of the objective aperture is imaged on the screen (this is usually near zero magnification.)

Under these conditions, an image of the objective aperture formed by the electrons scattered by the specimen, as well as a bright and smaller image of the condenser aperture formed by the undeflected electrons, will be seen on the screen (Fig. 1.12). The objective aperture is manipulated until the images of the two apertures are concentric. The objective aperture is now aligned.

The end result of alignment should be to make the beam parallel and then coincident with the optical axes of the lenses (which may differ from their geometric, i.e., mechanical axis).

When all the alignment adjustments are made, then

1. The illumination should be sufficiently bright to enable objects to be viewed and focused at the highest magnification for which the instrument was designed.
2. The illumination should remain centered as its intensity is varied by changing condenser focus.
3. The image should remain centered as the focus is varied, or as the voltage is varied.
4. The image should remain centered as the magnification is varied.

Detection and Correction of Lens Asymmetry

During the discussion of alignment it was assumed that lenses and, therefore, their magnetic fields, were perfectly axially symmetrical. In actuality, it is extremely difficult to machine lenses of short focal length. The tolerances required for the condenser and projector lenses are less

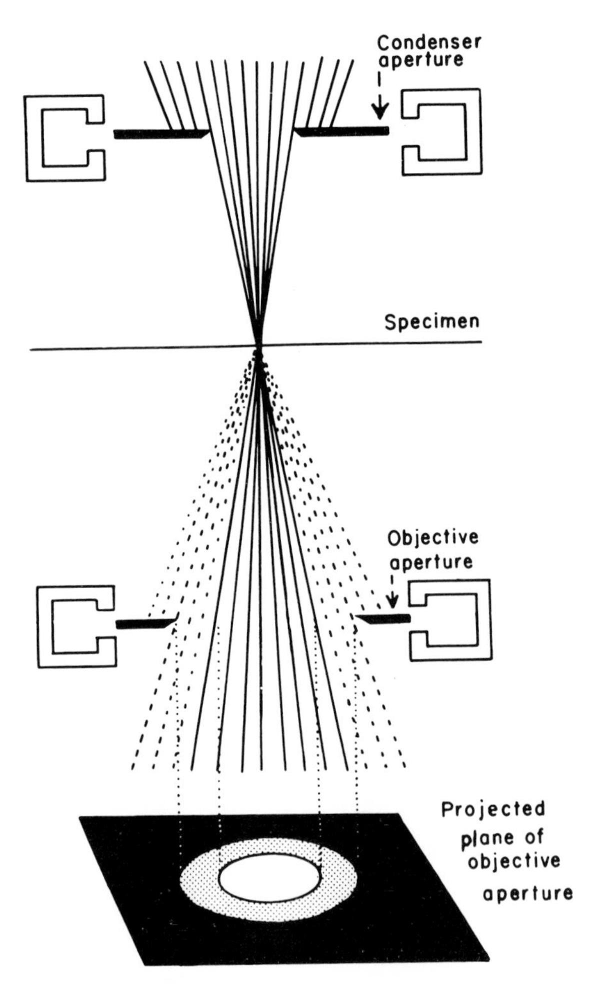

Fig. 1.12 Objective aperture alignment.

critical and those that come with the instrument usually need no compensation. For the objective lens where the mechanical tolerances are most critical, asymmetry of the pole pieces, whether due to the departure from true circularity of the bore or from inhomogeneities in the iron, always results in a slightly asymmetrical magnetic field being formed by this lens. A similar defect can also be induced by the presence of semiconducting contaminants, either on the lens surfaces or on its objective aperture.

An asymmetrical lens, in light optical terms, acts as though an addi-

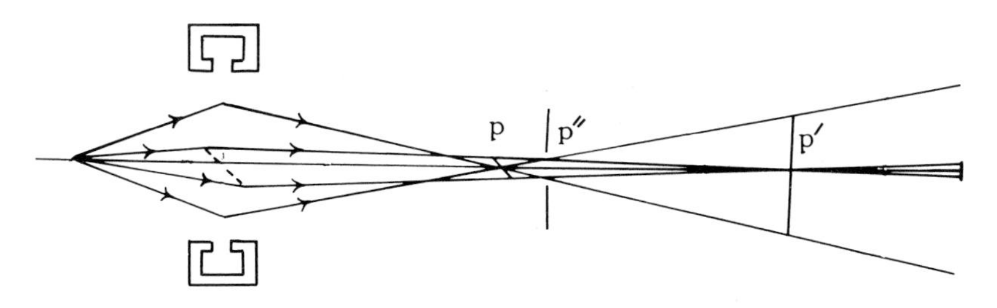

Fig. 1.13 Astigmatism. Rays parallel to the axis in two mutually perpendicular planes that pass through an astigmatic lens are brought into focus at two different points, p and p′. The disk of least confusion would be located between p and p′ and is designated as the "in-focus" position, p″.

tional weak cylindrical lens has been added to the objective of the existing convex (i.e., "spherical") lens (Fig. 1.13). The result is that the latter now has different focal lengths for rays entering in planes parallel to the optical axis and at right angles to one another. Thus instead of forming a point image of a point object, as is the case for rays passing through an axial symmetrical lens, two mutually perpendicular line images occur at different distances along the optic axis in image space. When the entering rays are at right angles to each other, rays in intermediate planes are focused at intermediate points along the axis and nowhere is there a "point" image of a "point" object; therefore, this aberration is known as *astigmatism.*

In the electron microscope, this defect can be detected by taking a through-focus series of spherical particles, such as carbon, and examining the rings that are the diffraction images at their edges (Fig. 1.14). Small holes in fairly thick films are even more suitable objects for visualizing these fringes. With a symmetrical lens, diffraction fringes are symmetrical both in-focus (the minimal *Frauenhoffer diffraction fringes*) and out-of-focus (*Fresnel diffraction fringes*). The presence of astigmatism apparent in the fringes at opposite poles of the test object is visible only in out-of-focus positions (Fig. 1.15). This is due to the fact that rays from one of the planes closer to focus give thinner diffraction fringes closer to the edge at one set of poles, while the rays from the other plane are farther from focus and thus wider fringes farther from the edge result. Astigmatism limits resolution since all image points appear as circles of confusion of a diameter related to the distance between the two focal planes for object detail in two mutually perpendicular directions.

Astigmatism can be corrected by superimposing on the lens magnetic field another field of asymmetric distribution and of variable magnitude that can be positioned so as to oppose and cancel the existing lens asym-

Fig. 1.14 Carbon particles (115,00X). The particles (from left to right) are over, under, and in-focus.

metry. The actual correction can be done most easily by means of a stigmator. Lenses equipped with stigmators are called *compensated objectives,* and in current instruments it is possible to vary the strength and angle of the compensating device by external controls so that the astigmatism can be eliminated during operation of the instrument.

There are essentially three basic types of stigmators currently in use to achieve the correction. The first type is a magnetic stigmator, which acts as an additional weak cylindrical lens of strength just sufficient to correct the cylindrical component of the objective lens. Its axis of asym-

Fig. 1.15 Diffraction pattern of a small hole in a thick carbon film demonstrating the presence of astigmatism in the lens systems.

metry can be oriented in a direction perpendicular to that of the objective lens' astigmatism. In practice, such systems are found in electron microscopes such as the Siemens Elmskop I and Philips 75 and 100B. In the Siemens instrument, it consists of two iron pieces on a ring of nonmagnetic material that are mounted concentric with the objective axis. The ring can be rotated about the objective axis as well as shifted along the direction of the axis. By these means it is possible to compensate independently for the influence of both the magnitude and the direction of the inherent astigmatism of the objective lens.

A second type of device consists of rod electrodes mounted symmetrically around the optical axis of the objective pole-piece. By changing the distribution of potentials on the electrodes, one can vary both the direction of the axis and the strength of the weak electrostatic cylindrical lens that is produced. The RCA EMU-3 and 4 and the AEI-EM 6 are instruments making use of electrostatic compensation. As an example, the former has the following system (Fig. 1.16): six nonmagnetic electrodes provide the electrostatic field and four adjustable iron shims vary

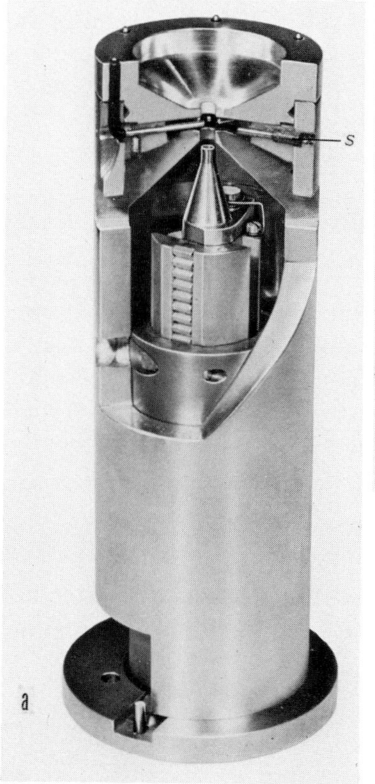

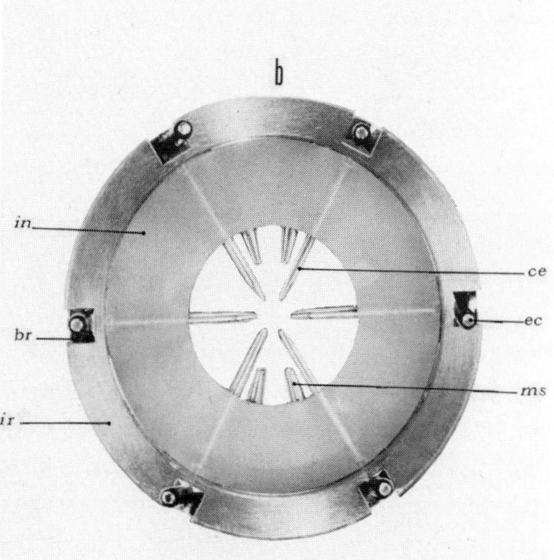

Fig. 1.16 RCA stigmator. a: A section through the pole piece of an objective lens showing the stigmator component (S). b: The stigmator seen from above. ir: Iron; ce: compensating electrodes; br: brass; ec: electrical connection; in: insulator; ms magnetic shims.

the magnetic field. The latter serve as a coarse adjustment to reduce the astigmatism to the point where the continuously variable electrostatic compensator is capable of operating at high sensitivity.

The Philips EM 200 and 300 and Hitachi HU 11E make use of electromagnetic compensation systems to correct for astigmatism. In these instruments, both the condensor and objective lenses are provided with electromagnetic stigmators where the magnitude and direction of the electromagnetic shims are controlling the current in their coils.

Compensation for lens asymmetry should be carried out after the other "external" sources of image defects have been eliminated. Correction for astigmatism is absolutely necessary where the desired level of resolution is below 50 Å.

Determination of Magnification

Upon installation of most microscopes, a magnification calibration chart is supplied by the manufacturer. Since the actual magnification will depend on such factors as the position of the specimen, voltage employed, image distortion, or asymmetry, it is not uncommon that manufacturer's calibrations may have an error of the order of $\pm$ 10%. When greater accuracy is necessary, such as in studies of particle size, special procedures must be used. For routine work, determination of magnification is made with simplicity and reasonable accuracy by the use of standard reference objects of known size, such as polystyrene latex spheres. Another means of calibration of magnification is by using a diffraction grating replica. These are now available with, for example, 50,000 lines per inch, and are used at a low range of magnification.

In making measurements of particles using low magnification micrographs, measurements should be made in the central region. This is due to the fact that a particle of well-defined size can appear larger or smaller in the periphery than in the center, depending on the kind of geometric distortion that the microscope's projector lens may produce. Since at high magnification only a very small central region of the image appears in the electron micrograph, this source of error is usually negligible.

Test of Resolution

A number of methods have been proposed to determine the resolution and limit of resolution of an electron microscope. The three most commonly used involve

a. Measurement of the minimum distance between particles that are recognizable as such.

b. Measurement of the minimum recognizable particle diameter; this method is much less reliable and meaningful than (a).
c. Fresnel fringe width at close to focus.

The use of partially graphitized carbon black as a test specimen for critical evaluation of resolution has recently been advocated. The value of such a specimen is that

1. The average distances have accurately been determined by X-ray and electron diffraction.
2. It is highly stable under electron beam bombardment and has a high degree of uniformity and reproducibility.
3. It is simple to prepare and has sufficient electrical conductivity to minimize electrostatic charging difficulties.
4. It has scattering characteristics not unlike those of the materials to be investigated by a high-resolution instrument.
5. It is finely divided and randomly oriented so that any field exhibits the structure.

Another advantage of carbon black is that it can serve for magnification calibration, to determine astigmatism, drift, and the contamination rate.

Operational Problems

Routine Maintenance

While it is the general policy of most microscopists to have their instruments serviced regularly by the manufacturer, certain operations fall within their province.

These include

1. Objective aperture cleaning (see p. 34), replacement (see p. 34), and alignment (see p. 25)
2. Filament replacement and alignment (see p. 32)
3. Detection and elimination of minor vacuum leaks (see p. 32)
4. Detection and elimination of mechanical and electrical disturbances (see p. 36)
5. Column alignment (see p. 19)
6. Detection and correction of lens asymmetry (see p. 25)
7. Cleaning the anode cap

When it is necessary to change the filament, the anode cap should be routinely cleaned. Cleaning may also be necessary in between filament changes when certain operational difficulties are encountered (see

Table 1.1). Cleaning of the cap can be done with levigated alumina powder or metal polish (e.g., Pikal Metal Polish, Nihon Maryo-Kogyo Co., Tokyo, Japan). Special attention during cleaning should be given to the region around the opening and its side walls. This can best be done with the aid of a cotton swab. The flattened surfaces can be cleansed with a wad of cotton. The cap should then be washed in a detergent and rinsed thoroughly using liberal amounts of distilled water for the last rinse. The cap should be blotted with towels and then dried with lint-free lens paper (e.g., Kimwipes).

The cleaning of objective apertures can best be avoided by the use of ultrathin metal apertures (see p. 34). Where these are not used, cleaning can be done by heating in vacuo in a tungsten boat. Chemical cleaning procedures are usually recommended in manufacturer's manuals.

Filament Replacement

The average life of a tungsten filament is 5–20 hr. To be replaced, the gun chamber should first be closed off, if such an option is possible, and air let in. If the gun chamber cannot be isolated, the entire column vacuum must be broken. The gun should be removed from the chamber and placed on a clean surface. The anode cap should be unscrewed and the old filament removed and replaced with a new straight and symmetrically bent one. A record can be kept of the life of the filament and the site of breakage.

Three adjustments are required upon replacement of the filament

1. *Centration.* The tip of the filament needs to be mechanically centered with relative accuracy with respect to the aperture of the anode cap. Careful centration will minimize the extent of column readjustment. A centration jig can facilitate this step.
2. *Depth adjustment.* The distance of the filament tip to the cap aperture should be carefully set with the aid of a depth gauge or jig.
3. *Column adjustment.* Reassemble the gun and check the alignment. If both the precisions and present filaments were carefully centered, then only minor translational adjustments of the gun may be needed.

Vacuum Leaks

The need for maintaining an adequate vacuum level in the column has already been emphasized. The mechanical and oil diffusion pumps that provide the vacuum need little routine care. The potential source of difficulty is an air leak of such magnitude that the capacity of the pumping system is exceeded. The presence of a serious vacuum leak will be indicated by the vacuum gauges or the activation of the vacuum failure switch.

A vacuum leak is located by the process of elimination. Since leaks may occur due to improper closure of the gun, specimen, or photographic chambers, these should be checked first to see if they are properly closed. More frequently, however, the leak is located in the vacuum system or microscope column. Thus if the aforementioned chambers are secure and the leak persists, the main valve between the pumps and column should be closed, and where the vacuum gauge is on the column, the pressure should be noted. If it rises rapidly, the leak is obviously in the column. A next step may be to determine if the leak is at the site of the external manipulative controls. Such leaks will become evident if, with the main vacuum valve open (i.e., normal operating condition), a pressure rise is noted when the specimen stage, fluorescent screen, or plate cassettes are manipulated.

If the leak is still not localized, then the reliability of the vacuum valves may be checked. This can be done by noting the vacuum at various stages of pumping from atmospheric pressure down. If the valves seem to be in order, a further search may require stepwise dismantling of the column. Prior to undertaking the rather laborious task of dismantling the column the search may be facilitated by using an organic solvent or gas as a "leak guide" and applying it to particular column junctions. Thus, where a large leak is present, wetting a junction may occasionally result in a drastic rise or drop in the pressure.

A useful method, feasible where a gas discharge tube is present, depends upon noting the color of the discharge after applying the solvent. For example, if the suspect joint is wet with alcohol and if alcohol passes into the column at the defective site, a striped whitish glow will be produced. The most sensitive method involves the use of a commercially manufactured leak detector. This is a simplified, sensitive mass-spectrometer that is connected to the vacuum system and adjusted to detect the helium line. Thus if helium gas is released near the site of the suspected leak and it passes into the column, an electric signal will be produced whose amplitude is proportional to the size of the leak and the closeness of the source of helium to the leak.

If dismantling of the column is unavoidable, it should be started at the top. The column is closed off at the level of the section removed (but left connected to the pumps). When the leak disappears, the joint at the level of the last section removed is implicated as the source.

Objective Aperture Replacement

The usual objective lens is provided with a physical aperture to permit the use of an optional aperture angle but, even more significantly, to provide usable contrast. While small apertures can improve contrast, if they are not kept clean they can degrade the image quality. If the edge of the aperture becomes covered with a thin, electrically insulating layer of contamination, then it becomes charged when subjected to bombardment by the wide-angle scattered electrons. The aperture then acts as a weak electrostatic lens that, however, is capable of exerting significant effects because of its proximity to the beam. Since the contamination process is not usually completely symmetrical, the "lens" ultimately formed will be asymmetric and thus introduce astigmatism.

With the aforementioned considerations in mind, the ease of cleaning was one of the three principal requirements for material selected for use for aperture discs; the other two being that they have a high electrical conductivity and be readily machinable to give openings that are round, smooth, and very small. These requirements were to a large extent met by platinum and molybdenum. Nevertheless, such apertures had to be frequently cleaned (e.g., after 6-hr exposure) because of the build-up of contaminants. This usually meant "down time" and interruption of the work schedule.

To overcome the need for frequent cleaning, multiple aperture holders have been developed for most microscopes. This type of holder permits not only quick change of several apertures without loss of the column vacuum but also the interchange of apertures of different sizes. Another, but not as yet widely adopted approach, is the use of *heated apertures.*

A more successful approach to overcoming the problem of frequent cleaning has been the use of ultrathin metal apertures. It has been found that such apertures have a negligible astigmatic effect. This may be explained as follows: The critical edge of the hole of the aperture has an area smaller than the typical aperture. The lens effects due to a charged layer of contamination on such an edge are, therefore, very much reduced. Because of its relatively low heat conductivity, the electrons impinging on such an aperture can be used to raise its temperature sufficiently so as to remove contaminating layers while it is still in the microscope. Such apertures are now commercially available. They can also be readily made by evaporating a metal, such as silver, over spheres of dextran resting on a collodion or Formvar membrane. Removal of the film support and spheres with solvent produces a thin (about 0.5μm) metal sheet with apertures corresponding in diameter to the dextran spheres. Such apertures are serviceable for long periods (e.g., months). They need to be replaced only if they close as a result of dust or excessive contamination build-up.

The diameter of the aperture opening used can be varied; most commonly used sizes being 15 to 50μm. Above 75μm, the presence of the aperture is of little consequence, since relatively few electrons are scattered at angles sufficiently large to be removed by such an aperture, while below 25μm the instrument's resolution may begin to be affected. This is due to the fact that at small aperture sizes we may be in the diameter range where the diffraction effect (rather than spherical aberration) limits the resolving power (e.g., at 50 kV, with 3.0 mm focal length objective lens, and a 6μm aperture in a rear focal plane, α is 10^{-3} and the resolution is 20 Å, while at 24μm, that is, at $\propto$ opt the resolution is 10 Å).

Thus, theoretically, the optimal aperture size should be determined from a resolution curve that will show the point where the effects of diffraction and spherical aberration are equal. In practice, where very thin, low-contrast specimens are being examined, use of an aperture substantially smaller than the theoretical optimum may be helpful, since the theoretical limit may, in any case, not be achievable in such specimens because of contrast limitations. Such a choice will, however, have another drawback in practice since the smaller the aperture the more rapidly it becomes contaminated and thus produces astigmatism.

The position of the aperture within the objective lens in relation to the specimen is a critical factor in determining image contrast, resolution, and rate of aperture contamination. Thus, for a particular desired $\propto$ if placed very close to the specimen, the aperture will be very small, the contamination rate will be high, and the field of view will itself be limited by the aperture. If placed close to the rear pole piece, the aperture can be larger, and the effects of contamination and field obstruction will be greatly reduced. Ideally, the aperture should be placed in the rear focal plane of the objective lens, but this often does not significantly improve its effectiveness over that in which the aperture is positioned just in front of the pole piece.

Disturbances

Under ideal instrument conditions, resolution of about 2 Å can be attained. In actual practice, however, a number of disturbances have to be avoided in order to insure that potential limitations to the achievement of high resolution are not present. Disturbances (other than vacuum leaks) can be classified into two different types depending upon the manner in which they are manifested on the screen. Thus there are those that produce a shift or "drift" of the image over a period of time that appears on the screen as image drift. Other disturbances produce a uniform or unidirectional blur in the image. Identification of the specific cause of the disturbance is possible by examining the electron image and correlating the change in appearance with a particular manipulation of the instrument.

The causes of image drift and image blur can be related to a number of different sources that induce one or both of these image defects.

STRAY MAGNETIC FIELDS

Electron microscopes are shielded against stray magnetic fields. Nevertheless, where such disturbance is intense the operation of the instrument may be seriously affected. Thus the presence of a slowly varying, strong, transverse magnetic field will make it difficult or impossible to align the column. When such alignment difficulty is found, a test should be made to detect the presence of such an external field.

MECHANICAL DISTURBANCES

To minimize possible erratic or systematic displacement of the specimen as a result of mechanical dislodgement of the specimen or its holder from its resting position by mechanical shocks or vibrations, the specimen chamber is usually designed so that the specimen holder is held rigidly when in its operating position. Nevertheless, since the specimen holder must be readily movable and replaceable, it cannot be permanently fixed in position and there may, therefore, be backlash or give-in of movable components of the stage mechanism. For this reason, external vibrations such as those coming from the passing of heavy trucks or trains can result in image drift, and thus the location of the electron microscope in a building should be where potential for disturbance is minimal. Where building vibration is known to exist, shock absorption mounts may be desirable. Also the rotary mechanical pump should be shock mounted and even mechanical controls on the microscope, such as the shutter, should be operated without erratic motion to avoid vibration and thus minimize image shift or drift. If the mechanical vibrations are intensive and continuous, image blurring will result.

ELECTRICAL DISTURBANCES

If the lens has been centered on the voltage axis, small drifts in lens current will produce image shifts. Slight variations in the objective lens current at frequencies below 5 cps, though causing slight oscillation of focus, may produce larger rotation of an image about the objective lens magnetic axis at large distances from that axis. Small fluctuations of the high voltage supply or lens supplies result in small changes in the effective strength of each of the imaging lenses so that the magnification and/or focus of each lens is altered slightly and this blurs the image.

A more common variety of this type of disturbance is due to a higher frequency alternating (e.g., 60 cps) magnetic field in the vicinity of the microscope. This may result from the proximity of a large AC carrying cable, transformer, or voltage stabilizer. The most sensitive part of the beam path is just on the image side of the specimen. Such an AC disturbance manifests itself as a blurring of the image in one direction, consistent from specimen to specimen. If, however, it is the region between the condenser and objective that is affected by the disturbance, then the presence of the transverse alternating magnetic field manifests itself by acting as a beam wobbler and is only detectable in out-of-focus planes.

SPECIMEN INSTABILITY

Image shifts may also be due to specimen instabilities (specimen drift). These may be due to thermal drift, e.g., the absorption of heat by the specimen grid so that thermal expansion results. This disturbance is evident even under normal operating conditions, when shortly after activation of the beam an image shift is apparent. Thermal equilibrium between the specimen grid and its holder is, however, quickly attained at constant levels of illumination, which results in improved image stability. If, however, the absorbed heat of the grid cannot be quickly dissipated by transfer to its surroundings, thermal equilibrium will not occur quickly and a slow image drift, detectable with the magnifying viewer, may persist for a long period of time. To insure image stability, good thermal contact must be provided for by fitting the specimen grid accurately into its cap, which in turn should be securely clamped onto the holder. The latter should then be placed in its mount so that good contact is made.

Another cause of specimen drift is broken or loose support film of the grid square under study. Under these conditions, thermal effects can produce changes in tension in the film that cause movements relative to the position of the tear. Also, the specimen and supporting membrane develop substantial positive charges (due to secondary electron emission) during exposure to the beam, and variations in intensity and/or position of the illuminating beam will cause variations in charge on the specimen and specimen holder. The resulting changes in repulsive forces between these charged elements can cause appreciable movements of the torn supporting film (similar to those operating in a gold-leaf electrometer).

CONTAMINATION

The slow irreversible deposition of organic matter from the residual gases or greased areas occurs on the beam-exposed surfaces of the column. Upon direct or indirect bombardment of such insulating layers by electrons, they acquire an electrostatic charge and can act as lenses or beam deflectors. Depending upon the magnitude, position, and rate of change of the charge, a varying degree of image shifting or blurring results. To minimize these disturbances, cleanliness of all the beam contacting surfaces, especially of the objective aperture, must be maintained.

To reduce contamination the following four steps are recommended

1. Replacing all "O" rings in the microscope column and vacuum manifold with ones made of Viton (obtainable from George Angus & Co., Wallsend-on-Tyne, England)
2. Using Apiezon "L" grease, but sparingly
3. Using D.C. 705 oil in the diffusion pump (obtainable from Dow Corning Corp., Midland, Michigan)
4. Handling of all internal components with nylon gloves

A summary of the most common microscope disturbances is listed in Table 1.1.

PHOTOGRAPHY WITH THE ELECTRON MICROSCOPE

Beam Intensity Level

In viewing the specimen image, or in its photography, there are three general considerations that together determine the working level of beam intensity during each of these two activities.

1. The beam intensity should be adjusted so as to avoid excessive electron bombardment of the specimen. Excessive irradiation can cause the effective temperature of the specimen to rise to hundreds of degrees Centigrade and results in considerable specimen damage. Therefore, to minimize specimen damage, both the maximum intensity of the beam and the length of exposure during visual examination and photography need to be carefully controlled.

TABLE 1.1 CHECK LIST OF COMMON MICROSCOPE DIFFICULTIES

Disturbances	*Common causes*
Alignment impossible	Stray magnetic fields
Image drift or blurring	External vibrations
Image shifts	External magnetic field, grid caps loose, support film loose, dirty objective aperture
Very short filament life	Small vacuum leak in gun area
Beam flickering with:	
stable image	Dirty anode cap
unstable image	Dirty anode cap and objective aperture
Image instability varying with:	
stage movement	Dirty specimen holder, grid cap or grid
beam intensity	Dirty objective aperture or pole-piece
Fixed and sharp image	Dirty projector-pole piece
Very out-of-focus image	Dirty objective and/or condenser aperture
Limited image distortion varying with stage movement	Dirty specimen holder, grid cap or grid
Unstable focus	High-voltage instability

2. The level of beam intensity must, however, be such that an image of sufficient brightness will be formed on the retina to permit location of the desired field of view, precise focusing, and examination of the electron image.
3. Short exposure of the specimen during photography is desirable because it reduces the possibility of obtaining a blurred picture. Exposures longer than about 3 sec may result in loss of resolution due to specimen drift and circuit instability.

Choice of Photographic Emulsion

With any given photographic emulsion, a longer exposure (to a given negative density) requires a lower beam intensity and, therefore, produces less net specimen damage but increases the risk of blurring due to specimen movement or instrument instability.

The smallest point discernible on a photographic plate is determined by the size of granules of the photographic emulsion. This means that a photographic emulsion of smaller grain will provide greater resolution. By permitting greater photographic enlargement, fine grain emulsions permit photography at lower electronic magnifications. This ought to permit the use of even lower beam intensities for the same exposure time. Fine grain emulsions, however, are photographically slower (i.e., require more exposure time and/or intensity) than coarser grain emulsions, and there is usually little net gain with respect to specimen damage when using fine grain emulsions.

At a low electronic magnification, the image on the screen or film is brighter than at a high magnification for a given illumination. This is a simple consequence of spreading the total energy over a smaller area and thereby obtaining a greater energy per unit area. The greater image illumination at low electronic magnification permits the use of the slower and higher resolution photographic emulsion for a fixed amount of specimen damage. Fine grain emulsion has an additional advantage. It usually permits greater contrast than does a coarse grain emulsion (fast photographic emulsion) and, therefore, permits increased resolution of low contrast specimens.

As a good approximation, contrast for electron-exposed emulsions is equal to about 2.3 times the developed density for all emulsions. Therefore, the greater the exposure the greater will be the contrast. Useful contrast requires the recording of a large range of densities with precision. Most of the noise or imprecision in the recording of density in an electron photomicrograph is due to the statistical fluctuations in the number of electrons falling on an elemental area of the emulsion. This noise increases as the square root of the number of electrons increases. The information bearing signal increases in proportion to the number of electrons and, therefore, the net signal to noise ratio, or precision of recording of density, increases as the square root of exposure increases. For this reason, slow, thin, and fine grain emulsions, which can record very high densities and collect the information from larger numbers of electrons to produce a given negative density, can provide more useful contrast.

The parameters involved in film choice are the electronic magnification, the optical magnification, the grain of the screen, and the grain of the film. The amount of optical magnification depends upon the necessary increase in retinal image size (without loss of brightness) to permit accurate focusing. It is usually convenient to choose a photographic emulsion with fine enough grain so that at minimum electronic magnification the required optical magnification approximately matches the resolution of the screen. Under these conditions, the brightest possible focusable image will be formed on the retina for a fixed level of specimen damage if the level of illumination during both focusing and photography is held constant.

In practice, it has been found that an emulsion suitable for spectroscopy is also satisfactory for electron microscopy. Specifically, the need is met by a slow, orthochromatic, fine-grained emulsion with good contrast. These criteria are met by the Kodak lantern slide plate and more recently introduced Electron Image Plates, as well as by the Ilford N. 40 process plate and the Gevaertt deapositive contrast plate.

Photography with the Microscope

The need for and the advantages of photographing the final electron image have already been discussed. The routine procedure involved in this operation can be summarized as follows:

a. The field of view is selected at a low magnification and at a low beam intensity, and preliminary centering and focusing is carried out.
b. The magnification is raised so that the desired level of resolution of detail in the electron image (visually near the level of resolution on the photographic emulsion) may be attained.
c. The intensity of the illumination must be raised but is kept to a level just adequate for critical focusing and/or reasonably short exposures.
d. The image is finally brought into critical focus by varying the current (i.e., focal length) of the objective.
e. The photograph is then taken without changing the condenser setting (i.e., intensity of illumination). To insure that it is in exact focus, especially at higher magnifications, a number of photographs are taken with the objective lens, and the current varied slightly in both directions from the assumed critical focus. Such a group of photographs are known as a *through-focus series.*

Until a few years ago it was the standard procedure to decrease the intensity (i.e., aperture angle, α_c) of illumination by about one order of magnitude by reducing the condenser lens current just prior to photography. This was usually done so as to increase the depth of field. However, since a change in the "in-focus" position of the condenser may secondarily introduce a number of possible kinds of image degradation, current preferred procedure requires that the condenser setting remain essentially unchanged throughout final focusing and photography.

One method, incorporated in some microscopes to aid focusing at low intensity of illumination (i.e., small illuminating aperture) for both visualization and photography, makes use of a built-in deflector type of focusing mechanism commonly known as a beam-wobbler. This involves the use of magnetic coils fitted between the condenser and specimen to deflect the electron beam from side to side over an angle of about 10^{-2} rad. This temporary increase in the aperture angle produces a marked decrease in the depth of field from the normal conditions. As a result, the in-focus (i.e., the range of variation of sharpness of the image with change in objective current) is reduced and departure from the exact focus is more readily recognized. The aperture angle can also be increased by inserting a larger physical aperture in the condenser lens. However, this results in a very large increase (often excessive) in the beam intensity on the specimen. The beam-wobbler, as an aid in focusing, is especially valuable at low magnifications, when a specimen of poor contrast is used or when the intensity of illumination must be kept low.

Photographic Processing

The following are a number of major factors to be considered in the handling of photographic material.

a. *Plate inspection.* It is a desirable precaution to inspect each plate before it is placed in the microscope to insure that it is dust- and lint-free. This is done by reflecting light from the recommended safelight on the emulsion surface and removing any particulate matter with a camel's hair brush or commercial "duster." If plates of a manufacturer are found to be consistently clean, inspection can be dispensed with and periodic checks of cleanliness will usually suffice. The plates should be handled by the edges to prevent scratches, preferably using lint-free cotton or polyethylene gloves.

b. *Plate degassing.* Emulsions are thinly spread on plates so that there is a minimum of gelatin. Nevertheless, prepumping the plates (loaded in holders) in a vacuum chamber before placing them in the casette magazine is desirable to insure maintenance of a good vacuum, and thereby a minimum interruption in picture taking.

c. *Plate processing.* The processing procedure for photographic plates, after exposure to the electron beam, consists of 5 steps.

1. Development. The developer selected should have a suitable balance between fine grain, convenience, and contrast, with emphasis on the latter. Besides exposure, four main considerations in development govern the density, contrast, and uniformity of the negative. These factors are: the kind of developer used, the temperature of the developer, the length of development, and the extent of agitation during development. Kodak recommends HRP (1:4) or D-19 (1:2) for use with its Electron Image Plates as having a great range of tonal values and causing less "chemical fog," thus resulting in a brilliant vivid image.

Development time is 3 min for HRP and 4 min for D-19 at 68°F in a tray with proper agitation.

2. Rinsing. This step should be carried out in running water for a full 1½ min at 65–70°F with continuous agitation. (The plates may become mottled if washing is not carried out properly or if stopbath solutions are used.)

3. Fixing. This step requires intermittent agitation. It should extend for twice the time it takes for the milky appearance of the plate to clear completely and this will vary with different emulsion types. Kodak fixer requires 10 min and its rapid fixer 5 min.

4. Washing. This should be carried out for 30 min using moderate agitation. This step can be drastically accelerated, if need be, by rinsing for 30 sec in water (to remove excess hypo) followed by 2 min in Kodak Hypo Cleaning Agent followed by a 5-min water wash.

5. Drying. This should be carried out in a dust-free area. To minimize drying marks, the washed plate can be rinsed for 1 min in Kodak Photo-Flo solution, or its surface wiped carefully with a chamois or a soft viscose sponge.

d. *Negative printing.* This is the final stage in securing an electron micrograph. This involves the use of a suitable enlarger, printing paper, and processing procedure.

A good quality enlarger, equipped with a standard enlarger lamp and a double condenser lens is essential, although for maximal resolution a point-light source is needed. The enlarger should be connected to the current via a timer that is calibrated into minutes and seconds, so that exposure time can be accurately controlled. Where line voltage fluctuations are suspected or known, the enlarger should also be connected up to a voltage control device (e.g., Variac unit). The enlarger should be positioned on a vibration-free base and its lens and condenser surface should be kept free of dust and fingerprints. Focusing the image should be carried out with the aid of a focusing magnifier.

Photographic printing papers are available in a wide selection of contrast, image tones, and surface properties. For scientific work a single weight, smooth, glossy or "F" surface provides best results.

Standardization of the enlarging procedure will insure the results of routine processing to be consistently of good quality. Thus one type of paper and developer should be selected. Kodak manufactures 2 types of suitable papers: Kodabromide, which comes in 5 contrast grades, and Polycontrast Rapid, which with the aid of filters provides 7 degrees of contrast.

The correct length of exposure of the paper to the image of the negative should be a routine goal. This will minimize the loss of tonal range that may result from attempts to compensate for incorrect exposure by altering the required development time. Illumination for the most commonly used grade of paper should be standardized. A densitometer with a suitable photocell pickup can facilitate this procedure.

Processing of exposed prints is also a 5-step process.

1. Development. For use with either of the aforementioned papers, Kodak Dektol (1:2) or D-72 (1:2) can be used for 1½ min at 68°F. Full development of the paper, i.e., maintenance in the developing solution until the picture reaches a plateau of inactivity, should be the standard processing procedure.

2. Rinsing. The prints should be rinsed for 5–10 sec, with agitation in a stopbath.

3. Fixing. This step requires constant agitation for 6–10 min, at a temperature between 65 to 70°F, in one of several fixers that are commercially available.

4. Washing. Fixed prints should be washed for 1 hr in a tray or washing tank where water exchange (at temperatures of between 65 and 75°F) is provided for. More rapid washing can be facilitated using Kodak Hypo Cleaning Agent as recommended on the package.

5. Drying. The prints can be freed of excess surface water by piling them on a smooth surface and passing a roller over their back surface. They then should be placed, face down, on a drying drum, squeezed into close contact with the clean surface of the drum, and kept there until they fall off by themselves.

In recent years automated processing of prints has been introduced. This "stabilization process" provides for rapid handling of prints within a single enclosed unit avoiding the need to set up trays for the develop-stop-fix-wash procedure. Such prints can be of good quality and provide for quick evaluation of the electron micrographs. However, because the chemical reactions within the emulsion have been stopped only temporarily, such prints are subject to deterioration on standing after a number of months. They can be converted into permanent prints by treating them in a standard fixing bath and then washing and drying in the usual manner. Automatic processing units are manufactured by Kodak, Ilford, and other companies.

e. *Photograph evaluation.* Unsatisfactory results are reflected by electron micrographs whose details are not sharp and/or which lack contrast. Such micrographs may be the result of the use of excessively thick tissue sections or supporting films, specimen drift due to torn specimens or support films, astigmatism, altered focus, microscope instability, poorly preserved or sectioned specimens, or poor photographic processing.

COMMERCIAL ELECTRON MICROSCOPES

The basic characteristics of currently available commercial electron microscopes are listed in Table 1.2.

Image Intensifiers

The last few years have seen the commercial introduction of image intensifiers as electron microscope accessories. These units are usually designed to:

a. Permit direct high magnification image observation at comfortable brightness levels.
b. Permit specimen observation at lower beam intensities (which would reduce the chances for specimen damage).
c. Permit the image to be electronically scanned and processed and, therefore:
   1. Extend the range of image contrast beyond that provided by direct visualization on the screen.
   2. Permit simultaneous specimen observation by groups of viewers, thus facilitating teaching and demonstration.
   3. Permit continuous recording (on magnetic tape) of transient phenomena.

There are a number of different electronic systems in current usage. As a group they are yet to fully meet the level of anticipated promise because of their limited resolution except at the highest electron microscope magnifications. Photographic recording still appears to be the best means of securing and recording most electron microscope information.

The image intensification system utilizes the effect of electron-bombardment induced conductivity. Some insulating semiconductors have electric resistance which decreases with the impact of high energy electrons. When this effect is applied, it is possible to get output conduction currents multiplied more than 1,000 times over the primary current of high-speed electrons, and also to obtain amplification of image brightness of approximately 1,000 times.

An image intensifier system (Fig. 1.17) consists of an electron microscope, a television pick-up system, and a television picture monitor. The TV-pick-up is mounted under the camera chamber of the electron microscope in vacuum. The unit includes an image converter and a scanning device for low-speed electrons. The image converter is bombarded directly with the electron beam. Output signal current from the converter is introduced to the TV receiver through the amplifier.

TABLE 1.2 COMMERCIAL ELECTRON MICROSCOPES

Company and model	*Price*	*Resolution capability (A)*	*Range*	*Type*	*Acceleration voltage (kV)*
JEOLCO (USA) INC.					
50	7,850	100	2,000 to 4,000	step	50
100B	49,000	2	190 and 500 to 5000,000	step	20, 40, 60, 80, 100
100U	43,000	3	150 and 300 to 3000,000	step	20, 60, 80, 100
JEM 120	39,000	4.5	120 and 600 to 250,000	cont.	25, 50, 80, 120
JEM 200	70,000	4.5	1,000 to 150,000	cont.	50, 100, 150, 200
JEM 500S	250,000	10	500 to 150,000	cont.; 1,000 to 1000,000	200, 350, 500
JEM 1000	457,000	7	1,000 to 150,000	cont.	500, 750, 1,000 std.
T7	23,000	7	300 to 80,000	cont.	60
PERKIN-ELMER CORP.					
HS-8	28,000	8	1,000 to 1000,000	12 steps	25, 50
HU-11E	44,000	3.5	250 to 300,000	cont. & 22 steps	25, 50, 75, 100
HU-125E	49,000	3.5	250 to 300,000	cont. & 22 steps	25, 50, 75, 100, 125
HU-200E	58,000	5	1,000 to 200,000	cont. & 22 steps	50, 100, 125, 150, 175, 200
HU-650	270,000	4	1,000 to 150,000	cont. & 40 steps	200 to 650
HU-1000	450,000	10	1,000 to 150,000	cont. & 4 steps	400, 500, 600, 700, 800, 900, 1000

TABLE 1.2 (Continued)

Company and model	*Price*	*Resolution capability (A)*	*Range*	*Type*	*Acceleration voltage (kV)*
PHILIPS ELEC. INSTR.					
EM 100C	25,000	10	0 to 90,000	cont.	40 to 100
EM 300	55,000	5 pt. to pt.	220 to 500,000	cont. & steps	20, 40, 60, 80, 100
FORGFLO CORP.					
EMU-4	48,000	8	1,400 to 200,000	cont. & 15 steps	50, 100
PICKER CORP.					
EM 8	NG		100 to 4000,000	cont.; 100 to 700	40, 60, 80
EM 801	NG		100 to 160,000	cont.; 100 to 700	40, 60, 80
EM 802	NG		150 to 160,000	cont.; 150 to 1,300 steps	40, 60, 80, 100
EMU-4	47,000 to 49,000	8	1,400 to 2000,000	cont.; 1,400 to 50,100; 2000,000 15 steps	50, 100
SIEMENS AMERICA INC.					
Elmiskop 101	51,600	3.5	285 to 280,000	cont.; 1,600 to 280,000; 5 steps	40, 60, 80, 100
Elmiskop 1A	37,800	5	200 to 174,000	900 to 174,000; 12 steps	40, 60, 80, 100 or 50, 75, 100, 125
CARL ZEISS INC.					
EM 9A	25,975	10	160 to 40,000	cont.; 4 steps	60

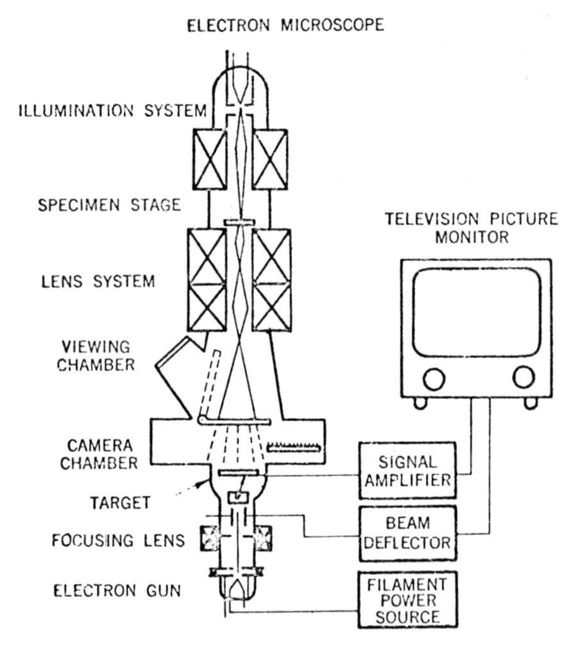

Fig. 1.17 Schematic diagram of an image intensifier.

Stereomicroscopic Accessories

Because of the small objective aperture that provides for a large depth of focus and because of the large depth of field, recording of stereoscopic views of objects is rather simple with the electron microscope. This is especially useful when viewing replicas of deep structures (which by conventional methods do not distinctly reveal the form of the surface) and with thick (0.1-0.2μm) biological sections of tissues.

To secure stereoelectron micrographs, the specimen holder is tilted through an angle of 5°–10° between 2 successive exposures.* In practice the specimen is centered, the first photograph is exposed, the specimen is tilted, re-centered, re-focused if necessary, and then re-photographed. If the specimen plane is rotated excessively, axial dimensions are exaggerated. When the two micrographs are viewed through a stereo-

*Tilting is accomplished either by rotation of the specimen holder around an axis in the plane of focus, by a built-in stage control, or by remounting the specimen support with a shim under one edge sufficient to produce the 5-10° tilt.

viewer, a 3-dimensional image of the object is seen. Prints for stereoscopic viewing need to be precisely mounted at correct orientation and separation to secure satisfactory results.

BIBLIOGRAPHY

Books

Adrenne, M. von (1940). "Electronen Übermikroskopie." Springer, Berlin.

Borries, B. von (1949). "Die Übermikroskopie." Aulendorf Württ., Cantor.

Brüche, E., and Scherzer, O. (1934). "Geometrische Elektronenoptik." Springer, Berlin.

Burton, E. F., and Kohl, W. H. (1946). "The Electron Microscope." Van Nostrand Reinhold Company, New York.

Busch, H., and Brüche, E. (1937). "Beitrage zur Elektronenoptik." Barth, Leipzig.

Cosslett, V. E. (1946). "Introduction to Electron Optics." Oxford at the Clarendon Press, Oxford.

———. (1952). "Practical Electron Microscopy." Academic Press, New York.

———. (1966). "Modern Microscopy or Seeing the Very Small." Cornell University Press, Ithaca.

De Broglie, L. (1950). "Optique Électronique et Corpusculaires." Herman, Paris.

Dupouy, G. (1952). "Eléments d'Optique Électronique." Colin, Paris.

Fischer, R. B. (1954). "Applied Electron Microscopy." Indiana University Press, Bloomington.

Gabor, D. (1948). "The Electron Microscope." Chemical, New York.

Glaser, W. (1952). "Grundlagen Der Elektronenoptik." Springer, Berlin and Vienna.

González-Santander, R. (1966). "Manual De Microscopía Electrónica. I. Elementos de Microscopía Electrónica." Monogra. de Ciencia Moderna, No. 75, Madrid.

Gregoire, C. (1950). Microscope Électronique et Recherche Biologique." Masson, Paris.

Grimstone, A. V. (1968). "The Electron Microscope in Biology." Arnold, London.

Grivet, P. (1965). "Electron Optics." Pergamon Press, New York and Oxford.

Haine, M. E., and Cosslett, V. E. (1961). "The Electron Microscope." Spon, London.

Hall, C. B. (1966). "Introduction to Electron Microscopy." McGraw-Hill Book Co., New York.

Hayat, M. A. (1970)."Principles and Techniques of Electron Microscopy: Biological Applications," Vol. 1. Van Nostrand Reinhold Company, New York and London.

———. (1972)."Basic Electron Microscopy Techniques."Van Nostrand Reinhold Company, New York and London.

Hayat, M. A., (editor). (1972). Principles and Techniques of Electron Microscopy: Biological Applications. Vol. 2. Van Nostrand Reinhold Company, New York and London.

Heidenreich, R. D. (1964). "Fundamentals of Transmission Electron Microscopy." Wiley, New York.

Jacob, L. (1950). "Introduction to Electron Optics." Methuen, London.

Kay, D. (ed.) (1965). "Techniques for Electron Microscopy." Davis, Philadelphia.

Klemperer, O. (1963). "Electron Optics." Cambridge University Press, Cambridge.

Magnan, C. (ed.) (1961). "Traité de Microscopie Électronique (2 vol.). Herrmann, Paris.

Mahl, H., and Golz, E. (1951). "Electronenmikroskopie," VEB Biographisches Inst., Leipzig.

Meek, G. A. (1970). "Practical Electron Microscopy for Biologists." Wiley-Interscience, New York.

Myers, L. M. (1939). "Electron Optics." Chapman & Hall, London.

Ockenden, F. F. J. (1946). "Introduction to the Electron Microscope." Williams & Northgate, London.

Palacios de Borao, G. (1950). "Supermicroscopia Electronica." Monografía de Ciencia Moderna, No. 21, Madrid.

Paszkowski, B. (1969). "Electron Optics." Elsevier, New York.

Pease, D. C. (1966). "Histological Techniques for Electron Microscopy." Academic Press, New York.

Picht, J. (1955). "Das Electronenmikroskop." Fachbuchverlag, Leipzig.

Portocala, R., and Ionescu, N. I. (1962). "Microscopia Electronica in Biologie si Inframicrobiologie." Acad. Republ. Populare Romine, Bucharest.

Reis, T. (1951). "Le Microscope Électronique et Ses Applications." Presses Univ. de France, Paris.

Ruhle, R. (1949). "Das Electronenmikroskop." Schwab, Stuttgart.

Selme, P. (1963). "Le Microscope Électronique." Presses Univ. de France, Paris.

Terriere, G. (1962). "Microscopie Électronique." Tech. Ingr. Mesures et Anal., Paris.

Wainrib, E. A., and Miliutin, W. I. (1954). "Elektronenoptik." EBV Verlag Technik, Berlin.

Wischnitzer, S. (1970). "Introduction to Electron Microscopy." Pergamon Press, Elmsford, New York.

Wyckoff, R. W. C. (1949). "Electron Microscopy: Technique and Applications". Interscience, New York.

———. (1958). "The World of the Electron Microscope." Yale University Press, New Haven.

Zworykin, V. K., Morton, G. A., Rauberg, E. G., Hillier, J., and Vance, A. W. (1945). "Electron Optics and the Electron Microscope." John Wiley, New York.

Review Articles

Hartman, R. E., and Hartman, R. S. (1965). "Elimination of Potential Sources of Contaminating Material." *Lab Invest. 14,* 1147.

Reynolds, G. T. (1968). "Image Intensification Applied to Microscope Systems." *Adv. Opt. Electron Micr. 2,* 1.

Robertson, J. D. (1959). The Electron Microscope. *In:* "Tools of Biological Research," Vol. 1 (Atkins, H. J. B., ed.), Thomas, Springfield, Illinois.

Seigel, B. (1964). The Physics of the Electron Microscope. *In:* "Modern Developments in Electron Microscopy." (Seigel, B. M., ed.), Academic Press, New York.

Valentine, R. C. (1966). "The Response of Photographic Emulsions to Electrons. *Adv. Opt. Electron Micr. 1,* 180.

Zeitler, E., and Bahr, G. F. (1960). The Interpretation of Contrast in an Electron Micrograph. *Sci. Instr. News* 5, 5.

2 Electron Microscopy of Selectively Stained Molecules

T. Koller, M. Beer, M. Müller, and K. Mühlethaler

Laboratory for Electron Microscopy I. Swiss Federal Institute of Technology, Zürich, Switzerland, and Department of Biophysics, The Johns Hopkins University, Baltimore, Maryland.

INTRODUCTION

The electron microscope has contributed extensively to our knowledge of the structure of biological systems, especially at the subcellular level. Electron microscopy has provided considerable information on the shape and size of molecules, but little on the distribution of chemical groups within them. Progress in this area, however, would provide an important approach to many problems in molecular biology. The purpose of this chapter is to summarize some of the work that has been directed toward this goal, especially the efforts devoted to the problem of recognizing the sequence of nucleotides of nucleic acids. First, some of the chemical problems of attachment of stain to specific sites on a macromolecule are reviewed, and then a discussion on the preparation of nucleic acid molecules for high-resolution electron microscopy and the questions related to the identification of small structural details and markers are presented.

INTRAMOLECULAR SELECTIVE STAINING

Nucleotide Selective Reagents

Guanine Specific Reactions

Moudrianakis and Beer (1965a and b) showed that diazotized 2-amino p-benzene disulfonic acid coupled with GMP (guanosine monophosphate) about 100 times more rapidly than with any of the other common nucleotides. They were able to show that it was possible to produce an altered DNA in which approximately 80% of the GMP residues were converted to the diazonium addition product while conversion of the other nucleotides was less than 5%. Although no detailed study of the breakdown of the molecules was carried out, it was clear from subsequent electron microscopy that the DNA retained a high molecular weight. Also, the staining of such molecules with uranyl acetate gave satisfactory visibility of single strands suggesting additional binding of UO_2 cations by the marked G sites.

Erickson and Beer (1967) examined the reactions of the same diazonium compound with RNA from the bacteriophage MS-2. In agreement with the results of Kössel (1965) for diazosulfanilic acid, they showed that the selective binding of the dye to the GMP residue was a

2-step process. Initially, at pH 9.0 the dye was bound extensively to both GMP and AMP (adenosine monophosphate) and to a lesser extent to CMP (cytidine monophosphate). After subsequent mild acid hydrolysis, the attachment remained only to GMP. The reagent was less selective for RNA than it was for DNA, presumably, because some coupling occured also with OH groups of the ribose moiety. In RNA polymers 60% of the GMP residues and ~15% of the other nucleotides were labeled. Erickson and Beer (1967) also examined the fragmentation accompanying the coupling reaction. When 30% of the GMP residues reacted, 5% of MS-2 RNA remained unbroken, but with 50% labeling only 1% remained intact.

Ulanov and Molysheva (1967) and Ulanov *et al.* (1967) proposed a new approach for making the guanine residues in DNA. DNA reacted with N-β-(chloroethyl)-NN-diethylamine, which is known to react in the free state with phosphotungstic acid (PTA) to form a precipitate. Correspondingly, it was found that DNA from the bacteriophage T_2 after treatment with the alkylating agent could be stained with PTA.

Cytosine Specific Reactions

Semicarbazide and certain substituted semicarbazides couple (at pH 4.2) to cytidine and not to the other nucleotides (Hayatsu and Ukita, 1964; Kikugava *et al.*, 1967). Similar coupling reactions occur with RNA in which the CMP residues have been converted to the expected addition product while the other nucleotides remain unaltered.

Gal-Or *et al.* (1967) examined a number of acyl hydrazides in an attempt to find an analogous coupling involving compounds possessing anionic groups for subsequent attachment of heavy metal cations. 3,4,-Dicarboxybenzoyl hydrazine (DCBH) reacted under conditions very similar to those necessary for the semicarbazide reaction. The authors were able to label 70% of the nucleotides in poly C and produced an altered macromolecular sample of RNA (MS-2 RNA) in which 80% of the CMP residues were involved in the coupling reaction and only 10% of the other nucleotides were modified.

Another approach is the incorporation of thiosubstituted nucleotides into polynucleotides synthesized *in vitro*. A new analogue of poly cytidilic acid, poly (s^2C) has been synthesized from 2-thiocytidine-5′-diphosphate (s^2CDP) by polynucleotide phosphorylase from *E. coli* (Scheit and Faerber, 1971). Certain heavy metal compounds (e.g., p-hydroxymercuriphenyl-sulfonate) react specifically with the thio groups (Faerber, personal communication). This procedure has the advantage of a very high specificity.

Thymine Specific Reactions

Bahr (1954), when testing osmium tetroxide with many biologically important materials, noticed that thymine formed a black compound, whereas the other nucleic acid bases did not. Burton and Riley (1966) and Beer *et al.* (1966) extended these studies and found that among the deoxynucleotides only dTMP (deoxythymidine monophosphate) reacted readily; dCMP (deoxycytidine monophosphate) reacted approximately 10 times more slowly than dTMP and dAMP (deoxyadenosine monophosphate), and dGMP (deoxyguanosine monophosphate) did not react at all. DNA reacted also to give a product in which primarily the dTMP residues were altered. Furthermore, Beer *et al.* (1966) showed that one mole of osmium reacted with one mole of nucleotide. However, they did not succeed in isolating an addition product, presumably because it was unstable.

3′,-5′-Diacetyl thymidine (DAT) treated with osmium tetroxide in benzene solution yielded an addition product containing 1 mole of osmium per mole of DAT (Highton *et al.*,1968). This osmate ester, combined with two moles of cyanide ions, gave a stable, water soluble product that in electrophoresis moved toward the anode. The same product could also be formed in an aqueous solution containing both osmium tetroxide and CN^-. The same procedure modified 80% of the TMP (thymidine monophosphate) residues of denatured DNA, and for each mole of reacted TMP there were bound 2 moles of cyanide and 1 mole of osmium. The extent of reaction of the other nucleotides was less than 10% (Highton *et al.*, 1968).

The physico-chemical properties and the morphology of bacteriophage ϕX-174 DNA were examined after reaction with osmium tetroxide and KCN by Di Giamberardino *et al.* (1969). It was shown by a density study in a CsCl or $CsSO_4$ gradient that the molecules became increasingly dense as the reaction proceeded, confirming the binding of the heavy osmium metal. The sedimentation velocity of the reacted DNA continued to be sensitive to ionic strength, suggesting that the molecules were not crosslinked. The latter was confirmed by direct electron microscopy of ϕX-174 DNA. The mildness of the thymine labeling reaction was shown by the fact that many of the molecules were unbroken and remained closed circles. Thus it appears that there is a set of reactions that can convert DNA into a macromolecule that remains largely unbroken and linear, but in which the majority of the thymine residues are converted to an addition product containing 1 osmium atom and 2 cyanide ions.

Recently, Subbaraman *et al.* (1971) studied the reaction of osmium tetroxide-pyridine complexes with nucleic acid components. They found that thymine and its derivatives react rapidly in aqueous solution to form stable bis (pyridine) osmate esters. The osmium tetroxide-pyridine system combined the advantages of rapid rates of reaction, satisfactory hydrolytic stability of the product esters and convenient flexibility with respect to the types of substituents that can be built into the ligand.

The binding of mercury to thiosubstituted nucleotides (see above) should also be applicable to the marking of T or U residues (Faerber, personal communication). Using *Bacillus subtilis* DNA polymerase, the base analogue 4-thiothymidine (s^4T) can be incorporated into the polynucleotide d(A-s^4T)·d(A-s^4T)(Lezius and Rath, 1971). Similarly, the polynucleotide phosphorylase from *E. coli* is able to polymerize poly 4-thiouridylic acid (poly s^4U) from 4-thiouridine-5′-disphosphate (s^4UDP) (Simuth *et al.*, 1970) and poly 2-4-dithiouridylic acid (poly s^2s^4U) from 2-4-dithiouridine-5′- diphosphate (s^2s^4UDP) (Faerber and Scheit, 1970).

Other Nucleotide "Staining" Reactions

Moshkovski *et al.* (1968) and Sakharenko *et al.* (1967), on the basis of spectroscopic evidence, concluded that K_2PtCl_4 reacts primarily with the adenine residues of DNA. When DNA molecules were stained with solutions of K_2PtCl_4 many dark clusters were found along the strand. These were interpreted to be the result of "physical development," as in the photographic process, catalyzed at the adenine sites where K_2PtCl_4 reaction was believed to have occurred.

Gibson (Gibson 1969; Gibson *et al.* 1971) examined the reactions of $HAuCl_4$ with the nucleotides, nucleosides, and deoxynucleotides. This ion in aqueous solution undergoes partial hydrolysis in which one or more chloride ions can be replaced by $(OH)^-$ ions. The corresponding equilibrium is dependent upon the concentration of chloride and $(OH)^-$ ions. The gold equilibrium system of ions reacted with all the nucleotides but at very different rates and stoichiometries. At pH near 5.6, AMP reacted more rapidly giving a product with a stoichiometry of 2 gold ions for 1 nucleotide.

GMP reacted three times more slowly and CMP 20 times more slowly than AMP. TMP was altered most slowly of all. The first three compounds gave a product containing 2 gold ions per nucleotide, whereas TMP contained 1. After a considerable length of time had passed, gold metal precipitated from the GMP and CMP reaction mixtures but not from the other two. The gold-dAMP, -dGMP, and -dCMP products were eluted early from G-50 Sephadex and had high sedimentation coefficients, whereas the gold-dTMP product was eluted late and sedimented more slowly. These results suggested extensive polymerization in the former products. With polynucleotides the reactions with gold (III) proved to be slower and less selective.

Fiskin and Beer (1965) studied the binding of Hg-p-disodium-p-hydroquinone diacetate (HgHDA) to the nucleotides. This binding was a reversible metal ligand interaction. They found that at pH values above neutral and at appropriate concentrations of HgHDA, G and U would extensively bind the metal, but A and C would not. Using polynucleotides at pH 8.5, poly U bound HgHDA to the extent of 62% when the concentration of the reagent was 1.7×10^{-3} *M*, while under the same conditions poly A and poly C bound negligible quantities. The binding of poly G could not be determined directly, but from the binding constant this could be estimated to be somewhat less than that for poly U. This reaction system involved the reversible binding of the reagent, and in this respect was less favorable than the earlier mentioned selective reagents.

Amino Acid Selective Reagents

Kühn *et al.* (1958) studied collagen fibrils stained with PTA. In such a system 2 types of binding occured. One was stable, could not be readily washed out, and was not blocked if the amino groups were acetylated, nor if the carboxyl groups were methylated. This binding was stoichiometrically equivalent to the number of arginine residues, increased by the formation of new guanidine bonds, and was attributed by these workers to arginine. The other binding was much weaker. It was readily washed out, was blocked when ε-amino groups were acetylated, and was attributed to binding by lysine and hydrolysine. Electron microscopy of PTA-stained collagen fibrils indicated that the tightly bound component lead to recognizable bands but the loosely bound component appeared to give only a diffuse increase in density.

Electron microscopy was also used to study the positions of the acidic and basic groups in collagen. To accomplish this, Hodge and Schmitt (1960) took advantage of the remarkable lateral aggregate that could be formed from the linear rodlike tropocollagen molecule. In these so-called SLS (segment long-spacing) bundles the molecules were adjacent to each other in such a way that the corresponding amino acids of the identical macromolecules lay in parallel planes that were perpendicular to the long axis of the component molecules. The planes containing stained amino acids appeared as dense bands in the electron microscope. Selective stains could thus indicate the position of the corresponding amino acids.

Hodge and Schmitt (1960) examined PTA-stained SLS bundles and found a number of dense bands of varying intensity. However, no blocking experiments had been done on this system and so it was not clear which bands were to be associated with ϵ-amino groups and which with guanidine groups. The band pattern generated by uranyl acetate was also examined by these workers. It had many similarities, as well as some differences, compared with the bands obtained with PTA. The uranyl acetate bands were attributed to negatively charged groups. However, it was not known if the ATP molecules involved in the formation of the SLS aggregate were also stained.

Formanek and Formanek (1970) stained the murein *sacculi* of bacterial cell walls. The murein *sacculi* of the cell walls of *Spirillum serpens* consist of a 2-dimensional network of polysaccharide chains covalently connected with oligopeptides. The single free amino groups of the oligopeptides were specifically stained in a 1-step reaction with stoichiometric amounts of the following heavy atom compounds: 3-hydroxymercury-4-hydroxy-5-carboxybenzenesulfonic acid, anhydro-3-hydroxymercury-4-toluene-sulfonic acid, Zeise's salt, platinum blue, and tribromoacetaldehyde.

In another group of experiments 2-step staining reactions were used. First, sulfur compounds were bound to the free amino or carboxyl groups of the oligopeptides. Propylene-sulfide, N-acetyl-DL-homocysteinethiolactone, or S-acetylmercaptosuccinic anhydride were used to insert the sulfur atoms, which then could be reacted specifically with stoichiometric amounts of 2, 4, 6,-tris-acetoxymercury-3-acetaminotoluene. Stained and unstained murein *sacculi* were examined by high-resolution electron microscopy. Unstained murein sacculi could hardly be recognized in bright field and did not expose any structural details. Stained sacculi, however, showed distinct contrast and revealed a network pattern in which the stain was localized as discrete particles of about 5 Å diameter in periods of 10 Å.

In these elegant studies two main difficulties remained unsolved. It is almost impossible to spread a single layer of a cell wall on a support film. Isolated membranes usually form vesicles in solution that collapse when deposited and dried on an electron microscope grid. Therefore, the pattern observed in the microscope is a superposition of the individual patterns of the two cell walls lying on top of each other. Second, the specifity of these staining reactions is limited to a special case such as the murein where the amino and carboxyl groups are the only probable reaction sites. For example, in lipid containing membranes, the heavy metal compounds should react also with the double bonds and should, therefore, also stain the lipid containing structures (F. Kopp, unpublished data).

PREPARATION OF NUCLEIC ACID MOLECULES FOR HIGH RESOLUTION ELECTRON MICROSCOPY

Since denatured filamentous nucleic acid molecules can be regarded as being two-dimensional, they are probably the most favorable system to develop methods for structural analysis by intramolecular selective staining. The following discussion pertains to studies related to nucleic acid molecules.

The ideal technique for specimen preparation should yield a uniform distribution of unbroken, unaggregated macromolecules over the whole grid in such a manner that randomly taken pictures could be analyzed quantitatively. Various methods have been devised, each of which has certain advantages as well as disadvantages. Hall (1956) proposed spraying the macromolecules on freshly cleaved mica. On the smooth and hydrophilic surface the molecules tend to spread. Spraying can also be performed onto hydrophilic carbon films mounted directly on grids (Bartl, *et al.*, 1966 and 1970). Although the latter method is more suitable for subsequent staining, its drawback is that large linear molecules, such as nucleic acids, are often broken. For preparing the above mentioned support films, the reader is referred to Hayat (1970) and Henderson and Griffiths (1972).

The most frequently applied methods for nucleic acids are the protein monolayer techniques (Kleinschmidt and Zahn, 1959; Lang *et al.*, 1964; Mayor and Jordan, 1968). A monolayer of basic protein is formed on the surface of water or a salt solution. The nucleic acid molecules, which are either spread together with the protein or diffused from solution to the surface, become fixed in the monomolecular layer. The mixed film (protein and nucleic acid) is then transferred to the grid. This method is very successful for demonstrating the shape and length of double- and single-stranded nucleic acids. The distribution of the molecules over the grid is regular and aggregation is easily recognized in a homogenous sample. The observed configuration of the molecules probably reflects their conformation in solution (Lang *et al.*, 1967). However, the fact that protein is always associated with the nucleic acid strands appears to be a limiting factor for resolving more detailed architecture of the latter.

The Streaking Method

Beer (1961) proposed a method that leads to unbroken straight single nucleic acid molecules. DNA is transferred to the support film as a grid is streaked along the surface of a solution. The films can be made either of a copolymer of styrene and vinylpyridine that has weakly basic ion exchange properties, or evaporated carbon (Hayat, 1970). The DNA is adsorbed as parallel straight molecules on the film as the grid is streaked along the surface of a nucleic acid solution for a distance of ~ 2 cm and at a speed of 2 cm/sec. The direction of streaking is noted on the grids. The excess liquid is immediately drawn off by touching the edge of the grid to absorbent filter paper so that the flow direction of receding excess liquid is the same as the direction of streaking. The abundance of strands transferred depends upon the pH, the ionic strength, and the concentration of the DNA solution. Satisfactory conditions are: DNA concentration 1 to 3 μg/ml and buffer 0.1 *M* ammonium acetate, pH 6.1.

Similar results can be obtained by putting a drop of double- or single-stranded nucleic acid (2–5 μg DNA/ml in 0.02 *M* salt) on a grid carrying a carbon film. The liquid is then removed by touching the edge of the grid with a piece of filter paper.

Using the streaking method, Highton and Beer (1963) demonstrated unbroken, single-stranded RNA molecules from tobacco mosaic virus (TMV) particles denatured in the presence of 4% formaldehyde. Grids covered with carbon films were streaked as described by Beer (1961) across the surface of 1 ml of RNA solution (25μg/ml of RNA in 0.01 *M* sodium phosphate buffer at pH 6.8 with 4% formaldehyde, 10^{-4} *M* EDTA). Separate, straight strands, oriented parallel to one another and to the direction of streaking were observed on the grids. The number of molecules deposited on the carbon films increased with increasing ionic strength and RNA concentration. Below 0.01 *M* ionic strength almost no strands were observed. Length measurements indicated that most of the molecules were either unbroken or suffered only one break. From the known molecular weight of TMV RNA the base spacing was calculated to be between 5 and 6 Å, indicating a certain amount of stretching of the strands during the deposition on the grid.

A similar technique was described by Bendet *et al.* (1962). The DNA strands were obtained by mild alkali treatment of T_3 virus particles. One or 2 ml of a solution containing 0.5 to 1 μg/ml T_3 DNA in 0.1 *M* ammonium acetate and 10^{-4} *M* EDTA was placed on a collodion covered microscope slide. The slide was then tilted to an angle of 45° and the excess liquid allowed to flow off onto bibulous paper. After shadowing, the films were floated on water and picked up on grids. With this procedure Bendet *et al.* (1962) found remarkably little variation in the length of T_3 DNA molecules.

The techniques described above suffer from three major drawbacks. (1) Since the strands are oriented in one direction, the study of the contour of the molecules is difficult; for example, circular nucleic acid molecules are seldom recognizable as ring shaped. Generally, they appear as two intertwisted, parallel strands or as a "double" strand. (2) When the concentration of nucleic acid is high, the molecules tend to form multistranded, longitudinal aggregates (Erickson and Beer, 1968). (3) In order to reduce the frequency of aggregated molecules, it is advisable to work at the lowest possible nucleic acid concentration (0.5 to 2 μg/ml in 0.01 *M* salt). Under these conditions, however, the distribution of molecules over the grid is irregular and it is difficult to find the individual strands.

The Adsorption Method

Nucleic acid molecules can be adsorbed very efficiently directly to carbon support films that have been positively charged by pre-treatment with quaternary ammonium salt (Koller *et al.*, 1969a; Koller *et al.*, 1969b). The method is quick and is suitable for the retrieval of nucleic acid molecules from a variety of conditions, such as varying ionic strength of the solvent and temperature of the sample. In the original procedure, DNA was adsorbed to hydrophobic quaternary ammonium salt treated grids by briefly touching the grids to the surface of the nucleic acid solution at an angle of $\sim$ 30°. The films remained hydrophobic and virtually none of the solution adhered to the grid. The method was tested using the double-stranded DNA of bacteriophage T_3 and the circular single-stranded DNA of bacteriophage ϕX-174.

There were no differences in contour lengths and configuration of T_3 DNA molecules, when the grids were pretreated with different concentrations (10^{-3} *M* to 10^{-8} *M*) of the quaternary ammonium salt solution. Below a concentration of $\sim 10^{-8}$ *M* the grids behaved essentially like untreated standard carbon support films. The appearance of ϕX-174 DNA, however, was highly dependent upon the quaternary ammonium salt concentration used during the pre-treatment of the grids. At high concentrations the molecules appeared heavily tangled even after prior melting of the DNA in the presence of formaldehyde at 60°C. When the concentration of the quaternary ammonium salt was lowered, clearly ring-shaped molecules appeared. Their contour lengths, however, did not correspond to the values reported for the Kleinschmidt spreading procedure by Freifelder *et al.* (1964). Contour lengths of only 1.0 to 1.3 μm were obtained. The pre-treatment of the films had to be carried out with $\sim 10^{-8}$ *M* solution of quaternary ammonium salt in order to obtain results comparable to the work by Freifelder *et al.* (1964). The strands then obtained were clearly ring shaped, but very thin and extremely difficult to detect after rotary shadowing. Better visibility was achieved by staining with hexatantalum dodecabromide diacetate (Koller *et al.*, 1969).

Although the mean contour lengths of the molecules tested agreed with the values reported in the literature, the variation in lengths was higher. It seemed from subsequent work that the time of adsorption was too short and that during attachment of the molecules stretching might be exerted on certain strands. Therefore, the procedure was modified and the following technique (Fig. 2.1) was adopted.

Standard carbon films produced in a conventional evaporator are hydrophobic. These grids are subsequently placed in a glow discharge to render them hydrophilic. In order to load the surface of the supporting membrane with positive charges, the grids are floated overnight on alkyl dimethyl-benzyl ammonium chloride (BAC), which is commercially available as a disinfectant under the name of Zephiran (Winthrop

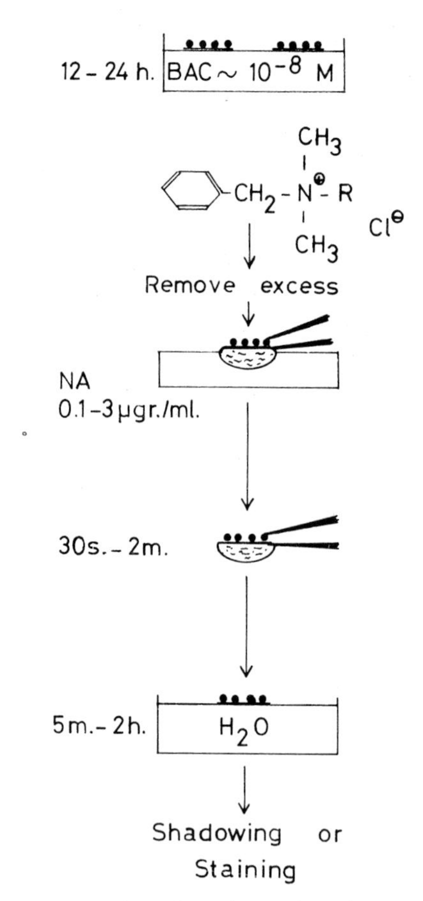

Fig. 2.1 Adsorption method: adsorption of nucleic acid molecules onto positively charged support films.

Laboratories). After the use of diluted solutions (in most cases $10^{-5}\% = \sim 10^{-8}\ M$), no washing of the grids is performed; the excess liquid is removed from the membrane with a filter paper. The grids are then ready to accept the nucleic acid and are used immediately. DNA is adsorbed to the films by briefly touching the grids horizontally to the surface of the nucleic acid solution.

On removing the grid from the surface, a droplet of the nucleic acid solution remains hanging on the hydrophilic grid. The grid is kept in this position and the molecules are allowed to attach to the carbon film for ½–2 min, depending upon the length of the molecule tested. The grids are transferred into clean water and washed for several minutes. After washing, the grids are either passed through ethanol, dried, and then rotary shadowed, or they are stained with uranyl acetate (Gordon and Kleinschmidt, 1968) or with dodeca-bromo-hexatantalum diacetate (Koller *et al.*, 1969).

The following nucleic acids were examined in our recent experiments: double-stranded circular polyoma virus DNA (16s) and single-stranded bacteriophage Qβ RNA.* The nucleic acid concentrations used varied between 0.1 and 2.5 μg/ml. Usually, the higher the ionic strength of the solvent, the lower was the required concentration of nucleic acid molecules. The composition of the solvent was up to 0.01 M salt (KCl or ammonium acetate), with or without 20% sucrose, and with or without 50% formamide.

Double-Stranded Nucleic Acid Molecules

Grids treated with $10^{-5}\%$ BAC solutions adsorbed an adequate distribution of DNA molecules when touched to the surface of a solution of 0.5 to 1.0 μg DNA/ml in 0.01 M salt. When the concentration of BAC was less than $10^{-5}\%$, grids behaved essentially as untreated carbon films, and ~ 10 times less molecules were adsorbed to the grids.

Generally the molecules were distributed and separated from each other. There appeared to be little of the lateral aggregation and branching of the strands found frequently on untreated carbon films. Overlapping molecules were easily recognized, and therefore the identification of a particular strand as being an isolated molecule was usually unambiguous. The strands were usually more tangled than expected from the appearance of double-stranded DNA prepared with the classical Kleinschmidt method (1959). Figure 2.2 shows a preparation of circular

*The polyoma DNA was a kind gift from Dr. R. Portmann and the Qβ RNA was generously supplied by Professor Ch. Weissmann.

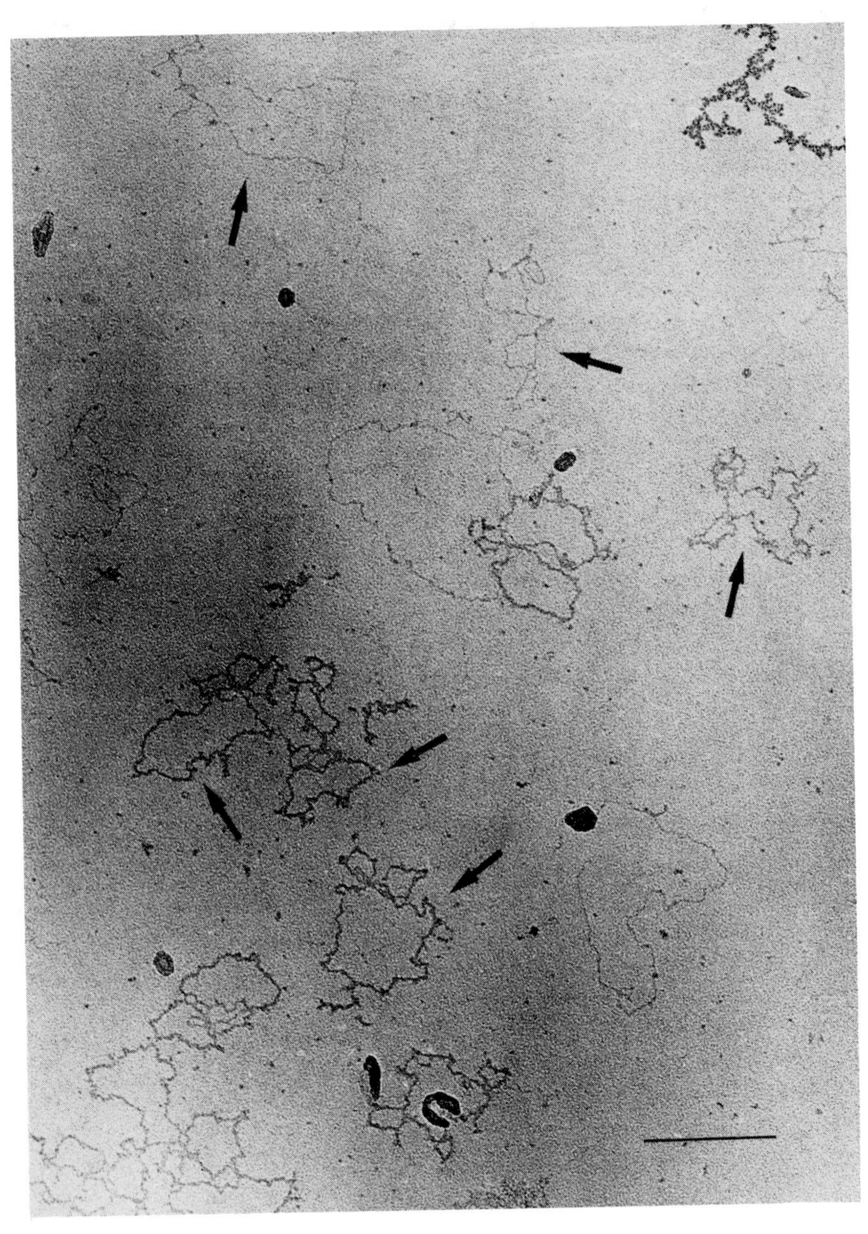

Fig. 2.2 DNA (16S) of polyoma virus particles prepared by the adsorption method from a solution containing 0.01 M ammonium acetate and 20% sucrose. Staining was accomplished with uranyl acetate in acetone. Several circular molecules are observable, some of which are tangled while others are unfolded. Note difference in the stainability of the molecules. Calibration equals 1.0 μ.

polyoma DNA (16s) stained with uranyl acetate and adsorbed from a solution containing 0.01 *M* ammonium acetate and 20% sucrose. The open circles that are quite frequently seen with the spreading technique (Crawford *et al.*, 1966) are not as often observed.

It appears that the basic protein used in the Kleinschmidt method (1959) helps to untangle the strands and leads to a conformation that is very convenient for analysis. Figure 2.3 shows a histogram of a prep-

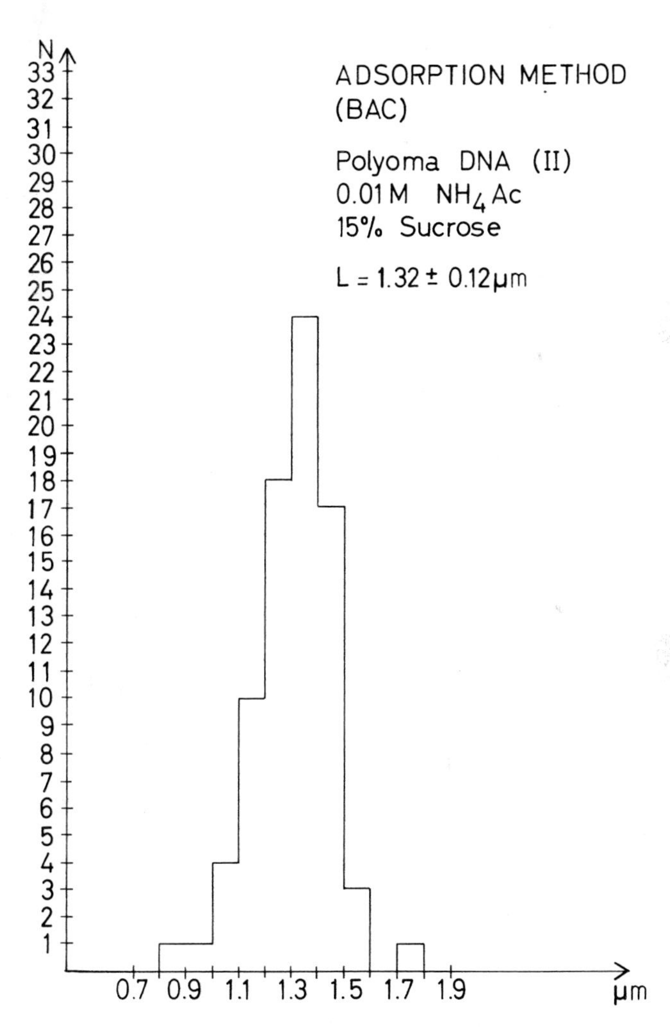

Fig. 2.3 Histogram of polyoma virus DNA (16S) prepared by the adsorption method from a solution containing 0.01 M ammonium acetate and 20% sucrose. The mean contour length found is 1.32 ± 0.12 μ.

aration of polyoma DNA (16s) in 0.01 *M* ammonium acetate and 20% sucrose. Sucrose was added to the solvent to untangle the strands. The mean length of the strands obtained was 1.32 $\pm$ 0.12 μm. The mean value and the spread correspond to the data reported in the literature employing the spreading technique (Crawford *et al.*, 1966). Barblan and Hirt (1969) report 1.52 $\pm$ 0.09 μm for a salt concentration of 0.01 *M* and 1.32 $\pm$ 0.15 μm for 10 *M* salt. From experiments carried out with the filter method (see following), it seems that the addition of sucrose to the solvent slightly reduces the contour length that might lead to conditions similar to the 10 *M* salt of Barblan and Hirt (1969).

Single-Stranded Nucleic Acid Molecules

Qβ RNA strands were adsorbed to grids pretreated with 10^{-5}% BAC from a solution containing 2.5 μg/ml RNA, 2 mM EDTA, and 50% formamide at room temperature. The formamide was added in order to reduce the extent of formation of secondary structure through hydrogen bonding. Figure 2.4 shows representative Qβ RNA strands prepared under these conditions. Untreated carbon films bound virtually no nucleic acid as was clear from the results of Highton and Beer (1963). Figure 2.5 demonstrates a histogram obtained measuring unselected molecules, adsorbed on hydrophilic BAC (10^{-5}%) grids. A mean value of 1.14 $\pm$ 0.12 μm was obtained, which corresponds to a mean internucleotide distance of about 3.15 Å. This spacing between the nucleotides agrees with the values found in the closely related RNA from bacteriophage R-17 (Sinha *et al.*, 1965).

Discussion

The contour lengths and their variations of double- and single-stranded nucleic acid molecules prepared by the adsorption method are close to what one usually expects with the spreading technique. Satisfactory results with little variation in length have been obtained with the adsorption method by Sheperd and Wakeman (1971) demonstrating the DNA of cauliflower mosaic virus particles. It is pointed out that the measurement of the contour lengths with the adsorption method is more difficult than with the Kleinschmidt procedure because the strands are more tangled (also reported by Gordon and Kleinschmidt, 1969, and Dubochet *et al.*, 1970 and 1971) and the width of the double-stranded molecules (also shown by Dubochet *et al.*, 1970 and 1971) is only 20 to 25 Å; whereas the width of the double-stranded DNA prepared with the

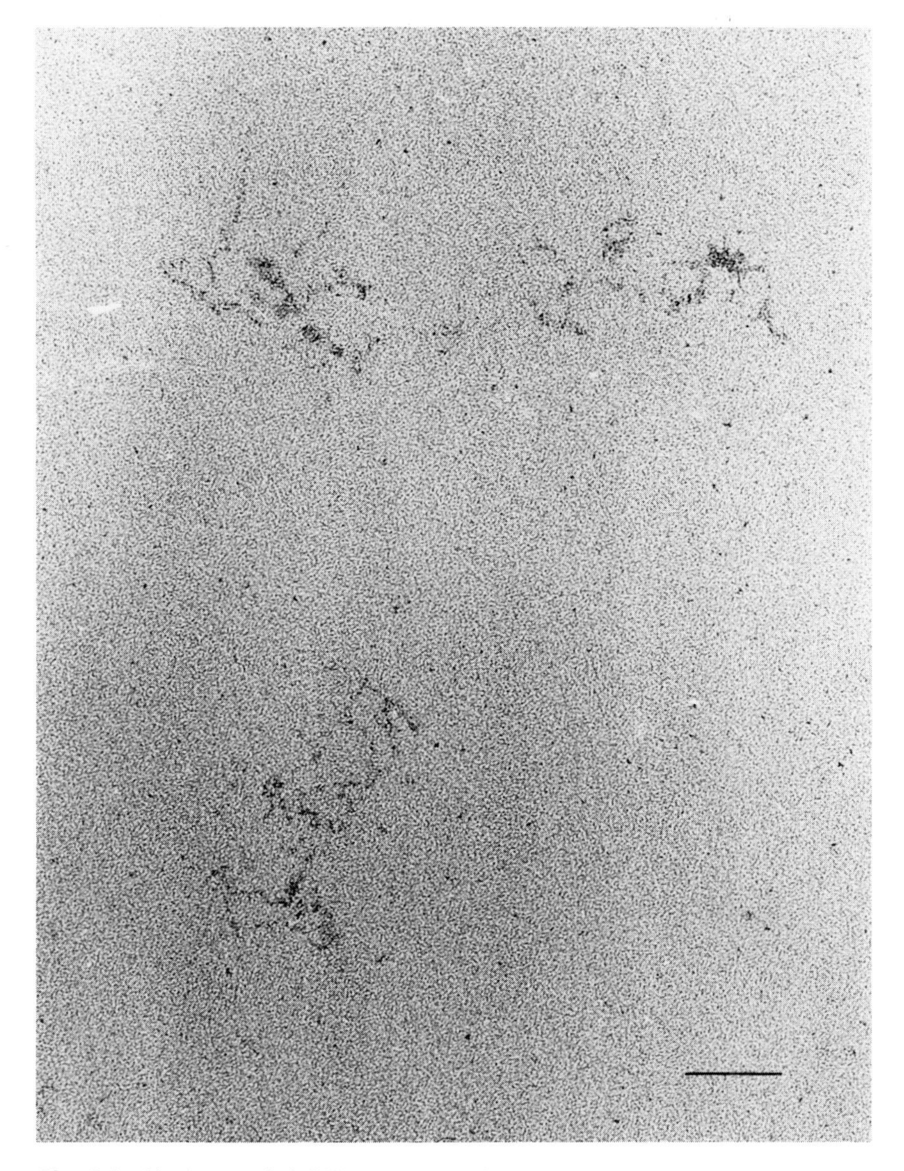

Fig. 2.4 Single-stranded RNA from bacteriophage Qβ prepared by the adsorption method from a solution containing 2 mM EDTA and 50% formamide. Calibration equals 0.1 μ.

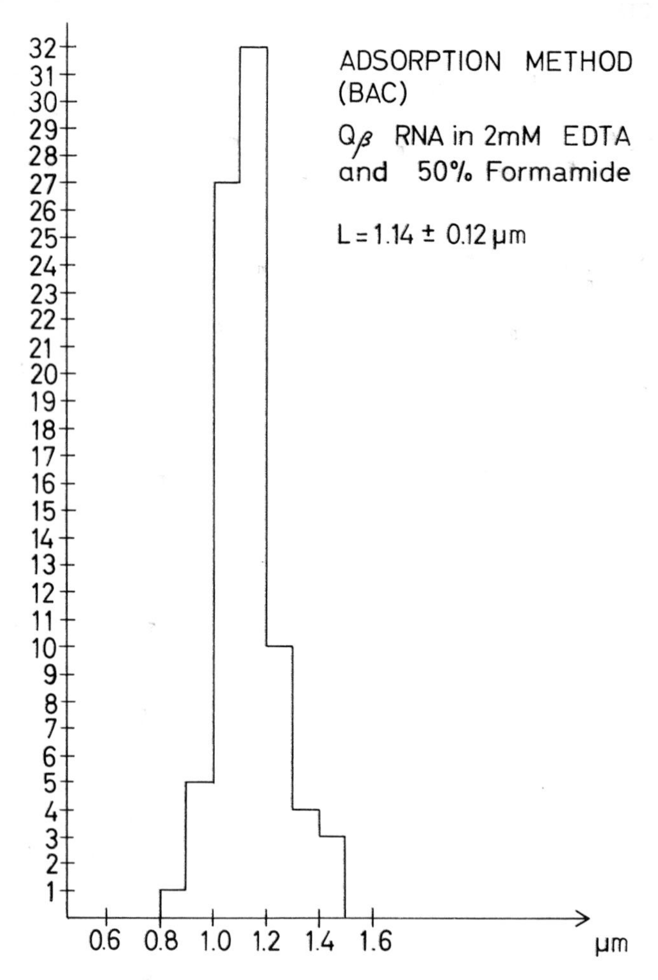

Fig. 2.5 Histogram of Qβ RNA prepared by the adsorption method from a solution containing 2 mM EDTA and 50% formamide. The mean contour length obtained was 1.14 $\pm$ 0.12 μ.

spreading procedure (Kleinschmidt and Zahn, 1959) is in the order of 100 Å (Dubochet *et al.*, 1970; Dubochet *et al.*, 1971).

Contrast enhancement and the visualization of nucleic acid molecules without a carrier protein after shadowing or staining is therefore much more difficult, especially on low magnification survey micrographs in bright field. It seems from the work by Ottensmeyer (1969) and Dubochet (this volume) that dark field microscopy might be of some help. To follow, on survey micrographs, the exact meandering of a strand prepared by the adsorption method is, therefore, not always unambiguous.

It appears that using hydrophilic grids with a carbon film to which the strands are allowed to attach slowly is more gentle and gives more reproducible results than the earlier procedure (Koller *et al.*, 1969) using hydrophobic grids. It is inherent in the use of hydrophobic films that virtually no liquid is left after the grid has been touched to the surface of the nucleic acid solution. Therefore, during the attachment of the molecules, stretching and breaking might be exerted on certain strands.

Since the report on the adsorption method (Koller *et al.*, 1969), different techniques based on the same principle have appeared in the literature. Gordon and Kleinschmidt (1969) utilized the ionic exchange properties of muscovite-mica. Upon cleavage of the mica, potassium ions are exposed and replaced by aluminum ions. The aluminum-mica was placed in a dilute DNA solution and allowed to remain undisturbed for about 30 min. After rinsing and drying of the mica, the contrast of the molecules was enhanced by rotary shadowing. Finally, carbon was evaporated onto the preparation. The replica was floated off onto water and picked up on grids.

As in the streaking method (Beer, 1961) the extent of adsorption was dependent upon the ionic strength of the DNA solution. Below 0.005 *M* salt, very little DNA was bound to the mica. At high ionic strength, however, the DNA had a convoluted, tangled appearance that was especially pronounced with longer DNA filaments. Unbroken circular single-stranded and double-stranded replicative (RF II) DNA molecules of bacteriophage ϕX-174 could be demonstrated, although there was an increased tendency toward breakage when compared with the spreading technique. The contour lengths of samples of selected circular and linear RF ϕX DNA molecules adsorbed to mica showed larger coefficients of variation (2 to 6%) than those of circular molecules examined by the protein monolayer technique (1 to 2%).

Dubochet *et al.* (1970, 1971 and this volume) deposited basic radicals onto thin carbon films in a reducing atmosphere produced by a glow discharge of amylamine vapors. The authors developed a sophisticated apparatus with a particularly clean vacuum for producing the glow discharge and controlling the inlet of amylamine into a bell-jar. A drop of the aqueous suspension of the particles to be observed was applied to the amylamine-carbon films. A period of ½ to 5 min was allowed for adsorption of the specimen. The excess liquid was removed by pressing the grid on blotting paper.

In contrast to the methods described by Beer (1961) and Gordon and Kleinschmidt (1969) the efficiency of adsorption increased from zero to a maximum when the salt concentration decreased from 0.1 *M* to 5×10^{-3} *M*. The efficiency is decreased by increasing the DNA concentration. The effect of this procedure on the contour length of DNA was examined, using the double-stranded replicative form of phage fd and the circularized form of phage lamda. The molecules prepared by this method were 10 to 20% shorter than after using the spreading technique (Kleinschmidt and Zahn, 1959). The variations were in the order of 3 to 7%, which compares well with the usual results by the Kleinschmidt procedure. With single-stranded nucleic acid molecules no reliable results could be obtained. In the preparation of phage fd, gamma globulins, and RNA polymerase, however, very good distribution of the particles was noticed.

The Centrifugation Method

A different approach, used mainly for visualization of nucleic acid-protein-complexes, was introduced by Miller, Jr., and Beatty (1969a–c), Miller, Jr., *et al.* (1970), and Miller, Jr., and Hamkalo (1971). The specimen, for example, the contents of an osmotically shocked cell, was immediately centrifuged in an ordinary table centrifuge through a 0.1 *M* sucrose plus 10% formalin (pH 8.5) cushion onto carbon coated grids. The grids were rinsed in 0.4% Kodak Photo-flo, dried, and stained with PTA and uranyl acetate (both stains in 70% ethanol). The grids were then washed and air-dried.

By using this method, Miller, Jr., and Beatty (1969a–c) were able to show DNA in the process of transcription. This work made use of a system in eukaryotic cells in which extrachromosomal DNA coding only for ribosomal RNA is engaged in very intensive transcription. Miller, Jr., and Beatty (1969a) also demonstrated nucleolar genes of *Triturus viridescens* oocytes engaged in transcription of ribosomal precursor RNA. The active genes, separated by inactive segments of DNA, were engaged in the simultaneous transcription of about a hundred RNA molecules; the shortest of its fibrils was attached to a polymerase molecule that had just initiated transcription, and the longest to one that had just completed synthesis of its RNA.

The above method was employed by Miller, Jr., *et al.* (1970) to visualize the morphology of active structural and putative ribosomal RNA genes after lysis of fragile *E. coli* cells. The diameter of the *E. coli* DNA strand was ~ 40 Å. This suggested that either the DNA was combined with some nonhistone proteins *in vivo* or proteins became attached rather uniformly to the DNA during or after isolation. Furthermore, it could be seen that the ribosomes were attached to the genome on mRNA molecules as monomers and in polyribosomes, and again genetically active and inactive portions of the *E. coli* chromosome could be distinguished.

The Filter Method

As will be shown, the features of "low noise" support films prepared from aluminum oxide are extremely sensitive to contamination. One source of contamination originates from the steps of the adsorption procedure for the specimen. Therefore, any carrier compound, such as the quaternary ammonium salt in the adsorption method, should be avoided, since it affects the noise level on high-resolution micrographs. We have developed a method that allows the preparation of unbroken and unaggregated nucleic acid molecules, and also does not eliminate the features of "low noise" support films.

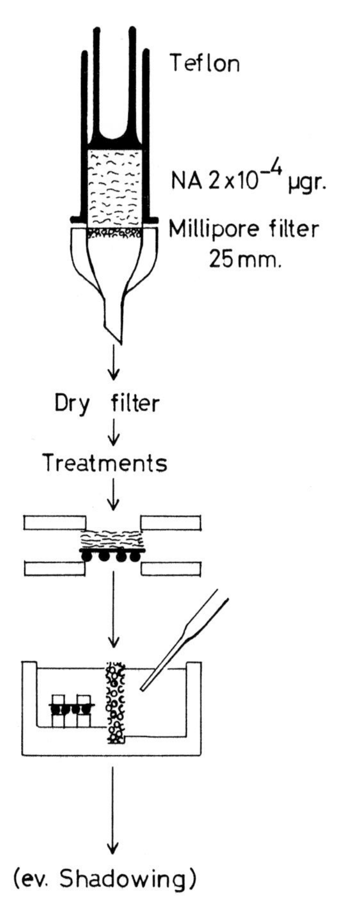

Fig. 2.6 Filter method. Macromolecules are collected and immobilized on a Millipore filter. Small pieces of the filter are pressed on a grid. The filter is then dissolved by diethyleneglycol dimethylether followed by acetone using the double chamber.

After modification of the original technique (Koller and Müller, 1970) the following procedure has been adopted (Fig. 2.6). A solution, containing between 10^{-3} and 5×10^{-4} μg total amount of DNA or RNA, is passed through a Millipore filter (VSWP 02500 or VSWP 05000) at a rate of ~ 0.7 ml/min. The strands are collected on the filter. We use a tightly fitting piston on top of the nucleic acid solution in order to reduce the air-water-interphase to a minimum. This point is of considerable importance, since nucleic acid molecules adsorb at the air-water-interphase and the amount adsorbed increases with increasing DNA or salt concentration (Frommer and Miller, 1968).

At the end of the process, the filter, which is still wet sticks to the Teflon piston. It is dipped into absolute ethanol, where the filter is dehydrated and released from the tip of the piston. The Millipore filter is dried and cut into small pieces of ~3 mm^2. For high-resolution work, staining of the molecules is performed at this stage with the strands immobilized on the filter. In the experiments presented below, the filter pieces were submerged into an aqueous solution of uranyl acetate (2%) for ~5 min and then washed in water for ~½ min and finally in ethanol for 1 min.

To dissolve the filter, the surface of the aluminum coated grid is wettened with diethyleneglycoldimethylether. The small filter piece is then laid on top of the grid, the side with the strands facing the film. The diethyleneglycoldimethylether dissolves the filter and leads, therefore, to a quick attachment to the surface of the aluminum film. Finally, the preparation is dipped into a large volume of acetone for complete removal of the Millipore filter. This process is best completed in the double chamber, a Teflon dish divided into two compartments; the other compartment is used for easy, extremely gentle addition or removal of solvents (Fig. 2.7). The procedure described leads to a regular distribution of molecules over the whole Millipore filter. Therefore, all grids prepared from a single filter can be regarded as almost identical.

Double-Stranded Nucleic Acid Molecules

The method was devised using T_3DNA and polyoma virus DNA (16s). The tangling of the filamentous molecules, which is already found with the adsorption method and which was also described by Gordon and Kleinschmidt (1969) and Dubochet *et al.*, (1970 and 1971), proved to be even more pronounced with the filter method. We found that the addition of 20% sucrose to the solvent unfolded the molecules to some extent. Since most of the sucrose available contained many contaminants,

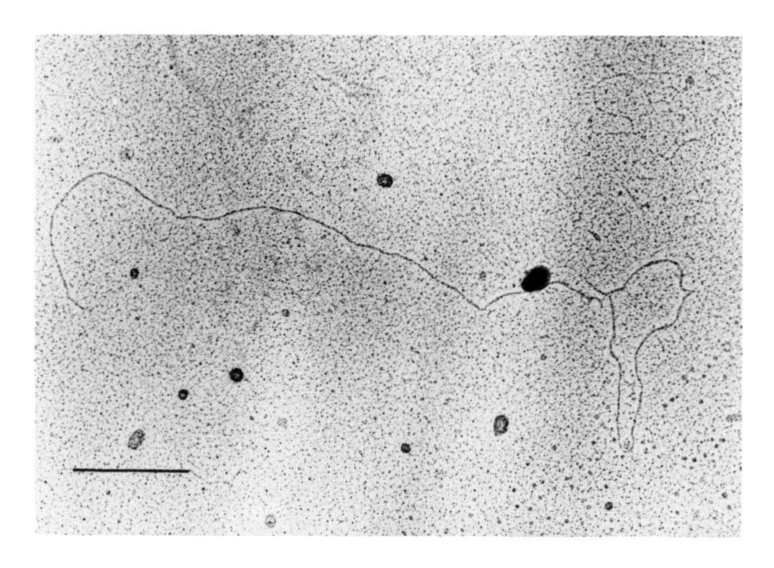

Fig. 2.7 Form II (circular) and form III (linear) of polyoma virus DNA prepared by the filter method from a solution containing 0.01 M ammonium acetate and 20% sucrose. Platinum was used for rotary shadowing. Calibration equals 0.25 μ.

the solutions had to be filtered through the same size Millipore filters before use. The circular conformation of polyoma virus DNA (16s) could now be recognized. This was even clearer in certain selected molecules as shown in Fig. 2.7. The closed circular molecule is what is called the form II and the linear strand is the form III polyoma DNA (Crawford *et al.*, 1966). In Fig. 2.8, a strand is shown that is believed to be a replicating molecule as described by Hirt (1969). More than half of the molecule consists of two parallel running strands of even thickness.

Contour length measurements were made under similar conditions as used for the adsorption method. Figure 2.9 gives histograms for 16s polyoma DNA in 0.01 *M* ammonium acetate without and with 20% sucrose. With sucrose the mean value of 1.35 μm $\pm$ 0.13 is equal to the results obtained with the adsorption method, and is again compatible with the data reported in the literature. Without sucrose, a mean value of 1.57 μm $\pm$ 0.17 is obtained, which corresponds exactly to the data reported by Barblan and Hirt (1969) with the spreading technique. However, the variation in length is somewhat greater, reflecting the tangling of the filaments, which leads to difficulties in interpretation. However, it should be possible to unfold the molecules by using ethidium

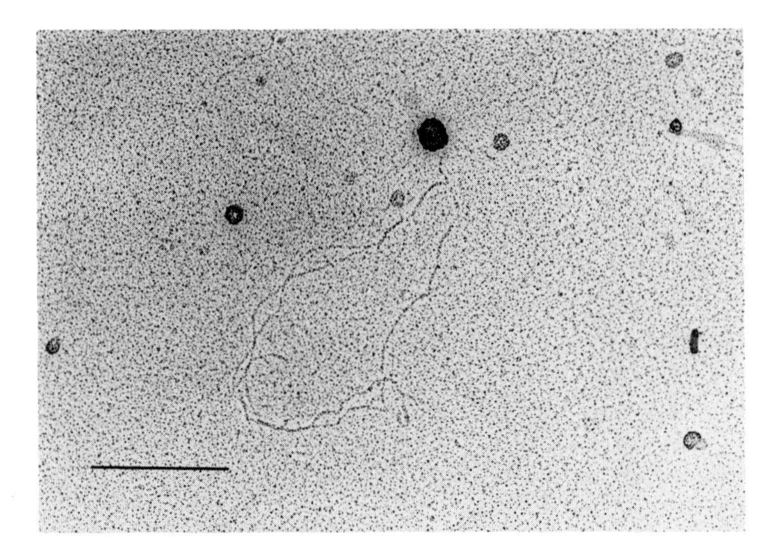

Fig. 2.8 Replicating molecule of polyoma virus DNA in 0.01 M ammonium acetate and 20% sucrose prepared by the filter method. Platinum was used for rotary shadowing. Calibration equals 0.25 μ.

bromide (Crawford and Waring, 1967; Wang, 1969). This work is in progress in our laboratory.

Single-Stranded Nucleic Acid Molecules

QβRNA molecules were used to test the method for single-stranded nucleic acids. Again, the conditions were kept similar to the ones used with the adsorption method. This time, the RNA was diluted in twice-distilled water. A total amount of 10^{-4} μg of RNA was collected on a Millipore filter of a diameter of 25 mm. Figure 2.10 demonstrated a representative QβRNA molecule stained with uranyl acetate according to Gordon and Kleinschmidt (1968). It is obvious that unfolding of single-stranded molecules is easy to obtain under conditions that do not favor hydrogen bonding. Although the deposition of the molecules seemed to be reproducible, the enhancement of contrast on survey micrographs was not consistent. This was true for the uranyl acetate staining as well as for platinum shadowing. We believe that this finding might be related to the fact that with the filter method nucleic acid strands, without any carrier substance, are deposited on the grid.

Figure 2.11 gives the histogram for QβRNA in distilled water. The mean length of 1.14 μm with a standard deviation between 8 and 9% corresponds to the findings with the adsorption method.

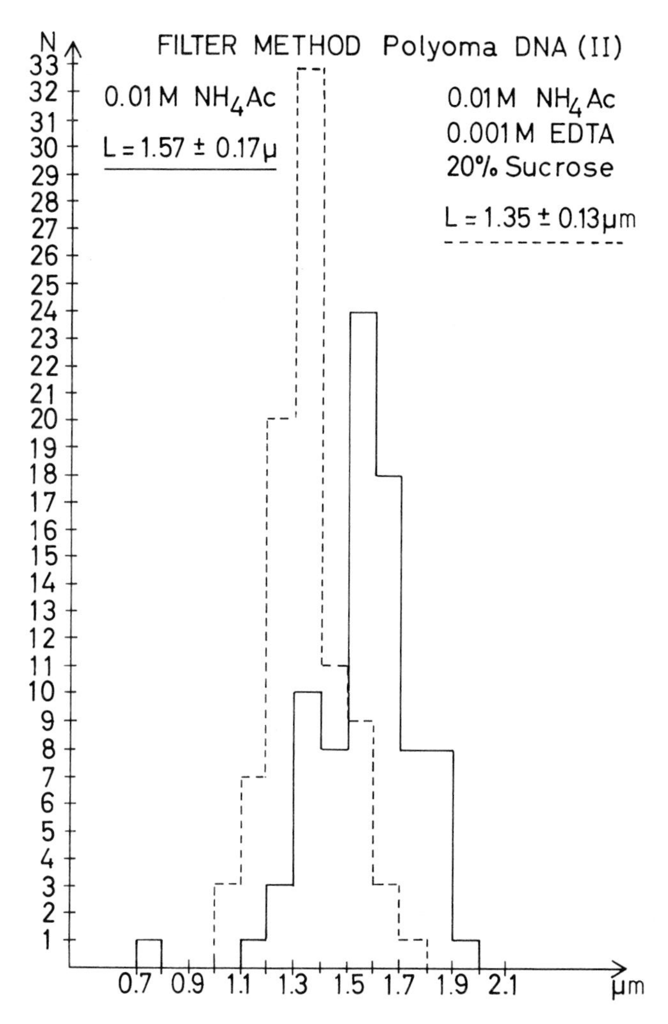

Fig. 2.9 Histogram of polyma virus DNA (16S) prepared by the filter method. The solution contained either 0.01 M ammonium acetate with a mean contour length of 1.57 ± 0.17 μ or 0.01 M ammonium acetate. 0.001 M EDTA and 20% sucrose with a mean contour length of 1.35 ± 0.13 μ.

Discussion

The filter method has certain advantages that have not yet been fully explored.

1. The usually practiced staining method for macromolecular systems with excess of heavy metal salts leads to contamination of the support films. This heavy metal contamination is one of the main sources of

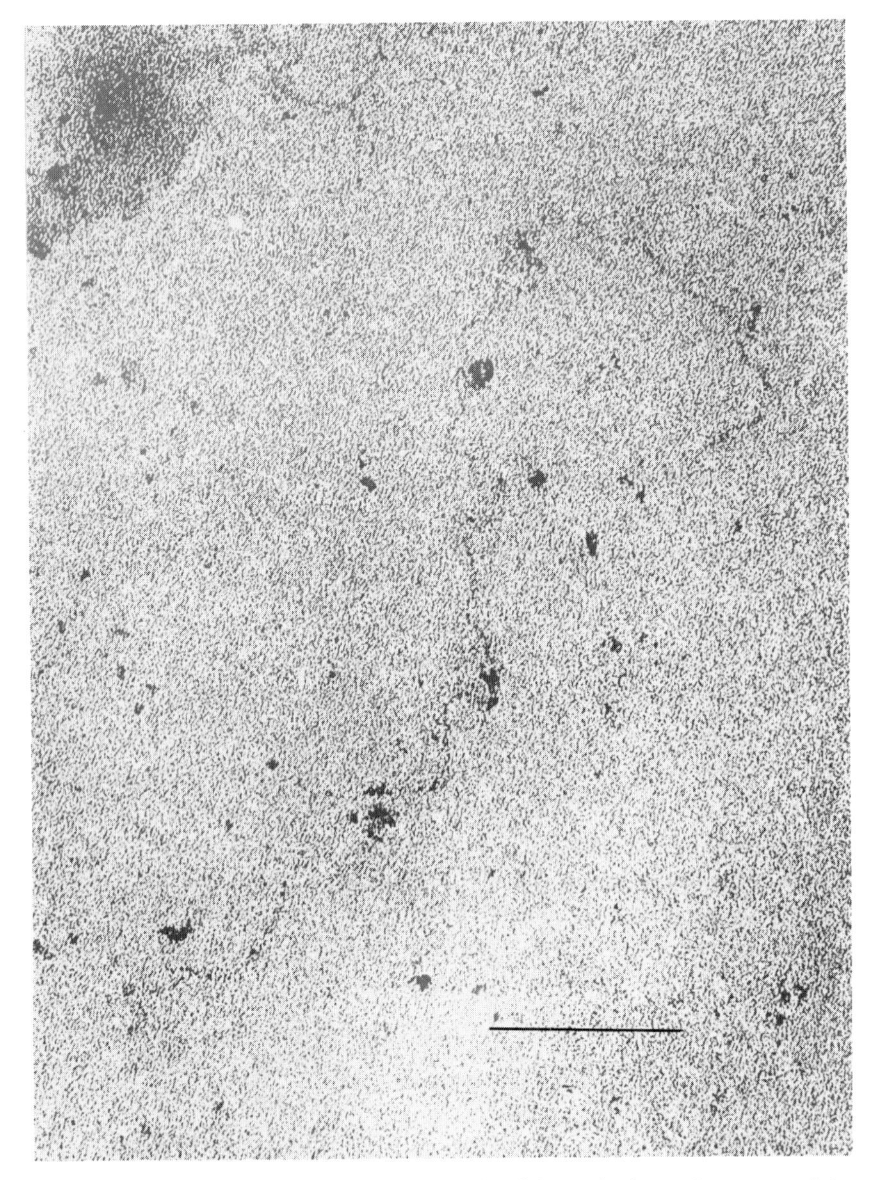

Fig. 2.10 Representative molecule from RNA of bacteriophage Qβ prepared by the filter method from a solution containing 2 mM EDTA. Staining was accomplished with uranyl acetate in acetone. Calibration equals 0.25 μ.

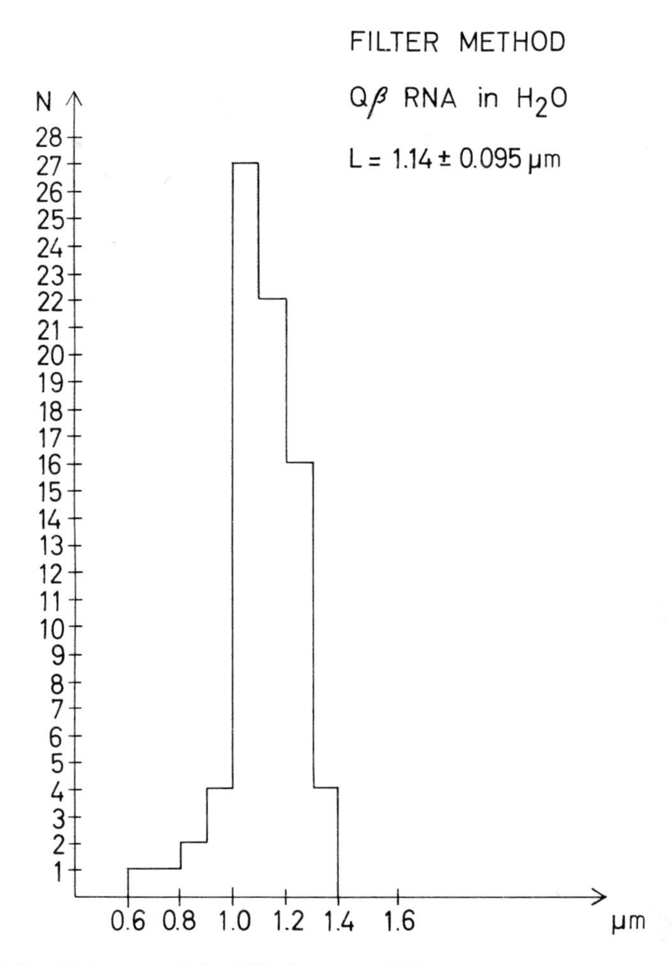

Fig. 2.11 Histogram of Qβ RNA in 2 mM EDTA prepared by the filter method.

noise in high-resolution micrographs. It also leads to irregular staining and to the formation of heavy metal clusters along the course of the nucleic acid filaments. We have tried to overcome this problem by staining the molecules while they are still on the filter. Contamination of the support film by the staining solution is thus avoided.

2. Hybridization experiments can be performed on Millipore filters. Single-stranded nucleic acid molecules are immobilized on a filter. This filter is then, under appropriate conditions, dipped into a solution of a different single-stranded nucleic acid. Base pairing and binding takes place for those polynucleotide segments that are complementary to

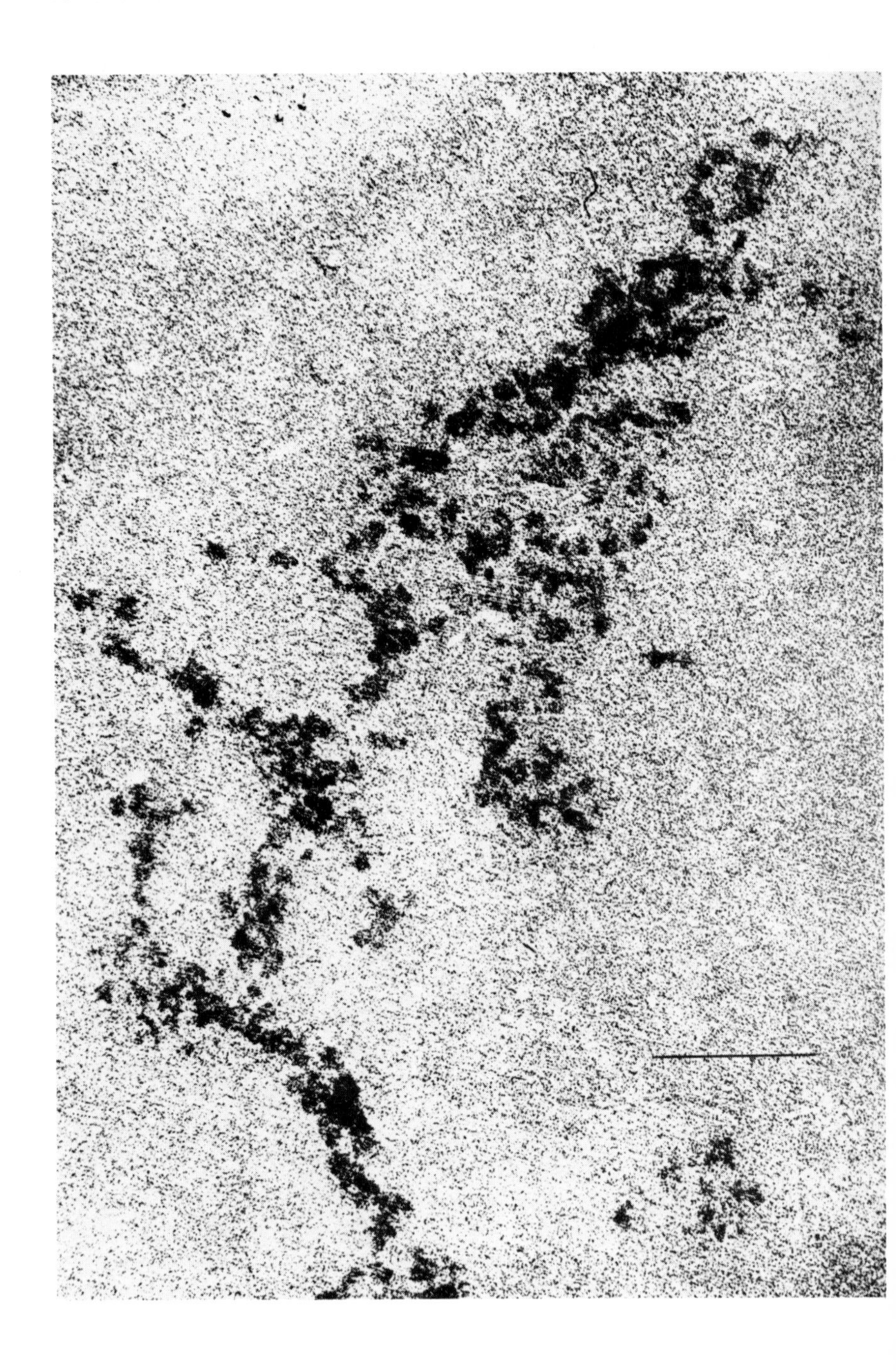

sequences of the immobilized strands on the filter (Gillespie and Spiegelman, 1965). It should by now be possible to examine such assay systems in the electron microscope, and it should be possible to use this technique for characterizing DNA sequences.

3. The mass of nucleic acid needed is extremely small, which might be important for chromosomal studies of higher organisms.

4. Macromolecules in very dilute solutions can be studied.

5. Different support films (carbon, graphite, aluminum oxide, mica) can be used.

6. Apparently the composition of the Millipore filters is very pure. The filter can be completely removed by dissolution in a large volume of solvent without leaving any disturbing contamination behind.

Application of the Filter Method for High-Resolution Microscopy of Nucleic Acid Strands

The application of selective staining for structural analysis of macromolecules requires that the conformation of the entire molecule can be visualized at a high-resolution level. The most suitable model system appears to be double- and single-stranded nucleic acid molecules. We concentrated our efforts on the morphology of double-stranded DNA. An attempt has been made to establish whether, at least over short segments, routine micrographs can be obtained that are compatible with the proposed double helical structure of viral DNA.

Figure 2.12 shows approximately half of a circular polyoma strand that was photographed as an entire molecule at 150,000 × magnification. The width of the strand is ~ 20 Å. Occasionally gaps of 20 to 60 Å are observed. These gaps have not been observed on the survey micrographs that served as a material for the histograms presented earlier. Since during drying and the initial phase of the dissolution of the filter, stress forces might be present, it seems conceivable that this leads to stretching and disruption of certain areas of the molecule. Figure 2.13 shows the same preparation at a higher magnification and about 200 Å under focus.

Fig. 2.12 Approximately half a polyoma virus DNA molecule (16S) prepared by the filter method from a solution containing 0.01 M ammonium acetate and supported by an aluminum oxide support film. Staining was carried out with 2% uranyl acetate in water with the DNA immobilized on the filter. The strand is tangled corresponding to the lack of sucrose in the solvent. Note the general appearance of a "pearl necklace," which is typical for all double-stranded DNA tested by this procedure. Calibration equals 250 Å. Electron optical magnification was 150,000 X.

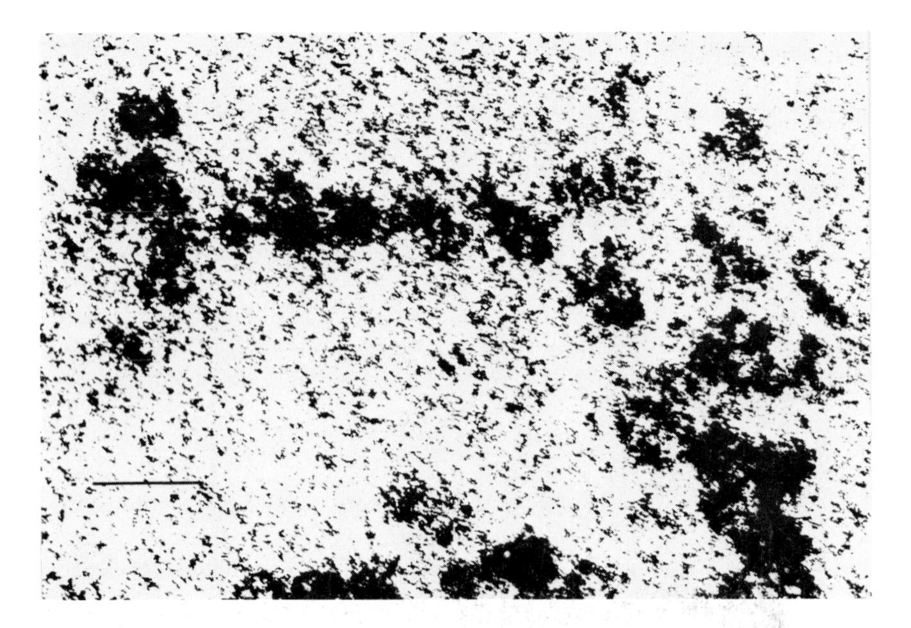

Fig. 2.13 Details of the same preparation as in Fig. 2.12, ~ 200 Å in the underfocus. The strand is ~ 20 Å wide. The distances between the "beads" range between 30 and 40 Å. Electron optical magnification was 150,000 X.

We have observed this characteristic appearance of a "pearl necklace" on all micrographs of double-stranded DNA. The width of the "bead"-like areas ranges between 18 and 25 Å. The distance between the "beads" varies between 30 and 40 Å. If it is assumed that one side of the strand is attached to a fiber of the Millipore filter and is, therefore, sterically hindered in binding the uranyl cations, this morphology seems compatible with the projection of a helical structure onto a plane. The "beads" would then represent the free surface of the molecule with the two single strands parallel to the grid plane and the small groove of the double helix in between. For the interpretation of these micrographs, one has to take into account that during drying, distortions of the three-dimensional structure may take place.

Figure 2.14 shows a different region on a close to focus micrograph. Again, the "bead"-like structure can be seen. One part of the loop is probably broken out and two gaps of about 25 Å are visible. One can almost clearly recognize two dense lines as a substructure along the course of the molecule (arrows). The center-to-center distance between the two strands is 12 Å. Possibly, these dense lines represent the stained backbones of the two single strands forming the double helix, since the dimensions fit reasonably well with the existing models of DNA.

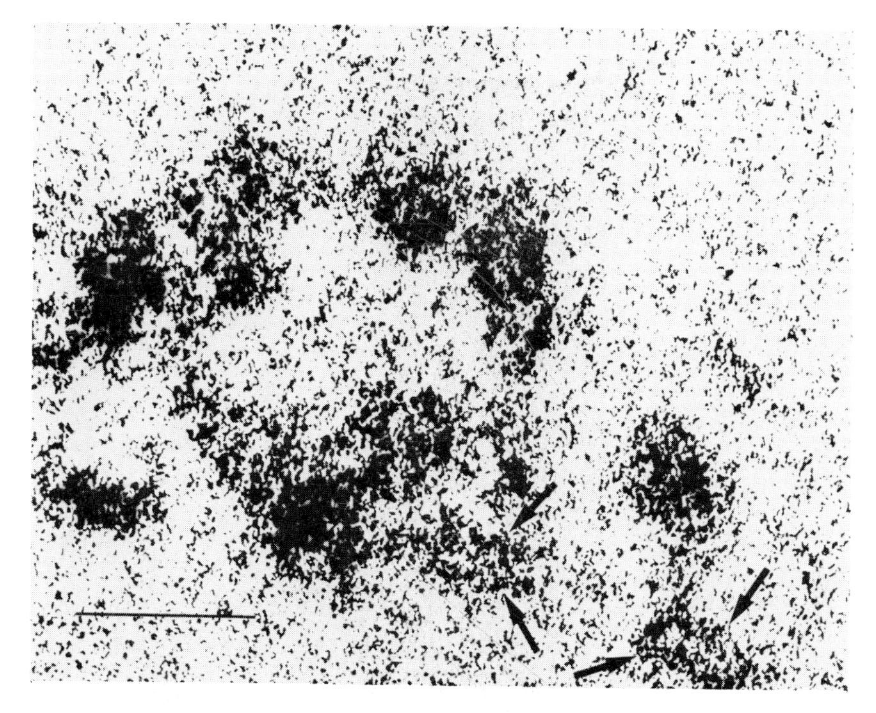

Fig. 2.14 Details of the same preparation as in Fig. 2.12, close to focus. One part of the loop is broken out and two gaps of ~ 25 Å are visible. Within the areas of certain "beads" one can recognize two dense lines along the course of the molecule that might correspond to the stained backbones of the two strands forming the double helix. Calibration equals 50 Å. Electron optical magnification was 150,000 X.

For comparison, Figure 2.15 shows a high-magnification micrograph in the parafocal region of a portion of a single-stranded QβRNA molecule. The strand can be followed as a dense line about 7 to 8 Å in diameter (arrows). There is a marked difference in appearance between this single strand and the double-stranded DNA (Fig. 2.12). It should be possible to distinguish morphologically at least short segments of single- and double-stranded nucleic acids at a resolution level of a few Ångströms.

IDENTIFICATION OF SELECTIVELY STAINED SITES

In the preceding sections it has been shown that nucleic acids and proteins can be stained with selective reagents and deposited in reproducible manners on the supporting films. An analysis of the structure of the macromolecules is possible if the selective reagents can be re-

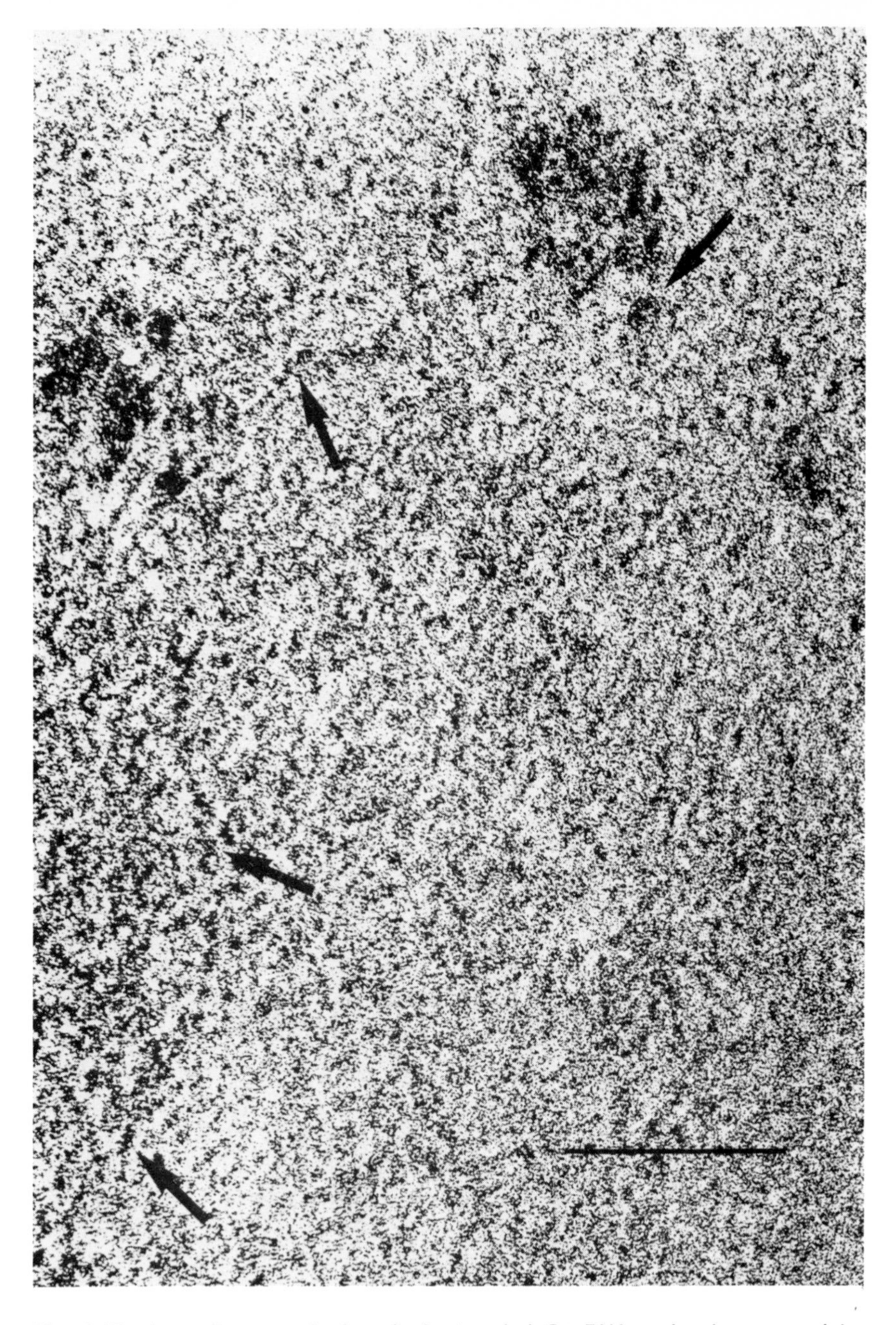

Fig. 2.15 A small segment of a single-stranded Qβ RNA molecule prepared by the filter method from a solution containing 2 mM EDTA. Staining on the filter was carried out with 2% uranyl acetate in water. Aluminum oxide support film was used. The micrograph is in the parafocal region. The course of the strand is indicated by arrows. Calibration equals 100 Å. Electron optical magnification was 150,000 X.

liably recognized. This problem is central in any attempt to obtain fine molecular information with the electron microscope. It is related to the apparent paradox that although electron microscopes today are capable of resolving as high as 2 Å, the finest detected molecular features are in the order of less than this resolution. An attempt has been made below to analyze this discrepancy.

An electron micrograph of a thin carbon film reveals image density variations at high resolution resulting from several causes. They are: (1) the photographic grain, (2) the electron noise, (3) the irregular structure of the support film, and (4) the phase and amplitude contrast transferred from the layered object.

Photographic grain depends upon the emulsion and the developing procedure. With appropriate choices the grain size can be kept down to the order of 30 μm. Overdevelopment generally leads to an increase in grain size. The electron noise, which is the statistical variation in electrons striking a fixed area, diminishes as the exposure increases. The importance of both electron noise and photographic grain are minimized at high magnification and long exposures. This problem has been discussed by Valentine (1964).

The irregularity resulting from the structures of the support film is not so readily circumvented, since it is fixed in the specimen and its relative importance is unaffected by changes in magnification. The appearance of the irregularities in the film is highly dependent upon the focal setting as was previously recognized by Sjöstrand (1956) and elegantly analyzed by Thon (1966). Thon showed, by using optical transforms of the images of carbon films, that granulation changes with the degree of defocusing in a predictable manner.

The information on the structure of a macromolecule is contained in the image density variations in the micrograph. This variation, however, is superimposed over the background density variations or noise mentioned above. Thus the minimum structural detail that can be readily detected must yield an image density recognizable in the presence of the noise, if a reliable identification of the structural feature is to be possible.

The Use of Thin Carbon Support Films

Until recently, the staining of the majority of the molecules has been accomplished with heavy metal cations. In the hope of working with aggregates of the excessively small cations, high pH or non-aqueous conditions have often been used (e.g., Gordon and Kleinschmidt, 1968). With these, frequently greater mass density has indeed been obtained but the results tend to be erratic (Erickson and Beer, 1968). O'Hara

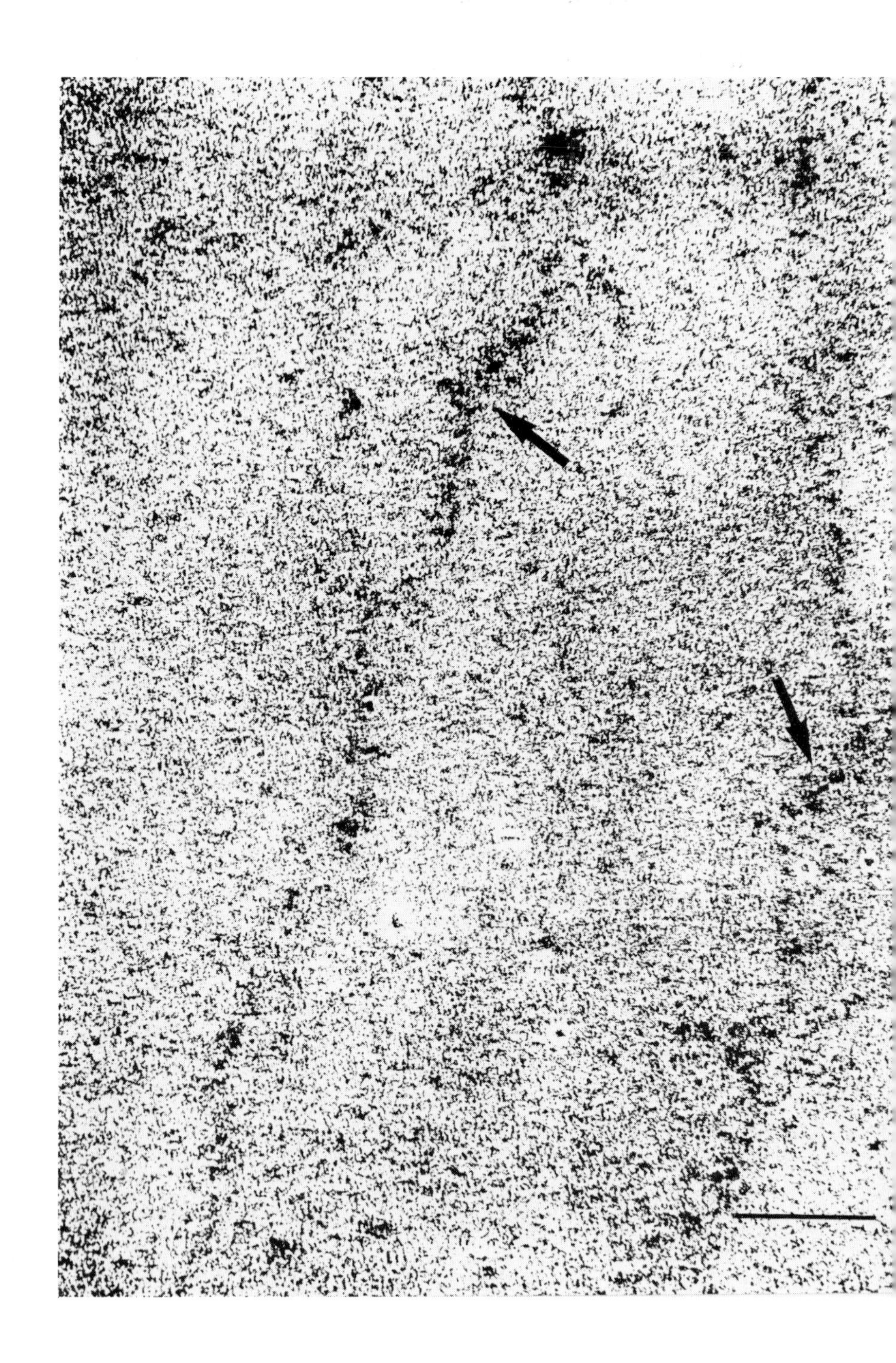

(1968) introduced the use of the large, stable yet compact cation $Ta_6Br_{12}^{2+}$. Koller *et al.* (1969) showed that with $Ta_6Br_{12}^{2+}$ the density of single-stranded nucleic acid molecules (ϕX-174 DNA) was indeed greater and more reproducible than that previously found.

However, high-resolution micrographs of such specimens stained on standard carbon films revealed that the density along the course of single strands (ϕX-174 DNA) was irregular and uneven. The strands were marked by clusters of 20 to 40 Å in size with gaps in between (Fig. 2.16). Only very small segments of evenly stained strands could be found and their appearance was strongly dependent upon the focal setting. Figure 2.17 demonstrates such a segment that is only visible close to focus and that disappeared on the following micrograph at 250 Å in the underfocus. Therefore, staining of nucleic acid molecules immobilized on Millipore filters, as described previously, appears to be more promising than this procedure.

Some discrete attempts to identify chemical groupings within a macromolecule using selective staining and electron microscopy have been reported in the literature. Beer and Moudrianakis (1962) coupled diazotized 8-amino-1,3,6,-naphthalene trisulfonic acid to DNA, streaked the altered macromolecules on standard carbon films, stained with uranyl acetate, and observed dense triangular spots that they attributed to the attached markers. Similarly, Moudrianakis and Beer (1965b) have claimed to recognize their guanine specific reagent after coupling to DNA and staining. In both studies there was considerable uncertainty in identification due to the noise of the support films.

Bartl *et al.* (1970) examined yeast alanine t-RNA after coupling with the guanine reagent. In the best micrographs linear structures of the expected length were indeed detected. The dense spots along the macromolecules were attributed to the stained reagent. Their abundance agreed with the expectation based on chemical studies. It was possible to correlate the observed patterns with the known structure by making reasonable assumptions regarding the number of unmarked bases between the marked ones. This number, however, was not completely reliable and the correlation was not unique. It was clear that unique sequence statements require improved visibility of both marked and unmarked bases.

Fig. 2.16 Two segments of a single-stranded circular DNA molecule from the coliphage OX-174 prepared by the adsorption method from a solution containing 0.01 M KC1 at 60°C. The staining was carried out with hexatantalum dodecabromide diacetate in ethanol according to Koller *et al.*, (1969). The staining is irregular with gaps and heavy metal clusters of different sizes. Calibration equals 250 Å. Electron optical magnification was 100,000 X.

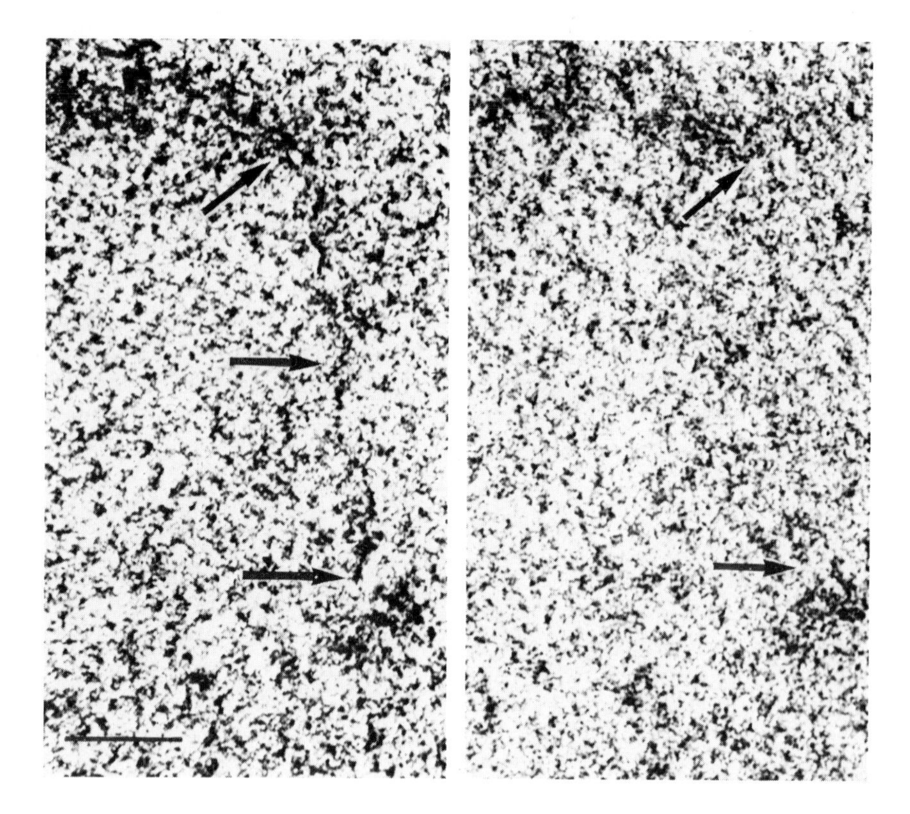

Fig. 2.17 A small segment of a single-stranded circular OX-174 DNA molecule prepared by the adsorption method and stained with hexatantalum dodecabromide diacetate. Only occasionally cleanly stained segments (arrows) can be found, and their visibility is dependent upon the focal setting. Calibration equals 50 Å. Electron optical magnification was 100,000 X.

The minimum mass that could be detected reliably was examined by Highton and Beer (1968) by attempting to measure the contrast given by a linear macromolecule. They used polyadenylic acid (poly A) stained with $HAuCl_4$. The contrast was measured by recording density changes along the molecule and in the adjacent background. Both densitometer tracings led to a noise pattern with similar distribution, but the trace over the molecule had a higher average density. The difference in mean densities was considered a measure of the scattering due to the stained molecules. Both the noise amplitude and the density difference resulting from the stained molecule varied with focal setting.

Best signal-to-noise ratio for such specimens was obtained at 0.1 μm under focus. The strand had a contrast of 1.5 $\pm$ 0.7%. Highton and Beer (1968) estimated that a nucleotide deposited on a carbon film and stained with two ions, each containing one heavy atom, had a 70% probability of detection.

Recently, considerable effort has been made to determine whether single heavy atoms can be visualized by present-day electron microscopy. It was Beer's idea (see Crewe *et al.*, 1970) to use chemically well-defined small organometallic compounds as a test substance. Crewe *et al.* (1970) sprayed the reaction product of tetracarboxylic acid and thorium nitrate or uranyl acetate on very thin carbon substrates. Upon examination in a high-resolution scanning electron microscope the thorium and the uranium product appeared as expected. The thorium compound showed chains of dots 13 and 23 Å apart, while the uranium complex appeared in pairs of dots 13, 23, and 33 Å apart. However, subsequent model building of the compounds tested (Formanek *et al.*, 1971) made it highly probable that the dots visible on Crewe's micrographs consisted of two heavy atoms or more. Furthermore, the models (Formanek *et al.*, 1971) were not consistent with the spacings observed.

The first evidence that single heavy atoms can indeed be localized by present-day microscopy came from two groups independently and simultaneously (Formanek *et al.*, 1971; Koller, 1971; Whiting and Ottensmeyer, 1971 and 1972). The latter used very thin carbon films treated with acetic acid (1 *M*) prior to specimen application. The authors, using dark field microscopy, tested several complexes containing palladium, iodine, platinum, and uranium and found the patterns expected from the models built. Their attempts to localize the T sites on ϕX-174 DNA, however, revealed the same ambiguity as in the earlier reports (Bartl *et al.*, 1970; Beer and Moudrianakis, 1962: Erickson and Beer, 1968: Fiskin and Beer, 1968: Moudrianakis, 1965b) on this subject.

Additional data concerning the micrographs presented by Whiting and Ottensmeyer (1972) are given in a paper by Henkelman and Ottensmeyer (1971). Dark field images are shown in which the configuration and spacing of dots in the image correspond to the configuration and spacing of atoms in the molecules (di-tetramethylammoniumdi-μ-iodotetraiodo-diplatinum). The observed scattering cross sections for atoms widely spaced in the periodic table of elements are found to be related to the theoretically calculated cross sections for these atoms. The results obtained by Formanek *et al.*, (1971), who by phase contrast visualized single mercury atoms on "low noise" aluminum support films, will be reported below.

The Use of Improved Support Films

Single Crystal Films

It is not surprising that considerable effort has gone into attempts to produce supporting films that are free of the granularity of the traditionally used carbon films. One of the most attractive suggestions for avoiding random granularity is to use single crystal supports (Fernández-Morán, 1960) and of these mica and graphite seem particularly attractive. For both, many claims have been made for successful cleavage by pulling the layers apart using Scotch tape and repeating this process until the thickness is decreased to a usable level. In the final step the Scotch tape is removed from the graphite or mica flakes by appropriate solvents. These procedures, however, are exceedingly slow. More recently, more convenient methods have been developed for the production of single crystal supporting films.

Karu and Beer (1966) developed a technique for pyrolitic growth of graphite on single crystals of nickel. Strips of high purity nickel were heated in a very clean vacuum chamber at 10^{-7} Torr (by passing current at a temperature of 900°C for a period of 8 to 10 hr). Such strips, when cooled and examined under a metallurgical microscope, gave clear evidence of single crystal domains in the size range of a few microns. When these nickel films were returned to the vacuum system, heated again to a temperature of 1100°C, and methane gas passed into the chamber to a pressure of 5×10^{-3} Torr, a dark surface began to cover the hot nickel. The pyrolitic reaction was allowed to continue for a period of 1 to 5 min after which time the heating current was switched off, the nickel strip was allowed to cool, and the graphite was floated off the nickel on the surface of HCl. The graphite was transferred to water and washed. Finally it was picked up on standard grids in a manner typical for evaporated carbon films.

The pyrolitic grown graphite did not yield very thin graphite films. Chemically cleaved graphite flakes, however, were much more successful (Beer *et al.*, 1971; Dobelle and Beer, 1968). Graphite crystals could be cleaved into extremely thin layers by drastic oxidation. Powdered graphite flakes, sodium nitrate, and concentrated sulfuric acid were thoroughly mixed. The flask was immersed in an ice bath and the contents allowed to cool. Potassium permanganate was added slowly while stirring and the temperature was maintained below 20°C. After the potassium permanganate was solubilized, the reaction mixture was removed from the ice bath, warmed to 35°C, and stirred on a magnetic stirrer. Then an equal volume of distilled water was slowly added and the temperature was maintained for 10 to 15 min at 98 ± 5°C. After cooling, 3% hydrogen peroxide was added.

The contents of the beaker were centrifuged at 500 rpm in a desk-top Sorvall centrifuge for 1 min. The precipitate was discarded, the supernatant was again centrifuged, and the precipitate washed by several re-suspensions and sedimentations. The last suspension was diluted in 10 volumes of ethanol and allowed to settle for 48 hr in the refrigerator. The yellow supernatant suspension was suitable for spraying on a holey carbon film. The ethanol wetted the carbon film and the suspended thin graphite flakes were stretched across the holes.

The most successful method for producing single crystal graphite membranes was described by White *et al.* (1969) and by Beer *et al.* (1971). It is known that at high temperatures in an inert atmosphere amorphous carbon can be converted to graphite. To withstand the high temperatures necessary, the standard grids had to be replaced by ones made of graphite. These were produced by cleaving thin slices from a graphitized carbon rod with the crystal C axis running along the cylinder axis, and drilling holes into this graphite disk by blasting with an abrasive drill through a mask of stainless-steel mesh. The grids were covered with holey carbon films. These films were again covered with a standard carbon film as commonly used. The heat treatment was carried out in a furnace at a temperature of 2500 to 2600°C for 1 to 5 min in an inert atmosphere of argon and nitrogen. It appeared that the absence of oxygen in the atmosphere surrounding the grids was of vital importance, since even small traces led to the loss of the films.

These graphite films appeared definitively less grainy than similar films of evaporated carbon. They showed extremely smooth "islands" several hundred Ångströms in diameter. Between the "islands," however, the typical carbon granularity was still apparent.

Riddle and Siegel (1971) have described an improved method for production of thin pyrolytic graphite films. The films were grown in an ultra-high vacuum chamber by exposing (III) epitaxial nickel films to carbon monoxide gas. The nickel served as a catalyst for the disproportionation of CO through the reaction $2\ CO \rightarrow C + CO_2$. The micrograph published in this paper is very similar to what has been observed on the graphite films produced by heat treatment of thin carbon films (White *et al.*, 1969; Beer *et al.*, 1971). It also shows extremely smooth "islands" 100–200 Å in diameter in between which are areas with characteristic carbon granularity.

Komoda *et al.* (1969) reported the use of beryllium single crystal flakes as substrates for high-resolution electron microscopy. These flakes were produced by evaporating beryllium in argon at low pressures. Apparently the flakes were very thin, and uniform single crystals were obtained, covered by a thin protective layer of BeO grown epitaxially on the metallic beryllium. No through-focus series micrographs were given by the authors and a comparison with aluminum oxide films described below is therefore not possible.

In a recent report, Heinemann (1970) commented on mica as an electron microscope specimen support film. He found that during the observation in the microscope, a relatively high rate of radiation damage occurred. However, even after complete decomposition due to radiation damage, a relatively low amount of background phase contrast was present. Furthermore, remarkable mechanical and electrical stability is reported.

Only rare applications of single crystal support films are described in the literature. Most likely these films have not proved to be more advantageous than carbon films because of specimen contamination occurring in conventional electron microscopes. However, very recently Hashimoto *et al.* (1971) used graphite thin crystals of ~100 Å in thickness and examined the possibility of visualizing thorium atoms with dark field microscopy. As employed by Crewe *et al.* (1970), these authors used the reaction product of thorium nitrate with benzenetetracarboxylic acid, and in addition they made micrographs of small crystals of thorium oxide. On the thorium-BTCA specimen linear arrays of spots with spacings of 10 Å and less could be seen. According to Formanek *et al.* (1971) these spots could be due to two thorium atoms on either side of the benzene ring. But in the micrograph of a thorium oxide crystal the arrangement of spots agreed with that of single thorium atoms in the (101) plane (cubic system).

Prestridge and Yates (1971) used silica supports ("Cabosil HS 5" from Cabot Corp., Boston) for the examination of a rhodium catalyst. Silica was impregnated with an aqueous solution of $RhCl_3$: 3 H_2O. The samples were reduced *in situ* in flowing hydrogen for 16 hr at 400°C. The degree of dispersion of the Rh after reduction was obtained by measurement of H_2 and CO chemisorption at room temperature. An average crystallite size of 12 Å was found. The value of 12 Å was also determined in the electron microscope using the principle of optimal sizing of the objective aperture for differential contrast in bright field. Within these rhodium clusters, spots ranging from 3.4 to 4.0 Å apart could be seen, which these authors considered to be single rhodium atoms.

Amorphous Films

An alternative approach to the use of single crystal support films would be to obtain very regular and perfectly amorphous films. Such films might be obtained by evaporation onto the coldest possible surface (Zinsmeister, 1964).

Müller *et al.* (1970) described a method for producing very thin aluminum support films. These films proved to possess very little inherent structure so that it became possible to visualize single mercury atoms by phase contrast (Formanek *et al.* 1971). The routine procedure for preparing these support films has been modified and improved considerably since the original publication, and will be described in detail.

Grids covered with a holey carbon-platinum film are rendered hydrophilic in a glow discharge. They are then placed onto or into a layer of glycerol that has been spread over a piece of a filter paper mounted on a cooling stage in a freeze-etching machine (Fig. 2.18). In order to prevent bubbling of the glycerol in the vacuum, the specimen stage has to be cooled before the bell-jar is evacuated. The cooling stage is protected against contamination by a liquid-nitrogen cooled trap. Only after the temperature of the glycerol has equilibrated at −50 to −100°C is the bell-jar evacuated. Usually a vacuum of 2×10^{-7} Torr is needed. Then, pure oxygen is injected through a needle valve and the pressure adjusted to 5×10^{-4} Torr.

Pure aluminum is evaporated from an electron gun, at an angle of 90°, onto the surface of the glycerol, where it condenses. The thickness of the film is estimated by a Balzers quartz thin crystal monitor. Films between 25 and 35 Å in thickness are suitable for high-resolution microscopy. The bell-jar is flushed with dry nitrogen while the glycerol and the cold trap are still cool. The aluminum coated preparation is placed on a pile of filter papers, which slowly absorbs the glycerol. As the glycerol is absorbed, the aluminum oxide film is gently deposited on the holey grids. Finally, the grids are washed with water and ethanol.

In the above procedure the cooling of the glycerol was necessary in order to reduce bubbling, which displaces the grids and creates a hilly surface on the glycerol. The cold specimen, however, easily becomes contaminated while the pressure in the bell-jar is still high. Therefore, a procedure that would allow the condensation of the aluminum on a surface of glycerol at room temperature could facilitate the large scale production of such films. Recently, Vollenweider and Koller (unpublished results) developed the following technique, which avoids almost completely the problem of bubbling of the glycerol kept at room

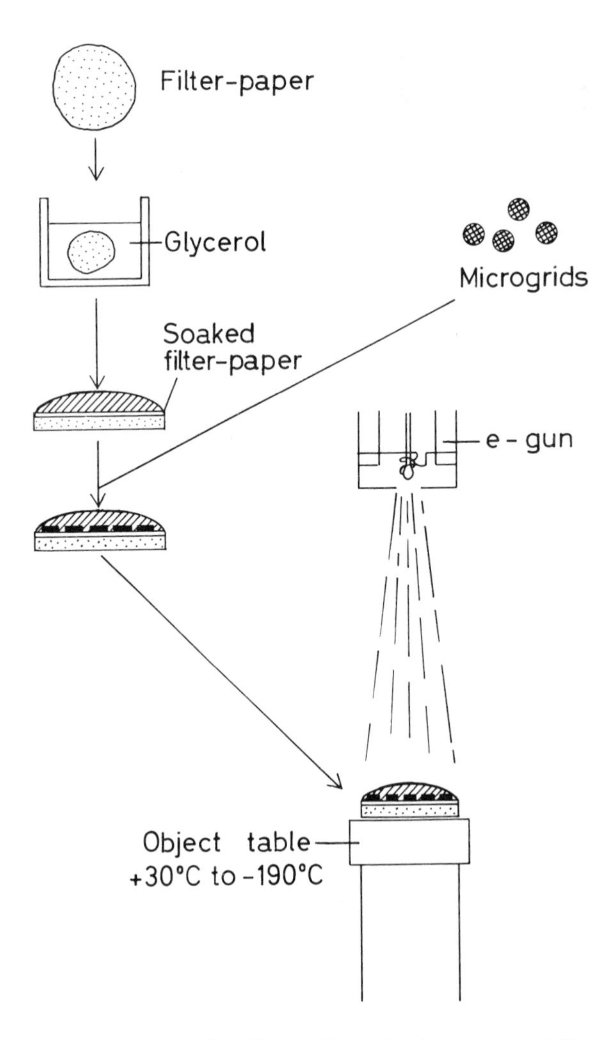

Fig. 2.18 Procedure for preparing "low noise" aluminum support films. Aluminum is evaporated from an electron gun under oxidizing conditions, and is condensed on the surface of cold glycerol.

temperature. Instead of a filter paper a copper mesh (mesh diameter ~1 mm) is used. This mesh is wettened with glycerol and the holey grids are placed inside the glycerol film formed on the mesh. The preparation is placed in the evaporator so that the space under the bottom side of the mesh is free and therefore accessible for evacuation. The evaporation of the aluminum and the washing procedure are carried out as described before.

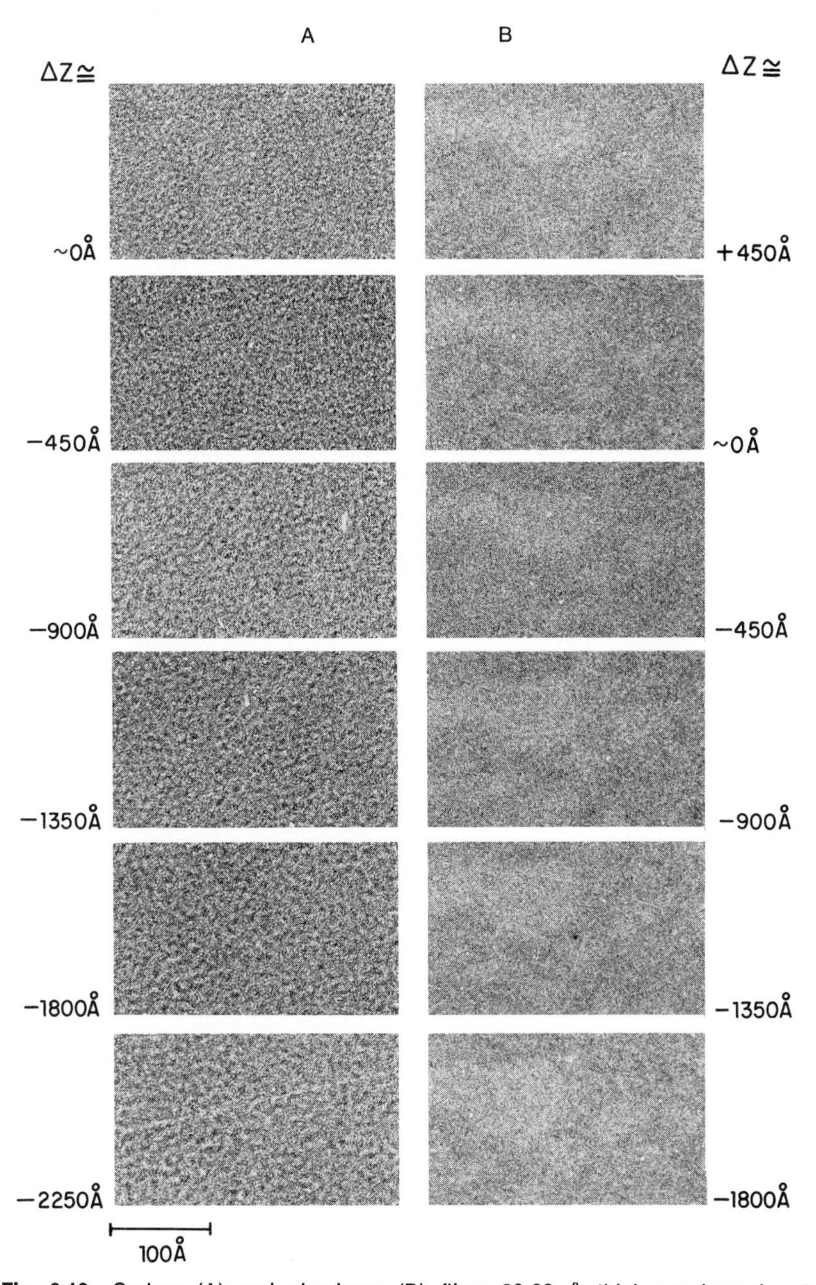

Fig. 2.19 Carbon (A) and aluminum (B) films 20-30 Å thick condensed onto glycerol at about —100°C. The focal settings of the micrographs range from close to focus to ~ 2300 Å. Calibration equals 100 Å. Electron optical magnification was 120,000 X.

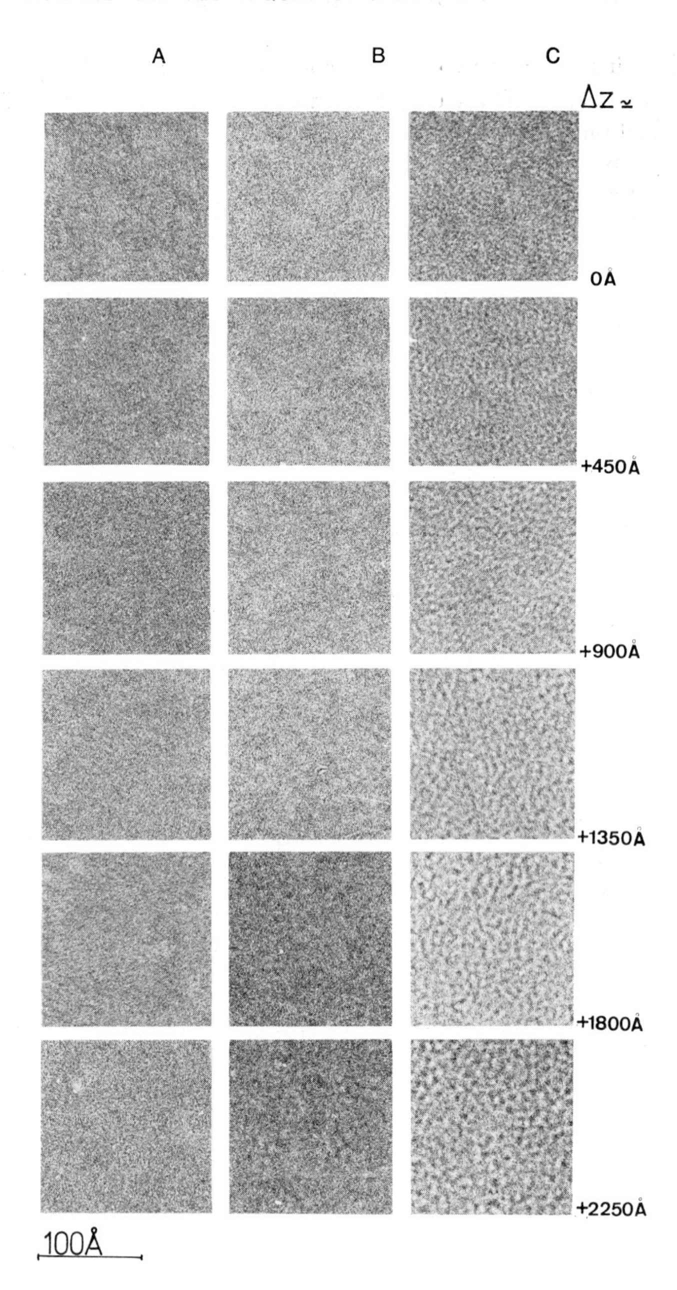
A
B
C
ΔZ ≃
0Å
+450Å
+900Å
+1350Å
+1800Å
+2250Å
100Å

Defocusing leads to imaging of fine structural details by phase contrast (Thon, 1966). Therefore, the "transparency" of a support film for high resolution microscopy can be judged from defocused micrographs. If by defocusing only little structure or "noise" appears within a support film, then imaging of a specimen on the film by phase contrast through defocusing is possible.

Electron micrographs were taken in a Siemens Elmiskop 101 at 100 kV (C_s = 3.2 mm). The illuminating aperture was usually 1.1×10^{-3} rad and mostly no objective lens aperture was used. The magnification was determined by the lattice spacings (3.44 Å) and the half-spacings (1.72 Å) of partially graphitized carbon black (Heidenreich *et al.*, 1968). Kodak Electron Image Plates were used and were developed for 3 min at 20° C in Kodak HRP developer (dilution 1:4).

The difference in appearance of carbon films and aluminum films is clearly shown in Fig. 2.19A. Carbon and aluminum films of similar thickness (~30 Å) and prepared in the same way have been compared by making through focus series micrographs. The defocus in both series ranges from near focus to about 2300 Å underfocus. The carbon film (Fig. 2.19A) shows the well-known focus-dependent granularity. The aluminum film (Fig. 21.9B), however, has the same smooth appearance throughout the focal range tested. From the comparisons made between aluminum films and graphite single crystals, it can be concluded that at present the former have two advantages: (1) the "smooth" and "low-noise" areas are often larger, and (2) the films are probably two to three times thinner.

The contamination during the observation in the microscope strongly affects the appearance of the aluminum films. In Fig. 2.20A, a fresh aluminum film is shown. The through focus micrographs were taken as rapidly as possible for minimum beam exposure. Figure 2.20C shows the same film after 10 min of irradiation. The contamination rate of the microscope was in the order of 2 to 3 Å/min. The appearance of the film (Fig. 2.20C) closely resembles a carbon film and has lost the features of the clean aluminum film. Some contamination may also be introduced

Fig. 2.20 Effect of contamination on the appearance of aluminum films. *A.* A fresh film. The micrographs were taken as rapidly as possible to minimize electron beam damage. *B.* "Clean" area of a preparation of 14 X recrystallized ferritin adsorbed onto an aluminum oxide film from the same preparation. *C.* As in *A* except after a 10-min exposure to the electron beam. The appearance of film *C* is indistinguishable from a carbon film. On film *B* increased noise is apparent below 1800 Å underfocus, probably due to contamination that originates from the deposition of a specimen. Calibration equals 100 Å. Electron optical magnification was 120,000 X.

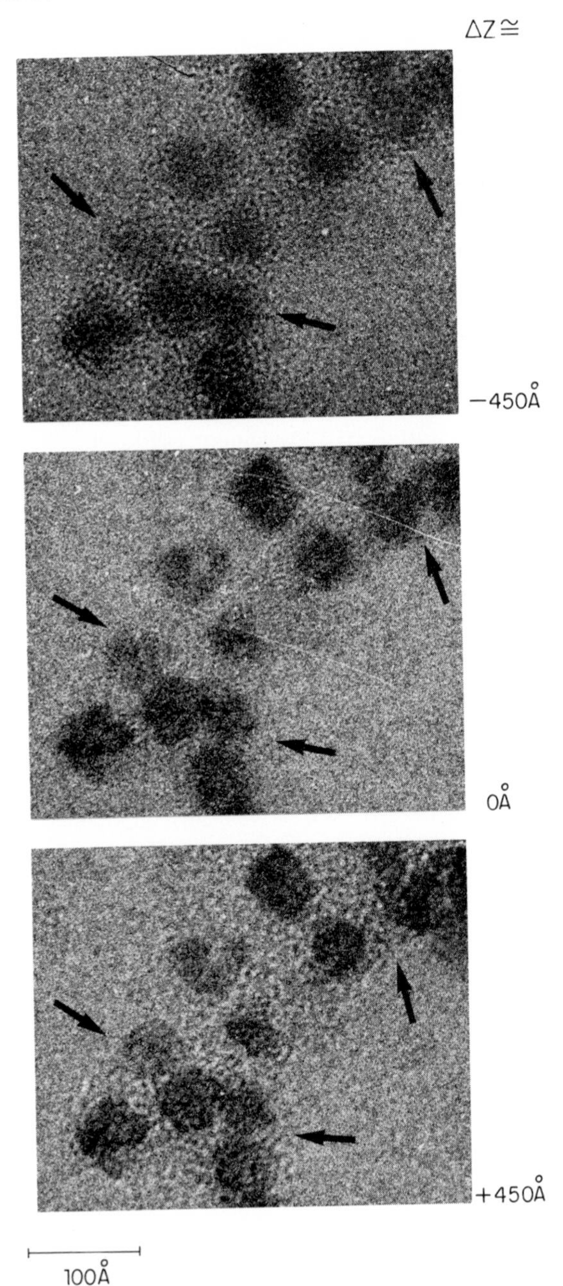

Fig. 2.21 Ferritin molecules deposited on aluminum support films. Micrographs range from ~ 450 Å overfocus to ~ 1700 Å underfocus. Close to focus the iron-free apoferritin is almost invisible. Out of focus, however, the structure of the apoferritin appears in phase contrast. The ultrastructure of the apoferritin is

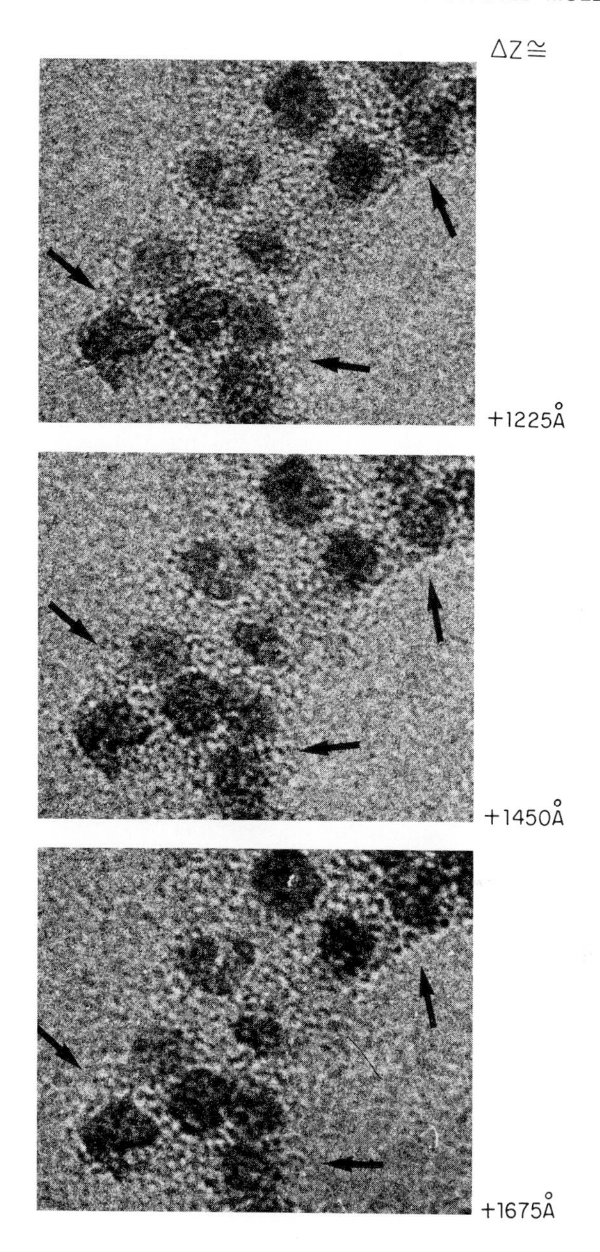

different from that of the support film. The arrows indicate the clear boundary between the edge of the molecules and the area of the film. Calibration equals 100 Å. Electron optical magnification was 120,000 X. Objective lens aperture was $\sim 5 \times 10^{-4}$ rad.

by contact of the aluminum film with the specimen. Figure 2.20B shows a "clean" area of an identical aluminum film onto which ferritin from a very clean solution (14 × recrystallized) was adsorbed. Below 1800 Å in the underfocus, some "noise" is apparent.

The signal-to-noise ratio determines the probability with which a fine structural detail, ultimately a single atom, on a support film can be recognized. Although good contrast as in dark field microscopy is very helpful, the information content is entirely dependent upon the signal-to-noise ratio. In order to obtain a qualitative estimate of the signal-to-noise ratio that can be obtained using aluminum films, ferritin and triacetoxy-mercury aurine were utilized as test specimens.

The iron core of the ferritin molecule is very electron dense and can be easily recognized on the screen in the microscope. The iron-free apoferritin, however, which by negative staining can be seen as a halo around the core, is barely visible on unstained preparations. The pertinent questions are: Is it possible to image the apoferritin without staining, and are there fine structural details which can be distinguished from the background of the support film? Figure 2.21 shows through focus micrographs of ferritin molecules on aluminum films from ~450 Å in the overfocus to ~1800 Å in the underfocus. Close to focus, as suggested by theory, the halo of the apoferritin is barely visible. Out of focus, however, where phase contrast appears, fine structural details appear and a fairly clear boundary can be seen between the edge of the molecules and the support film structure (arrows).

For comparison, through focus micrographs of ferritin supported by very thin carbon films are shown in Fig. 2.22. Out of focus, the apoferritin can be recognized as a halo. The fine structural details, however, cannot be distinguished from those of the support film, and it is supposed that most of them are carbon film structures superposed over the ferritin molecule. This has already been pointed out by Haydon (1969).

The employment of heavy metal atoms as markers on specific sites of organic molecules has been proposed by Beer and Moudrianakis (1962) as a method for improving contrast and reducing the effects of radiation damage to the specimen (see also Breedlove and Trammell, 1970). Therefore, it has been of interest to determine whether single heavy atoms can be visualized with contemporary electron microscopy using aluminum oxide support films. In order to examine the contrast of a single heavy atom, a sterically well-defined organo-metallic compound was needed. Triacetoxy-mercury aurine (TAMA) is a fairly rigid disk-like molecule with three mercury atoms being located at the corners of an equilateral triangle (Fig. 2.23). According to this model, the distances between the three mercury atoms vary between 8.2 and

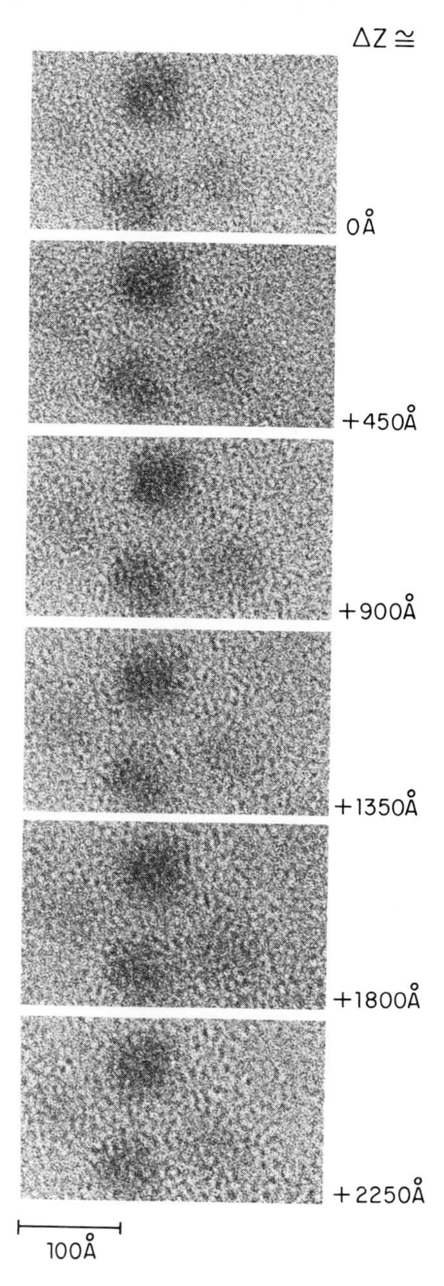

Fig. 2.22 Ferritin molecules deposited on a very thin carbon film (probably below 20 Å in thickness). The ultrastructure within the area of the apoferritin is indistinguishable from the structure of the support film. Calibration equals 100 Å. Electron optical magnification was 120,000 X. Objective lens aperture was $\sim$ 1 X 10^{-3} rad.

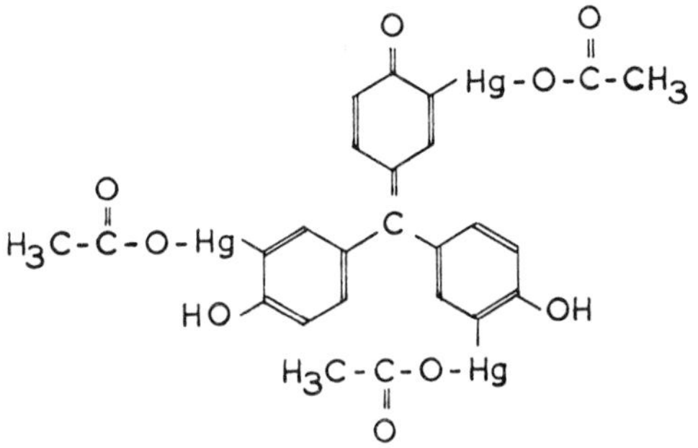

Fig. 2.23 Formula of triacetoxy-mercury aurine (TAMA).

8.9 Å. If such a molecule could be visualized in a plane projection, one would expect to see three dense dots separated by 8 to 9 Å.

Formanek *et al.* (1971) have reported that the mercury atoms of this molecule can be localized. Fig. 2.24a shows aggregates of TAMA with many sharp electron dense dots on a micrograph close to optimum defocus for single heavy atoms (Reimer, 1969). Presumably some of these dots are the mercury atoms of the rigid molecules deposited in all orientations. Among them, at the edge of aggregates, a certain number of isolated dots in a triangular arrangement can be recognized on equivalent focal settings. These are believed to be due to the three mercury atoms of a single TAMA molecule. Contrast measurements were carried out from dense dots protruding into a clear area. Values between 5 and 6% were obtained, which agree well with the contrast calculations for heavy atoms by Reimer (1969). The signal-to-noise ratio due to quantum noise was in the order of 8, which was greater than the usually assumed readability threshold of 5. Figure 2.24b shows a corresponding focal setting of a control film without molecules. No dots of comparable contrast are apparent.

Fig. 2.24 A preparation of TAMA at ~ 1000 Å underfocus is shown in *a*. At the edge of the aggregate toward the clean support film; triangular arrangements of three dense dots 8 to 9 Å apart can be seen. The individual dots presumably present the sites of single mercury atoms within a TAMA molecule. An aluminum oxide film at ~ 1000 Å underfocus is shown in *b*. No dots of similar density as in *a* can be recognized. Magnification calibration with partially graphitized carbon black and lattice spacing was 3.44 Å (insert). Electron optical magnification was 467,000 X. Illuminating aperture was 1.1 X 10^{-3} rad. Objective lens aperture was 1.0 X 10^{-2} rad and spherical aberration constant was Cs = 2.26.

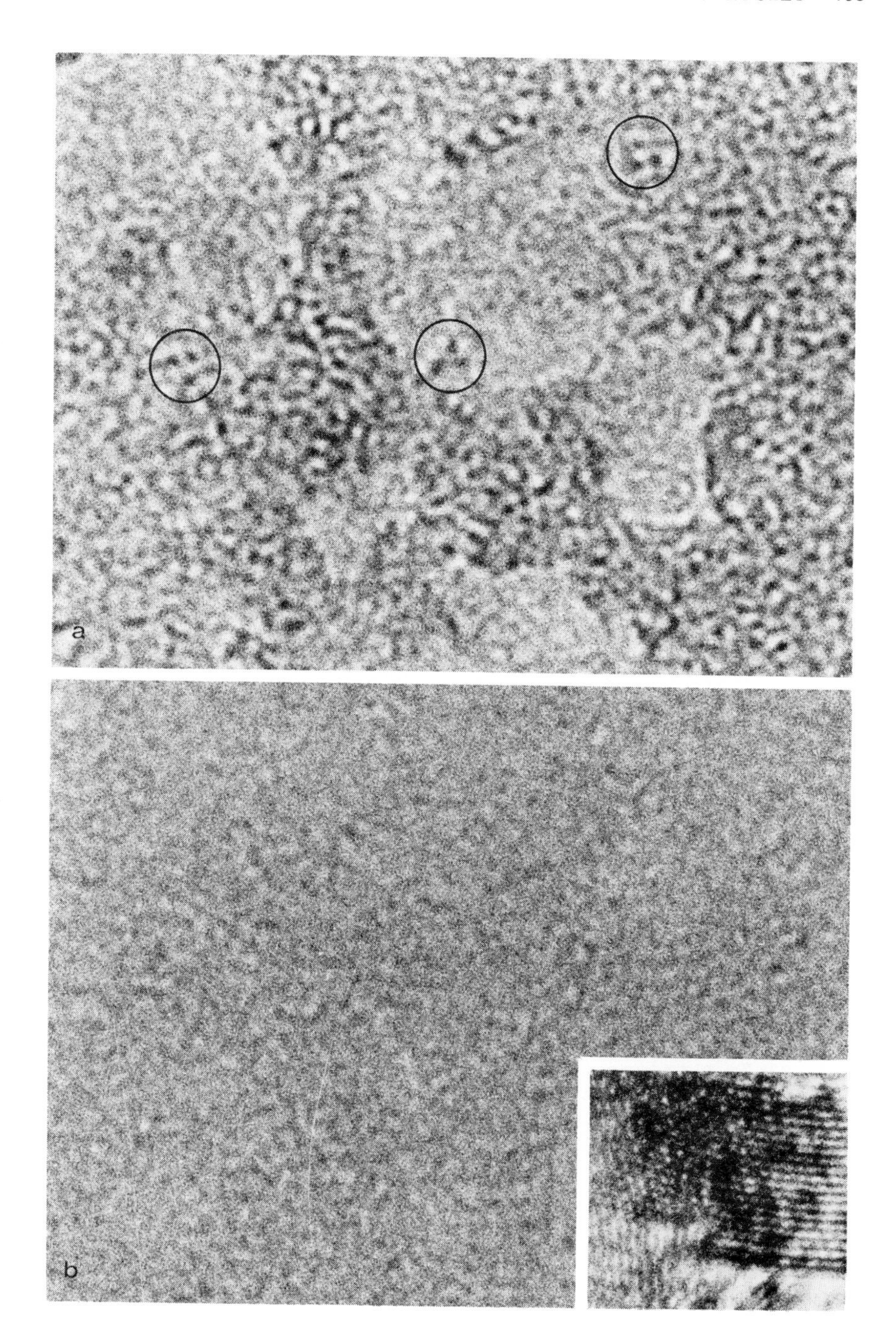
a
b

Recently attempts have been made to stain selectively synthetic polymers of known base sequences. One of the most promising nucleotide markers available is osmium for the T residues (Di Giamberardino *et al.*, 1969); and for this staining reaction the most obvious model polymer is poly d(A-T), a copolymer composed of alternating dA and dT units. If the data from Qβ-RNA (see p. 70) can be extrapolated to poly d(A-T), an internucleotide distance in the order of 3 Å is to be expected. Since osmium reacts much faster with the T residues than with A (Beer *et al.*, 1966; Subbaramen *et al.*, 1971), osmium dots are to be expected about every 6 Å.

Figure 2.25 shows a representative strand of osmium stained poly d(A-T); staining conditions are similar to those employed by Di Giamberardino *et al.* (1969) and are given in the figure legend. The strands are seen on micrographs of images defocused to give phase contrast. On micrographs of images close to the Scherzer focus, the strands are ~ 8 Å in diameter and their meandering can be followed over short stretches (100-300 Å) at this resolution level. The density variations within the strands, however, show no regular pattern compatible with regularly stained and spaced osmium atoms. This result is different from the

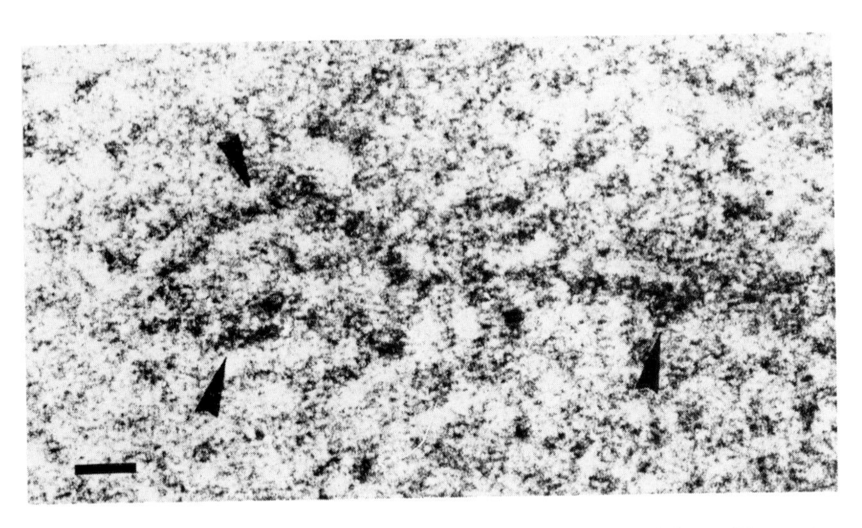

Fig. 2.25 A strand of a preparation of poly d(A-T) after staining with osmium. Reaction of poly d(A-T) (from Miles) for 6 hr at 65°C in a solution containing 0.1 M osmium tetroxide, 0.1 M KCN and 0.1 M NaH_2PO_4, pH 7.0; dialysis against distilled water at 60°C for 30 min, quenched on ice, strands collected on a Millipore filter at 4°C and transferred onto aluminum support films. Electron optical conditions as described in Fig. 2.24. Micrograph is ~ 1000 Å underfocused. Magnification calibration equals 20 Å.

results of osmium stained poly uridylic acid (poly U) presented by Whiting and Ottensmeyer (1972), in which one can see a row of dots with somewhat variable spacings of about 7 Å. The discrepancy between these results can probably be explained by the fact that the reaction rate of osmium is several times faster with thymine, thymidine, and TMP than with uracil, uridine, deoxyuridine, and UMP (Beer *et al.*, 1966; Subbaraman *et al.*, 1971). In our study the osmium reaction was taken to completion, but in the study of poly U probably only a fraction of the U residues were stained with Os atoms.

CONCLUSION

The investigations dealing with the visualization of single heavy atoms (Crewe *et al.*, 1970; Formanek *et al.*, 1971; Henkelman and Ottensmeyer, 1971; Hashimoto *et al.*, 1971; Prestridge and Yates, 1971; Whiting and Ottensmeyer, 1972) are complementary. They indicate, although do not prove, that present-day electron microscopy is capable of resolving single heavy atoms, independent of the mode of operation of the microscope. The crucial factor appears to be the supporting film, the structure of which is the main source of noise. This result is of considerable importance because it opens the way for the use of intramolecular selective staining of biomacromolecules. Presently, it is obviously much more difficult to localize a single heavy atom within an unknown, flexible molecule than within a rigid, well-defined compound. Problems such as specimen damage in the electron beam, variable internucleotide distance in a flexible single stranded polynucleotide deposited and dried on a support film, and insufficient specificity of the staining reaction still must be overcome.

SUMMARY

Progress has been made in selectively "staining" certain reactive groups within a molecule and identifying the marked groups by high-resolution microscopy. Among the intramolecular staining reactions, the reactions of diazotized compounds with guanine, acyl hydrazides with cytosine, and osmium tetroxide with thymine may be applicable to the study of nucleic acids. A stain with less specificity, but one that binds strongly to nucleotides is gold (III) chloride. With regard to selective staining of proteins, the amino and carboxyl groups of murein have been marked with mercury compounds, and the guanidine bonds in collagen have been localized with PTA.

In order to obtain unbroken and untangled single nucleic acid molecules for high-resolution microscopy, two principal procedures for preparation have been explored. (1) The surface of the support film is positively charged, which increases the binding capacity of these films for nucleic acids and other acidic molecules. (2) The solution of macromolecules is filtered through a Millipore filter that collects the strands, which are then transferred to the support film by dissolving the filter. High-resolution micrographs of DNA prepared according to these procedures suggest that double-stranded viral DNA has an appearance compatible with the proposed double helical structure.

The identification of selectively stained sites is hindered largely by the structure of the support films, which obscures the structure of the specimen. Two support films, however, seem to be very promising in overcoming this problem: (1) single crystal graphite films which are produced by heat treatment of standard amorphous carbon films, and (2) amorphous aluminum oxide films made by evaporating aluminum on the surface of glycerol under oxidizing conditions. Studies carried out with well-defined organo-metallic compounds on improved support films suggest that individual heavy atoms can be localized within a simple organic molecule either by phase contrast or by dark field microscopy.

ACKNOWLEDGEMENTS

The contents of this chapter have been published in Cytobiologie *4*:369, 1971 with the exception of Fig. 2.25 and some minor additions to the text. This work was supported by the Swiss National Foundation for Scientific Research Grant No. 3.434.70. The authors thank K. LaFountain for correcting the manuscript.

References

Bahr, G. (1954). OsO_4 and RuO_4 and their reactions with biologically important substances. *Exptl. Cell Res.* 7, 457.

Barblan, J. L., and Hirt, B. (1969). Influence of salt concentration on the length of polyoma virus DNA. Personal communication.

Bartl, P., Erickson, H. P., and Beer, M. (1966). Studies of the s-RNA and λ-DNA molecules labeled with guanine specific reagent. *Proc. 6th Intern. Congr. Electro. Micr.*, Kyoto, p. 119.

———, (1970). Electron microscopic study of base sequence in nucleic acids. VI. Guanine sites in yeast alanine transfer RNA. *Micron 1,* 374.

Beer, M. (1961). Electron microscopy of unbroken DNA molecules. *J. Mol. Biol.* 3, 263.

———, and Moudrianakis, E. N. (1962). Determination of base sequence in nucleic acids with the electron microscope. III. Visibility of a marker.

Chemistry and microscopy of guanine-labeled DNA. *Proc. Natl. Acad. Sci. 48*, 409.

———, and Highton, P. J. (1962). A simple preparation of graphite-coated grids for high resolution electron microscopy. *J. Cell Biol. 14*, 499.

———, Stern, S., Carmalt, D., and Mohlhenrich, K. H. (1966). Determination of base sequence in nucleic acids with the electron microscope. V. The thymine-specific reactions of osmium tetroxide with deoxyribonucleic acid and its components. *Biochemistry 5*, 2283.

———, Gibson, D. W., and Koller, T. (1970). Some Metal Ion DNA Reactions in Electron Microscopy. *In:* "Effects of Metals on Cells, Subcellular Elements and Macromolecules" (Maniloff, J., Coleman, J. R., and Miller, M. W. eds.), p. 131. Charles C. Thomas, Springfield, Illinois.

———, Bartl, P., Koller, T., and Erickson, H. P. (1971). Electron Microscopy of Nucleic Acids. *In:* "Methods in Cancer Research," Vol. 6 (Busch, H. ed.), p. 283. Academic Press, New York and London.

Bendet, I., Schachter, E., and Lauffer, M. A. (1962). The size of T_3DNA. *J. Mol. Biol. 5*, 76.

Breedlove, J. R., and Trammell, G. T. (1970). Molecular microscopy: fundamental limitations. *Science 170*, 1310.

Burton, K., and Riley, W. T. (1966). Selective degradation of thymidine and thymine deoxynucleotides. *Biochem. J. 98*, 70.

Crawford, L. V., Follett, E. A. C., and Crawford, E. M. (1966). An electron microscopic study of DNA of three tumor viruses. *J. Microscopie 5*, 597.

———, and Waring, M. W. (1967). Supercoiling of polyoma virus DNA measured by its interaction with ethidium bromide. *J. Mol. Biol. 25*, 23.

Crewe, A. V., Wall, J., and Langmore, J. (1970). Visibility of single atoms. *Science 168*, 1338.

Di Giamberardino, L., Koller, T., and Beer, M. (1969). Electron microscopic study of the base sequence in nucleic acids. IX. Absence of fragmentation and crosslinking during reaction with osmium tetroxide and cyanide. *Biochim. Biophys. Acta 182*, 523.

Dobelle, W. H., and Beer, M. (1968). Chemically cleaved graphite support films for electron microscopy. *J. Cell Biol. 39*, 733.

Dubochet, J., Ducommun, M., and Kellenberger, E. (1970). A new method for the observation of biomacromolecules in dark-field electron microscopy. *Proc. 7th. Int. Cong. Electron Micros.*, Grenoble, Vol. 1, p. 601.

———, Ducommun, M., Zollinger, M., and Kellenberger, E. (1971). A new preparation method for dark-field electron microscopy of biomacromolecules. *J. Ultrastruct. Res. 35*, 147.

Erickson, H. P., and Beer, M. (1967). Electron microscopic study of the base sequence in nucleic acids. VI. Preparation of ribonucleic acid with marked guanosine monophosphate nucleotides. *Biochemistry 6*, 2694.

———, (1968). Visibility of single nucleic acid molecules and implications for the study of nucleotide sequence. *Proc. 4th. Europ. Reg. Conf. Electron Micros.*, Rome, p. 87.

Faerber, P., and Scheit, K.-H. (1970). Synthesis and properties of poly-2,4-dithiouridylic acid, a new analog of poly uridylic acid. *Febs. Letters 11,* 11.

Fernández-Morán, H. (1960). Single-crystals of graphite and of mica as specimen supports for electron microscopy. *J. Appl. Phys. 31,* 1840.

Fiskin, A. M., and Beer, M. (1965). A guanosine and uridine specific reagent for study of base sequence in nucleic acids. *Biochem. Biophys. Acta 108,* 159.

———, (1968). Autocorrelation functions of noisy electron micrographs of stained polynucleotide chains. *Science 159,* 1111.

Follett, E. A. C., and Crawford, L. V. (1967). Electron microscope study of the denaturation of human papilloma virus DNA. I. Loss and reversal of supercoiling turns. *J. Mol. Biol. 28,* 455.

Formanek, H., and Formanek, S. (1970). Specific staining for electron microscopy of murein *sacculi* of bacterial cell walls. *Eur. J. Biochem, 17,* 78.

———, Müller, M., Hahn, M. H., and Koller, T. (1971). Visibility of single heavy atoms with the electron microscope. *Naturwiss. 58,* 339.

Freifelder, D., Kleinschmidt, A. K., and Sinsheimer, R. L. (1964). Electron microscopy of single-stranded DNA: circularity of DNA of bacteriophage ØX-174. *Science 146,* 254.

Frommer, M. A., and Miller, J. R. (1968). Adsorption of DNA at the air-water interface. *J. Phys. Chem.* 72, 2862.

Gal-Or, L., Mellema, J., Moudrianakis, E. N., and Beer, M. (1967). Electron microscopic study of base sequences in nucleic acids. VII. Cytosine-specific addition of acyl hydrazides. *Biochemistry 6,* 1909.

Gibson, D. W. (1969). Gold(III) reactions with nucleotides and polynucleotides. Thesis, The Johns Hopkins University, Baltimore (Md).

Gibson, D. W., Beer, M., and Barrnett, R. J. (1971): Gold (III) complexes of adenine nucleotides. *Biochemistry 10,* 3669.

Gillespie, D., and Spiegelman, S. (1965). A quantitative assay for DNA-RNA hybrids with DNA immobilized on a membrane. *J. Mol. Biol. 12,* 829.

Gordon, C. N., and Kleinschmidt, A. K. (1968). High contrast staining of individual mucleic acid molecules. *Biochim. Biophys. Acta 155,* 305.

———. (1969). Adsorption of DNA on aluminum-mica. *Proc. 27th Ann. Meet. Electron Micros. Soc. Amer.,* p. 266.

Hall, C. E. (1965). Method for the observation of macromolecules with the electron microscope illustrated with micrographs of DNA. *J. Cell Biol.* 2, 625.

Hashimoto, H., Kumao, A., and Hino, K. (1971). Images of thorium atoms in transmission electron microscopy. Japan. J. Appl. Phys. *10,* 1115.

Hayat, M. A. (1970). "Principles and Techniques of Electron Microscopy: Biological Applications," Vol. 1. Van Nostrand Reinhold Company, New York.

Hayatsu, H., and Ukita, T. (1964). Selective modification of cytidine residue in ribonucleic acid by semicarbazide. *Biochim. Biophys. Res. Comm. 14,* 198.

Haydon, G. B. (1969). Visualization of substructure in ferritin molecules: an artifact. *J. Microscopy* 89, 251.

Heidenreich, R. D., Hess, W. M., and Ban, L. I. (1968). A test object and criteria for high resolution microscopy. *J. Appl. Crystallogr. 1,* 1.

Heinemann, K. (1970). A comment on mica as electron microscope specimen support film. *28th Ann. Proceedings EMSA,* p. 526.

Henderson, W. J., and Griffiths, K. (1972). Shadow Casting and Replication. *In:* "Principles and Techniques of Electron Microscopy: Biological Applications," Vol. 2 (Hayat, M. A. ed.). Van Nostrand Reinhold Company, New York.

Henkelman , R. M. and Ottensmeyer, F. P. (1971). Visualization of single heavy atoms by dark field electron microscopy. *Proc. Nat. Acad. Sci. 68,* 3000.

Highton, P. J., and Beer, M. (1963). An electron microscopic study of extended single polynucleotide chains. *J. Mol. Biol. 7,* 70.

———. (1968). The minimum mass detectable by electron microscopy. *J. Roy. Micros. Soc. 88,* 23.

———, Murr, B. L., Shafa, F., and Beer, M. (1968). Electron microscopic study of base sequence in nucleic acids. VIII. Specific conversion of thymine into anionic osmate esters. *Biochemistry 7,* 825.

Hirt, B. (1969). Replicating molecules of polyoma virus DNA. *J. Mol. Biol. 40,* 141.

Hodge, A. J., and Schmitt, F., (1960). The charge profile of the tropocollagen macromolecule and the packing arrangement in native-type collagen fibrils. *Proc. Natl. Acad. Sci. 46,* 186.

Karu, A. E., and Beer, M. (1966). Pyrolytic formation of highly crystalline graphite films. *J. Appl. Phys. 37,* 2179.

Kikugawa, K., Hayatsu, H., and Ukita, T. (1967). Modifications of nucleosides and nucleotides. V. A selective modification of cytidylic acids with Girard-P reagent. *Biochim. Biophys. Acta 134,* 221.

Kleinschmidt, A. K., and Zahn, R. K. (1959). Über Desoxyribonucleinsäure-Molekeln in Protein-Mischfilmen. *Z. Naturf. 14b,* 770.

Koller, T. (1970). Probleme der Hochauflösungselektronenmikroskopie an Nuclein-Säuremolekülen. *Hoppe-Seyler's Zeitschr. Physiol. Chemie 351,* 117.

Koller, T. (1971). Suitability of aluminum oxide support films for high resolution electron microscopy. *Proc. 15th Ann. Meeting Biophys. Soc., New Orleans. Biophys. J. 11,* 216a.

———, Lee, Y. K., and Beer, M. (1969a). New methods for the study of nucleic acids and nucleoproteins with the electron microscope. *Proc. 13th Ann. Meeting Biophys. Soc., Los Angeles. Biophys. J. 9,* A-173.

———, Harford, A. G., Lee, Y. K., and Beer, M. (1969b). New methods for the preparation of nucleic acid molecules for electron microscopy. *Micron 1,* 110.

———, and Müller, M. (1970). Preparation of nucleic acid molecules for high resolution electron microscopy. *Proc. 7th Intern. Congr. Electr. Micr., Grenoble,* Vol. 1, p. 605.

Komoda, T., Nishida, J., and Kimoto, K. (1969). Beryllium single crystal flakes as substrates for high resolution electron microscopy. *Japan. J. Appl. Phys.* 8, 1164.

Kössel, H. (1965). Zur Reaktion von Mononucleotiden mit Diazoniumsalzen. *Z. Physiol. Chem. 340,* 210.

Kühn, K., Grassmann, W., und Hoffmann, U. (1958). Die elektronenmikroskopische "Anfärbung" des Kollagens und die Ausbildung einer hochunterteilten Querstreifung. *Z. Naturf. 13b,* 154.

Lang, D., Kleinschimdt, A. K., und Zahn, R. K. (1964). Konfiguration und Längenverteilung von DNA-Molekulen in Lösung. *Biochim. Biophys. Acta 88,,* 142.

———, Bujard, H., Wolf, B., and Russell, D. (1967). Electron microscopy of size and shape of viral DNA in solutions of different ionic strength. *J. Mol. Biol. 23,* 163.

Lezius, A. G., and Rath, U. (1971). Synthesis of poly d[(A—s^{4}T) · d(A—s^{4}T)] by *Bacillus subtilis DNA polymerase. Eur. J. Biochem. 24,* 163.

Mayor, H. D., and Jordan, L. E. (1968). Nucleic acid molecules: new microdiffusion technique for visualization. *Science 161,* 1246.

Miller, Jr., O. L., and Beatty, B. R. (1969a). Visualization of nucleolar genes. *Science 164,* 955.

———. (1969b). Extrachromosomal nucleolar genes in amphibian oocytes. *Genetics 61,* 133.

———. (1969c). Portrait of a gene. *J. Cell. Physiol. 74,* (Suppl. 1), 225.

Miller, Jr., O. L., and Hamkalo, B. A. (1971). Electron microscopy of active eukaryotic and prokaryotic genes. *Proc. 15th Ann. Meeting Biophys. Soc. New Orleans. Biophys. J. 11,* 219a.

———, Hamkalo, B. A., and Thomas, Jr., C. A. (1970). Visualization of bacterial genes in action. *Science 169,* 392.

Moshkovski, Y. S., Malysheva, L. F., Mirlina, S. Y., and Seytanidi, K. L. (1968). Selectivity of interaction of potassium chloroplatinite with deoxyribonucleic acid. *Biofizika 13,* 320.

Moudrianakis, E. N., and Beer, M. (1965a). Determination of base sequence in nucleic acids with the electron microscope. II. The reactions of a guanine-selective marker with the mononucleotides. *Biochim. Biophys. Acta 95,* 23.

———. (1965b). Base sequence determination in nucleic acids with the electron microscope. III. Chemistry and microscopy of guanine-labeled DNA. *Proc. Natl. Acad. Sci.* 53, 564.

Müller, M., Koller, T., and Moor, H. (1970). Preparation and use of aluminum films for high resolution electron microscopy of macromolecules. *Proc. 7th Intern. Congr. Electr. Micr., Grenoble,* Vol. 1, p. 633.

O'Hara, D. S. (1968). The selective coupling of diazosulfanilic acid with amino acids in proteins: Chemistry and applications to electron microscopy. Thesis, The Johns Hopkins University, Baltimore (Md.).

Ottensmeyer, F. P. (1969). Macromolecular fine-structure by dark field microscopy. *Biophys. J.* 9, 1144.

Prestridge, E. B., and Yates, D. J. C. (1971). Imaging the rhodium atom with a conventional high resolution electron microscope. *Nature 234,* 345.

Reimer, L. (1969). Elektronenoptischer Phasenkontrast. II. Berechung mit komplexen Atomstreuamplituden für Atome und Atomgruppen. *Z. Naturf. 24a*, 377.

Riddle, G. H. N., and Siegel, B. M. (1971). Thin pyrolytic graphite films for electron microscope substrates. *Proc. 29th Ann. Meet. Electron Micros. Soc. Amer.*, (Arceneaux, C. J. ed.). Claitor's Pub. Division, Baton Rouge. La.

Sakharenko, E. T., Malysheva, L. F., and Moshkovski, Y. S. (1967). The influence of potassium chloroplatinite on physical chemical properties of native DNA. *Molekulyarnaya Biologiya 1*, 830.

Scheit, K.-H., and Faerber, P. (1971). Synthesis and properties of poly(s^2C), a poly(C) analog. *Eur. J. Biochem. 24*, 385.

Sheperd, R. J., and Wakeman, R. J. (1971). Observation on the size and morphology of cauliflower mosaic virus deoxyribonucleic acid. *Phytopathol. 61*, 188.

Sinha, K., Fujimura, K., and Kaesberg, P. (1965). Ribonuclease digestion of R 17 viral RNA. *J. Mol. Biol. 11*, 84.

Simuth, J., Scheit, K.-M., and Gottschalk, E. M. (1970). The enzymatic synthesis of poly 4-thiouridylic acid by polynucleotide phosphorylase from *Escherichia coli. Biochim. Biophys. Acta 204*, 371.

Sjöstrand, F. S. (1956). A method to improve contrast in high resolution electron microscopy of ultrathin tissue sections. *Exptl. Cell Res. 10*, 657.

Subbaraman, L. R., Subbaraman, J., and Behrman, E. J. (1971). The reaction of osmium tetroxide-pyridine complexes with nucleic acid components. *Bioinorg. Chem. 1*, 35.

Thon, F. (1966). Zur Defokussierungsabhängigkeit des Phasenkontrastes bei der elektronenmikroskopischen Abbildung. *Z. Naturf. 21a*, 476.

Ulanov, B. P., and Molysheva, L. F. (1967). Possibilities of electron microscopic determination of the position of guanine in DNA. *Biofizika 12*, 325.

———, Molysheva, L. F., and Moshkovski, Y. S. (1967). Determination of the sequence of bases in nucleic acids. *Biofizika 12*, 326.

Valentine, R. C. (1964). Graininess in the photographic recording of electron microscope images. *Nature 203*, 713.

Wang, J. C. (1969). Degree of superhelicity of covalently closed cyclic DNA's from *Escherichia coli. J. Mol. Biol. 43*, 263.

White, J. R., Koller, T., Beer, M., and Bartl, P. (1969). High resolution electron microscopy of DNA. *Proc. 27th Ann. Meet. Electron Micros. Soc. Amer.*, p. 268.

Whiting, R. F., and Ottensmeyer, F. P. (1971). Dark field electron microscopy of atoms and molecules. *Proc. 15th Ann. Meeting Biophys. Soc., New Orleans. Biophys. J. 11*, 215a.

Whiting, R. F., and Ottensmeyer, F. P. (1972). Heavy atoms in model compounds and nucleic acids imaged by dark field transmission electron microscopy. *J. Mol. Biol. 67*, 173.

Zinsmeister, G. (1964). Theory of thin film condensation. Part C: Aggregate size distribution in island films. *Thin Solid Film 4*, 363.

3 High Resolution Dark-Field Electron Microscopy

Jacques Dubochet

Biozentrum,
Basel, Switzerland

INTRODUCTION

In electron microscopy, as in optical microscopy, the dark-field image is obtained when unscattered rays are intercepted by an opaque mask. Intentionally or unintentionally, every electron microscopist has come across a dark-field image, either by misalignment of the objective aperture or by making observations outside of the field of this aperture at low magnification. It is highly probable that the originators of electron microscopes had clearly in mind the possibility of dark-field.

v. Borries and Ruska (1935) first discussed dark-field using conical illumination. The next year, Boersch (1936a and b) published a paper proving his ability to obtain micrographs of a gold film at a magnification of 20 diameters. In spite of the simplicity of his microscope, Boersch was able to show that it is more difficult to obtain a good image in dark-field than in bright-field. He also showed the importance of the spherical aberration and anticipated a loss of image quality due to electrons scattered inelastically within the specimen. The same author showed, in an indirect way, that a sufficient contrast may be obtained in dark-field by a monoatomic layer of chlorine. As early as 1937, the major possibilities and limits of dark-field were known.

v. Borries and Ruska (1938) studied the effects of spherical and chromatic aberration. A short time later (1940) they demonstrated the effect of accelerating voltage of electrons on the intensity and contrast of the image. v. Ardenne (1940a) built an electron microscope specially adapted to dark-field using conical illumination. Using a very small objective aperture, he was able to obtain a resolution almost equivalent to that obtained in bright-field (1940b). Boersch (1943a) studied the problem of intensity distribution obtained when a crystalline specimen was observed in dark-field by tilted illumination. This same system of image formation was indirectly studied when a second image was observed on out-of-focus micrographs of crystals (v. Ardenne, 1940c). These micrographs were explained as being dark-field images displaced by spherical aberration of the objective (Hillier and Baker, 1942; v. Ardenne, 1943; Boersch, 1943b). During the early forties, dark-field was also being developed in the United States (Anderson, 1942; Levy, 1944).

Schiff (1942a and b) has discussed the theoretical basis of the conditions under which a single atom is made visible in dark-field. Boersch (1947), working on the same problem, introduced interference contrast and indicated that interference contrast did not play any role in dark-field.

The papers by Hall (1948a and b) are in some way a turning point, for they distinguished between specimens producing crystalline diffraction and those producing diffused scattering. The former group includes all crystalline specimens, particularly those of metallurgy. For crystalline specimens a resolution, nearly as good as that obtained in bright-field, is generally possible in dark-field. The method has been often used in metallurgy. By using certain conventional electron microscopes it is possible to obtain an immediate dark-field image by tilted illumination (Hale and McLean, 1964). The advantages and application of this approach can be found in general books on electron microscopy (e.g., by Magnan, 1961). The application of this method has been demonstrated by Pashley and Presland (1959). The possibility of obtaining a very high resolution has been indicated by Yada and Hibi (1967). They showed in both bright- and dark-fields the 1.02 Å spacing fringes due to (200) and (200) reflections of a gold crystal.

Amorphous specimens are primarily biological and noncrystalline, but some are formed from very small crystallites, such as an evaporated carbon film. According to Hall, it is not possible to obtain a high resolution image of these specimens in dark-field. The work published prior to 1960 confirmed the poor resolving power in dark-field.

Locquin (1956), using conical illumination, and Faget *et al.* (1960), using axial illumination, showed micrographs obtained in dark-field on biological specimens while studying the control of the phase in the microscope. Contributions to the theory of image formation were made by several authors (Scherzer, 1949; Langer and Hoppe, 1967; Hanszen *et al.*, 1964 and 65; Lenz, 1965; Lenz and Wilska, 1966/67). Thon (1968) obtained the first high-resolution picture of an amorphous specimen with axial illumination dark-field.

Dupouy *et al.* (1966a and b) and Dupouy (1967) had already shown that the excellent contrast obtained in dark-field may be useful for the observation of biological specimens. These workers also indicated that this technique would become useful when used with a very high-voltage electron microscope. Johnson and Parsons (1969) explained further the contrast in dark-field in relation to the thickness of the specimen.

Dupouy *et al.* (1969) prepared unstained bacterial flagella and DNA according to the method of Kleinschmidt *et al.* (1962) and Kleinschmidt (1968), and showed that micrographs of the specimens invisible in bright-field may be obtained in dark-field. Ottensmeyer (1969) showed convincingly, in addition, that the resolving power in dark-field is at least as good as in bright-field when biological macromolecules are used.

Dubochet *et al.* (1970a and b; 1971) described a method of preparing biomacromolecules specially adapted for dark-field observation, and defined some conditions allowing high-resolution observation.

The use of a transmission-scanning electron microscope allows a different approach to dark-field image formation. In the early days of electron microscopy, such instruments were already built (v. Ardenne, 1938). Thirty years later Crewe *et al.* (1968) succeeded in achieving high resolution. They were successful in obtaining a workable field emission gun that gave sufficient brightness when the spot size was demagnified to the level of a few Ångströms. The dark-field micrographs they obtained were comparable to the best ones obtained with a conventional microscope (Crewe and Wall, 1970a). In addition, the scanning microscope provides additional information picked up by the beam on the specimen. Taking into account the information by the inelastically scattered electrons, Crewe *et al.* (1970b) were able to give good evidence for single atom visibility on dark-field images. The problem of dark-field with a transmission-scanning microscope will not be considered further in this chapter.

THEORY

An electron microscope can generally be described in the following schematic way: An electron beam picks up information while crossing the specimen; the whole or part of this information is translated by the optical system that projects it in a usable form on the screen or on photographic film.

In many cases, dark-field imaging takes more of the information contained in the beam than does bright-field. The study of this information transfer from the specimen to the image may be carried out with the help of two different physical models. The first, a corpuscular model, considers the electrons of the beam to be independent particles. By crossing the specimen, some of these particles are scattered with an eventual energy loss. The second, a wave model, considers the electrons of the beam to be independent waves with a spatial extension equal to the coherence length. The interaction with the specimen scatters the electron wave and modifies the phase.

The two models will be considered in succession. No general description of dark-field image formation has been published in the literature, although elementary details can be found in several publications, some of which are cited in the text. The discussion of mathematical formulation has been avoided; this topic can be found in the references.

Image Formation of an Amorphous Object According to the Corpuscular Theory*

Electrons may be divided into two categories on the basis of their interaction with the specimen: (1) those that are not deviated and (2) those that are deviated by an angle Θ. The deviation can be with or without an energy loss. A bright-field image is obtained by eliminating part of the deviated electrons. By eliminating all the undeviated and eventually a part of the deviated electrons, one obtains the dark-field image. Several methods can be used to realize this. Only the method by axial illumination is considered here.

The electrons not scattered from the beam are eliminated by an axial screen located in the back focal plane of the objective lens. With a good approximation, this is the image plane of the objective aperture for most microscopes. Figure 3.1 defines the characteristics of this optical system. The electron beam has a small half angle of illumination α_1. The electrons scattered in the specimen by an angle smaller than Θ_1 are stopped by the axial screen (radius R_s) on which the image (radius r) of the condenser aperture is formed. The electrons scattered by an angle between Θ_1 and Θ_0 will go through the objective aperture (radius R_a), and thus, participate in the image formation.

Electrons scattered by an angle bigger than Θ_0 will be stopped by the objective aperture. The focal length of the objective is f. This objective is considered to be a thin lens. This method has been used by Dupouy *et al.* (1966a and b), Dupouy (1967), Thon (1968), Johnson and Parsons (1969), among others. This method is also called "Strioscopy." The historical reasons for this term are no longer important; the term "dark-field by axial illumination" is preferred.

During the interaction between the specimen and the beam, the following characteristics, which are illustrated in Fig. 3.2, will be considered. The beam of uniform intensity I_o traverses a film of uniform density ρ and thickness t. At depth x, the undeviated beam has the intensity I_x. From the initial beam, the specimen scattered the intensity I_s in an angle bigger than Θ_1. This intensity I_s may be divided into one intensity I_T scattered by an angle between Θ_1 and Θ_0, and an intensity I_L scattered by an angle bigger than Θ_0. I_T is transmitted to the image in dark-field. I_L is taken away from the bright-field image. Considering that there are no electrons absorbed in the specimen, which is an excellent approximation for a thin specimen, it is then possible to write the con-

*This is mainly based on the work by Hall (1951; 1966); some definitions have been modified in order to simplify the explanation of dark-field.

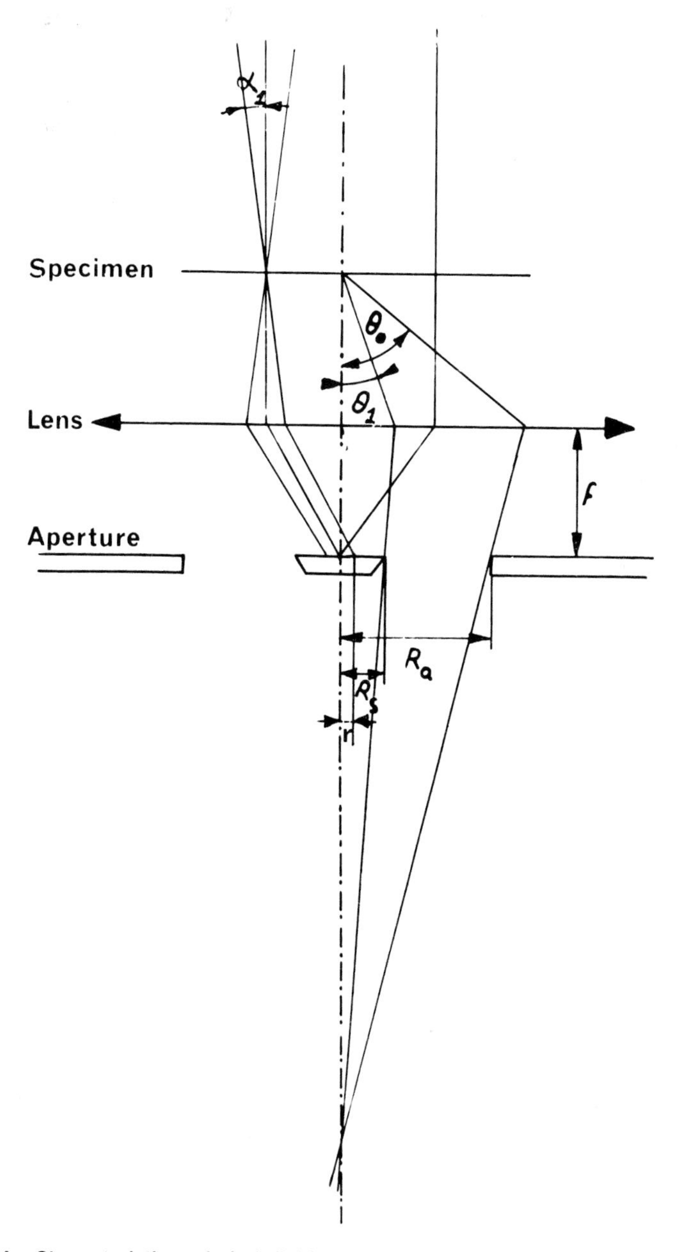

Fig. 3.1 Characteristics of dark-field by axial illumination. α_1: illumination half angle; θ_0: objective aperture angle; θ_1: objective aperture screen angle; R_a: objective aperture radius; R_s: objective aperture screen radius; r: radius of the image of the condenser aperture; f: focal length of the objective lens.

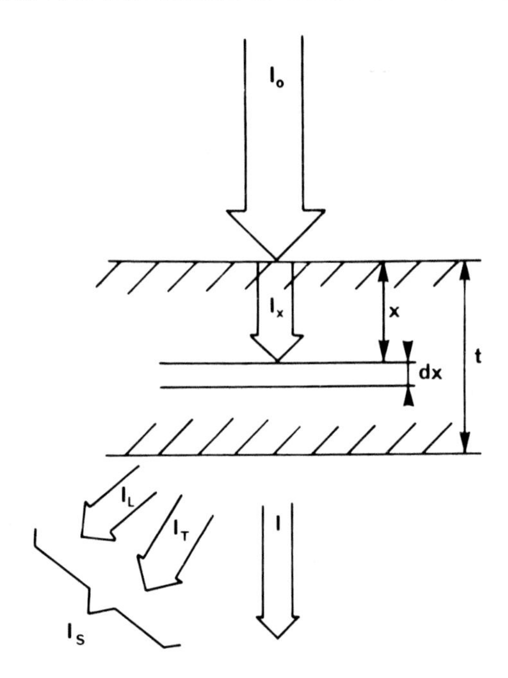

Fig. 3.2 Characteristics of the interaction between the electron beam I_0 and the specimen of thickness *t*. At the depth *x*, the unscattered beam is I_x. After having crossed the specimen, the unscattered beam is *I*; the scattered beam is I_s. I_s is divided into a transmitted intensity I_T and a lost intensity I_L.

servation equations

$$I_0 = I_x + I_s$$

$$I_s = I_T + I_L$$

The Thin Specimen

Let us consider the thin specimen as described in Fig. 3.3. It could, for example, be two thin perforated carbon films placed one on top of the other. Figures 3.4a and 3.4b are micrographs of such a specimen visualized in bright- and dark-field respectively. In the hypothesis of the thin specimen, it is considered that I_s is much smaller than I_0. This means that I_x may be considered equal to I_0 throughout the thickness of the specimen. In that case, I_s *is a linear function of the mass thickness* ρx*;* that is to say, the amount of material the beam traverses.

$$I_s = S(\Theta_1)\rho x$$

Another coefficient of scattering $S(\Theta_0)$ gives the value for I_L.

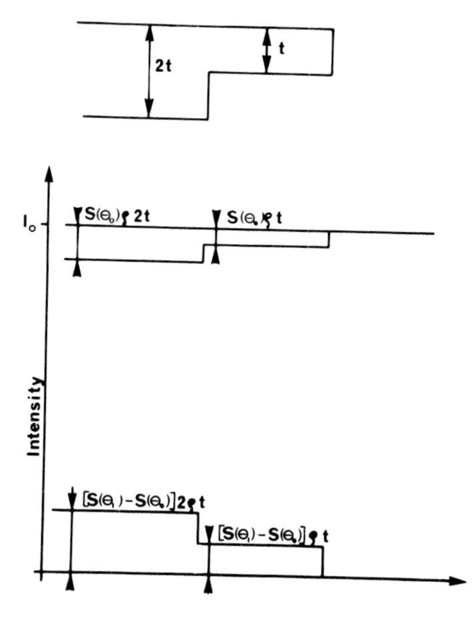

Fig. 3.3 Intensity in bright- and dark-field on the image of a thin film of thickness *t* and 2*t*. For explanation see text.

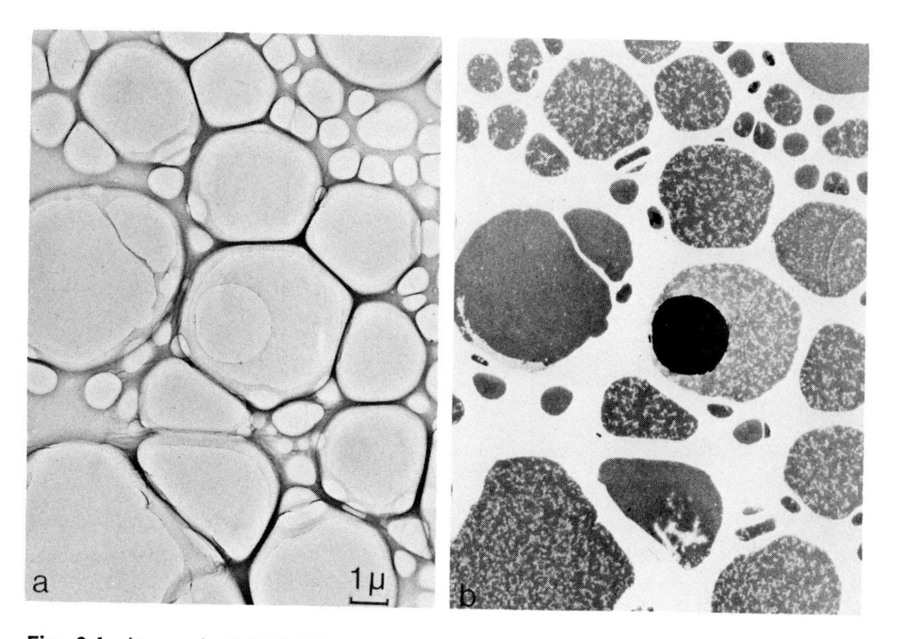

Fig. 3.4 Image in bright (a) and dark-field (b) of a thin carbon film deposited on a perforated carbon film. The hole in the thin film is barely visible in bright-field. The dirt is invisible in bright-field. It makes this film useless for dark-field observation. The thickness is in the order of 30 Å. Dark-field by conical illumination; objective aperture diameter: 25μ; objective focal length: 2.2 mm; acceleration voltage: 80 kV. AEI Electron Microscope EM 6B.

$$I_L = S(\Theta_0)\rho x$$

Figure 3.3 gives the intensity on the image of this specimen in bright- and dark-fields. In bright-field, the electrons crossing the holes are not scattered. They will reach the image, which receives the intensity I_0, disregarding the factor $1/M^2$. M is the magnification of the microscope and, as it is a constant factor, will not be considered. In dark-field all the electrons are stopped by the axial screen and the intensity on the image is zero. The object of thickness t scatters the intensity

$$I_L = S(\Theta_0)\rho t$$

outside of the aperture. This intensity will not take part in the image formation in bright-field. The object of thickness $2t$ scatters the intensity

$$I_L = S(\Theta_0)\rho 2t$$

In dark-field, only the intensity

$$I_T = I_s - I_L = S(\Theta_1)\rho t - S(\Theta_0)\rho t = [S(\Theta_1) - S(\Theta_0)]\rho t$$

will take part in the image of the region of thickness t. Twice as much intensity will form the image of thickness $2t$.

This linear relationship between intensity and thickness of the specimen has been confirmed using between 100 and 600 Å carbon film and a large objective aperture angle Θ_0 at 80 kV (Johnson and Parsons, 1969). Below 100 Å, linearity is no longer verified. The intensity seems to vary with the square root of the thickness.

The Thick Specimen

The linear relationship between mass thickness and intensity of the image in dark-field is not valid for thick specimens. Furthermore, when the object is sufficiently thick, the specimen is no longer transparent and the intensity on the screen fades to nothing. The curve relating intensity on the screen to thickness of the specimen has a maximum and takes a value of zero when the thickness is naught or infinite. This curve was first described qualitatively by v. Borries and Ruska (1938; 1940). Hall (1951; 1966) made the quantitative study of it.

With reference to the characteristics of the beam defined in Fig. 3.2, in the layer dx the contribution to I_s will be:

$$dI_x = -I_x S(\Theta_1)\rho dx$$

and then by integration it will be:

$$I_x = I_0 e^{-S(\Theta_1)\rho x}$$

From dI_x the amount kdI_x is scattered by an angle smaller than Θ_0 and contributes to I_T, that is, to the image in dark-field. The amount $(1-k)dI_x$ is scattered by an angle larger than Θ_0 and thus contributes to I_L. The introduction of this parameter k takes into account the electrons being scattered more than once. In dx, I_T is increased by kdI_x due to electrons scattered from I_x. In this same layer part of the electrons of I_T are scattered into I_L and then lost. It is possible to approximate this loss by

$$(1 - \kappa)S(\Theta_1)\rho dx \cdot I_T$$

Therefore, the differential equation will be

$$dI_T = -kdI_x - I_T(1 - k)S(\Theta_1)\rho dx$$

whose solution is

$$I_T = I_0 e^{-S\rho t}(e^{-kS\rho t} - 1)$$

The above results correspond to the experimental results obtained by Hall (1951) and more recently by Johnson and Parsons (1969). In order to give a numerical value to I_T, the knowledge of $S(\Theta)$ is necessary. This highly important parameter has been studied by several workers (Marton and Schiff, 1941; Lenz, 1954; Burge and Smith, 1962). On the basis of the work by Burge and Smith (1962), Dupouy *et al.* (1966b) and Dupouy (1967) calculated the curves that relate intensity on the screen to thickness of the specimen (Fig. 3.5).

The intensity on the image is given as a function of the thickness of the graphite film (ρ = 2.25g/cm^3). The objective aperture angle Θ_0 is 5.10^{-3} rad and the curve is calculated for several values of the parameter Θ_1. Figure 3.5a is valid for electrons accelerated at 75 kV and Fig. 3.5b at 1000 kV. For comparison with biological specimens, one has to take into account that the interaction of the beam is weaker with biological materials than with graphite. For the same mass thickness, a biological specimen is three or four times thicker than a graphite film (this may not seem consistent with the simple density ratio between graphite and biological specimens, but this is based on experimental observation and is justified because graphite is denser, crystalline and less subject to evaporation by the beam). Furthermore, the approximation of the thin specimen (defined in the previous paragraph) is valid under these conditions only for very thin specimens, e.g., 50 Å for graphite at 75 kV if Θ_0 is limited at 5.10^{-3} rad.

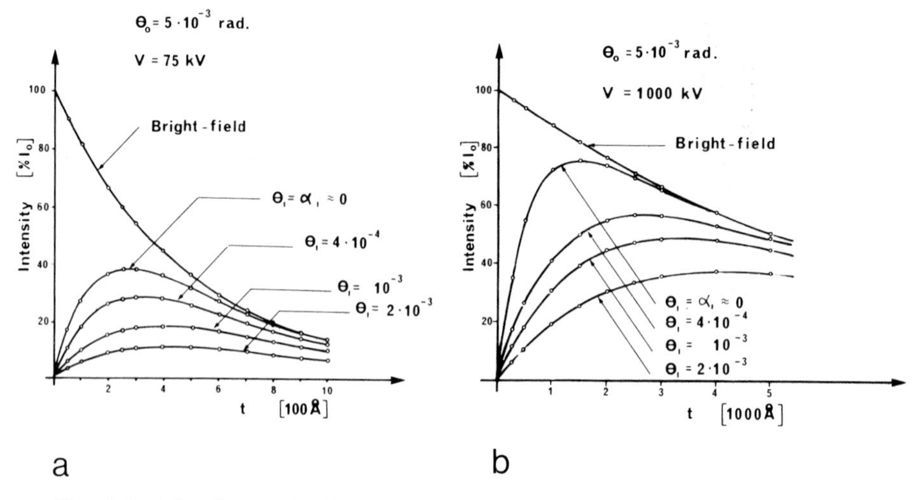

Fig. 3.5 Intensity on the image of a graphite film in bright- and dark-field by axial illumination for $\theta_0 = 5.10^{-3}$ rad and different value of θ_1. Figure 3.5a is adequate for a microscope with 75 kV accelerating voltage. It has been calculated from the data of Dupouy *et al.* (1966b). Figure 3.5b is adequate when using a microscope with 1000 kV accelerating voltage. It is redrawn from the same paper.

Contrast

Contrast C between two image points is defined by

$$C = \frac{\Delta I}{I_{\max}}$$

where ΔI is the intensity difference between the two points and $I_{\max}$ is the larger of these two intensities. The contrast may then vary from zero, for two points that have the same intensity, to one, if one of these intensities is zero. This definition is seldom used in electron microscopy but is useful for dark-field. It should be noted that the physiological perception of contrast of the film negative is related to its optical density when observed. Due to the linear relationship between the optical density of the film and the electron intensity (Valentine *et al.*, 1965), the above definition of contrast corresponds to the physiological perception. From this definition and the intensity curve of Fig. 3.5, it is not difficult to determine the contrast produced by a layer Δt laying on a film of thickness t. Dupouy *et al.* (1966b), and Dupouy (1967) have proposed a geometrical construction that will now be described.

In Fig. 3.6, if $I(t)$ is the intensity on the image produced by a film of thickness t and $I(t + \Delta t)$ is the one produced by the layer and the film, then the contrast will be

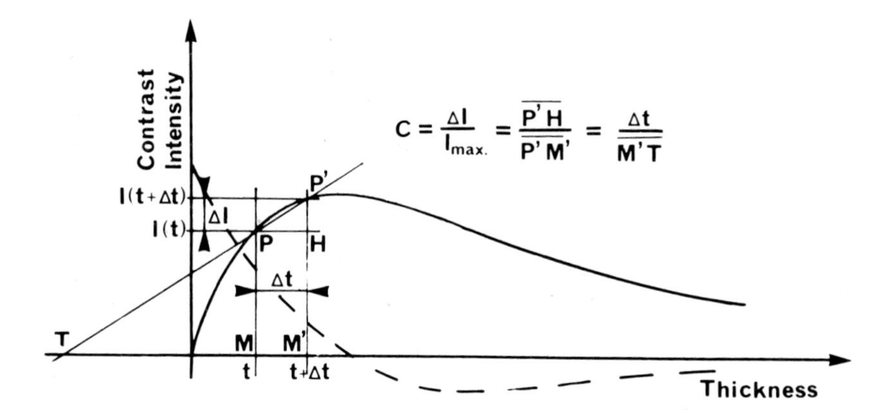

Fig. 3.6 Dark-field contrast of an object of thickness $\triangle t$ deposited on a supporting film of thickness t. If $\triangle t$ tends toward zero the contrast is expressed by the dotted curve. (After Dupouy, 1966b).

$$C = \frac{\Delta I}{I_{\max}} = \frac{\overline{P'H}}{\overline{P'M'}} = \frac{\Delta t}{M'T}$$

If Δt tends towards zero, the contrast produced by the small layer will be related to the derivative dI/dt of the curve $I(t)$.

Figure 3.7 shows the inversion of contrast for large thickness. In *A* the increment of thickness due to the folding of the film produces an increase of intensity of the image (left side of the curve in Fig. 3.6). The thickest part of the image of the specimen *B* is less illuminated than the thinner *C* (right side of the curve). Further comparison of the contrast of a thin specimen in bright- and dark-field is given below.

In bright-field, the intensity on the image of a film of thickness t will be

$$I(t) = I_0 e^{-S(\theta_0)\rho t} \simeq I_0(1 - S\rho t)$$

if

$$S\rho t \ll 1$$

The intensity on the image of the layer is

$$I(t + \Delta t) \simeq I_0[1 - S\rho(t + \Delta t)]$$

The contrast is then

$$\frac{\Delta I}{I(t)} \simeq \frac{S\rho\Delta t}{1 - S\rho t} \simeq S\rho\Delta t$$

The minimum thickness of a given material visible on a micrograph can be calculated, taking into account that the contrast has to be at

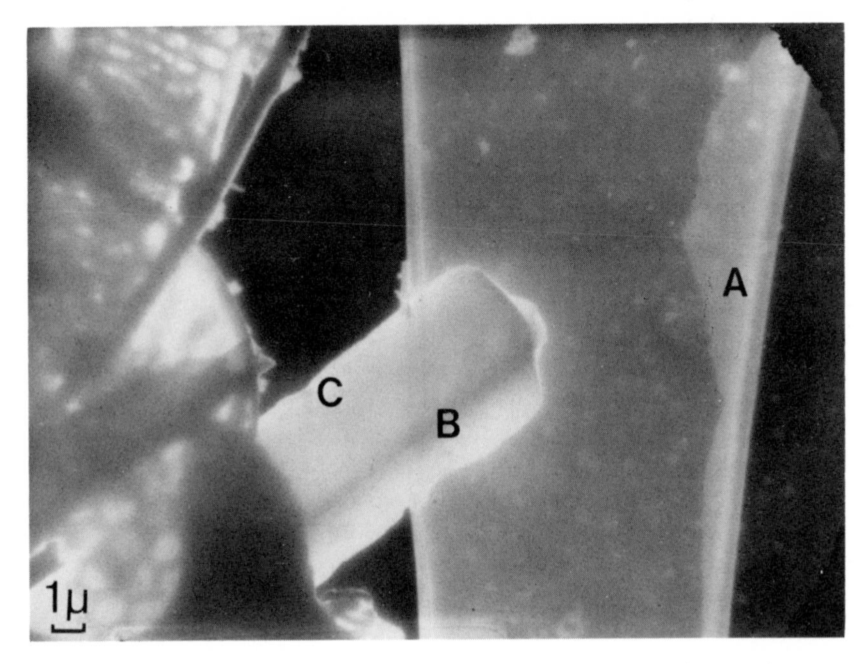

Fig. 3.7 Diatomaceous earth celite in dark-field. In the thin region, an increase of thickness by folding the film (A) produces an increase of intensity in the image. The thicker region (B) is less illuminated than its neighbor (C). Axial illumination; objective aperture diameter: 750μ^{m}; the central screen is a wire of 10μ^{m} diam.; acceleration voltage: 100 kV. Siemens Elmiskope I Electron Microscope. (After Johnson and Parsons, 1969).

least 5% in order to be seen by the human eye. From the results of Hall (1951) and De (1962), if the operating voltage is ~ 100 kV, the minimum thickness is less than 10 Å for graphite and more than 30 Å for biological carbon (what remains after exposition of biological material to the electron beam).

In dark-field, the intensity is

$$I(t) = I_0 e^{-S\rho t}(e^{kS\rho t} - 1)$$

If $S\rho t \ll 1$ the above equation becomes

$$I(t) \simeq I_0 kS\rho t$$

and

$$I(t + \Delta t) \simeq I_0 kS\rho(t + \Delta t)$$

The contrast is then

$$C = \frac{\Delta I}{I(t + \Delta t)} \simeq \frac{\Delta t}{t + \Delta t}$$

The minimum mass thickness of a layer of matter visible in dark-field may be theoretically decreased by decreasing the thickness of the supporting film. For instance, Boersch (1937) obtained a dark-field image of a gas stream by injecting CCl_4 vapors in the microscope. The density of this stream corresponded to a monoatomic layer of chlorine placed in the object plane without supporting film.

The last formula qualitatively describes the contrast of thin specimens. It seems that the corpuscular theory is no longer valid for small dimensions, and the numerical values that can be calculated from it should be considered with care. In addition, at these dimensions, it is difficult to know what the state of the observed material is and therefore its density.

Image Formation of Amorphous Objects According to the Wave Model*

As illustrated in Fig. 3.8, a plane wave falling on a grating of constant Λ^{-1} is divided. One part goes straight through while another is diffracted by an angle Θ so that

$$\frac{\lambda}{\Lambda} = \sin\Theta \simeq \Theta$$

if Θ is small. λ is the wave length of the electron.

The wave theory considers the amorphous specimen as the superposition of every possible grating, each of them diffracting an electronic

*The wave theory of the microscope may be found in any general book concerning optics, in particular by Born and Wolf (1959). In the case of the electron microscope, the reader is referred to Johnson and Wischnitzer in this volume.

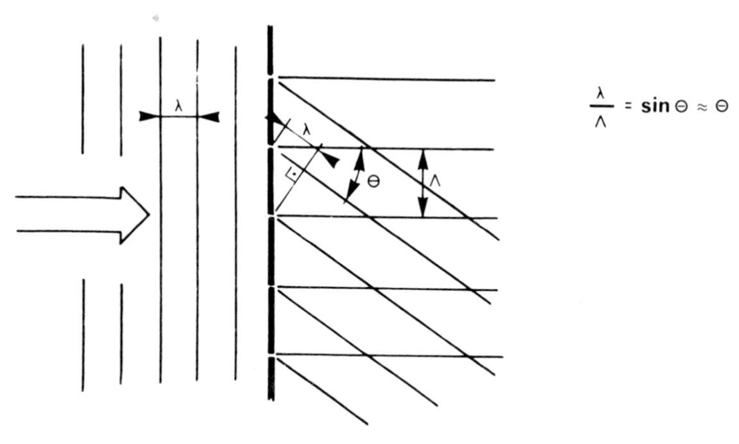

Fig. 3.8 Diffraction of a plane wave by a grating of constant Λ^{1}. λ is the wavelength. θ is the diffraction angle for the first order; it is considered to be small.

amplitude by the characteristic angle Θ. This means that all the infor tion contained in the beam, which is relative to the dimension the specimen, will be contained in the electronic amplitude diffracte angle Θ.

In the previous discussion of the thin specimen, the scattering efficient $S(\Theta)$ is defined as the electron intensity scattered by the mass thickness by an angle larger than Θ. The "differential cross sec $D(\Theta)$ can be defined as the intensity scattered by an angle Θ.*

$S(\Theta)$ and $D(\Theta)$ are related by the equation

$$S(\Theta_0) = \int_{\Theta_0}^{\pi} D(\Theta) d\Theta$$

Area Contrast

A relationship exists between each scattering direction and each poir the objective image focal plane. The intensity of this plane can be scribed, for an amorphous specimen, by $D(\Theta)$. A point on the in focal plane can be determined by its distance R to the optical axis o the diffraction angle Θ of the wave going through this point. R an are related by

$$R \simeq \Theta f$$

where f is the focal distance of the objective lens. This relation is valid if Θ is small. The position of the point can be further definec the "reciprocal spatial frequency" Λ, which is related to Θ by Bragg formula

$$\frac{\lambda}{\Lambda} \simeq \Theta$$

These three scales are shown in Fig. 3.9 for an electron microsc with a 2.5 mm objective focal length and an accelerating voltag ~60 kV. The wavelength of the electron is then 0.05 Å. In every moc electron microscope, the objective aperture plane is near the image f plane of the lens. The 25 μm aperture will cut out the information r tive to object spacings smaller than 5 Å. Resolving power will be lim at that value. The shape of the intensity curve is different from any small area to a second small area on the specimen. In bright-field, every specimen point, the objective aperture will cut off a vari amount of intensity. This is the method by which area contras created. In dark-field, the whole axial region is eliminated too should contain all information relative to the large dimensions

*With a rigorous definition $D(\Theta)$ would be an amplitude and not an intensity.

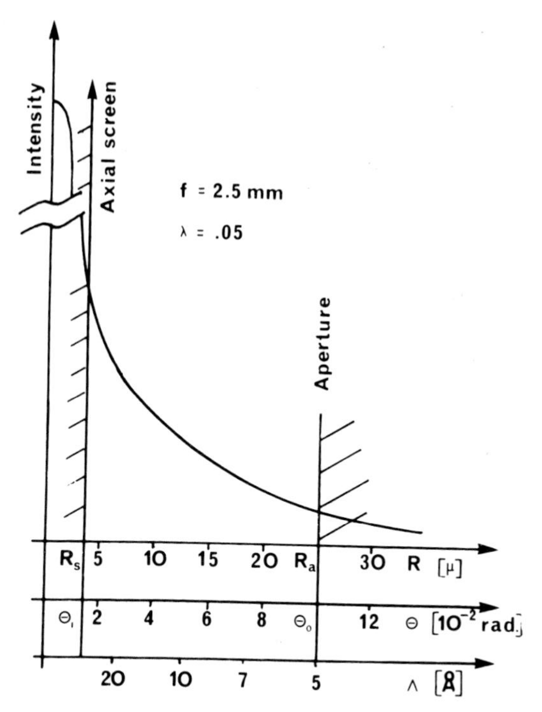

Fig. 3.9 Intensity distribution in the image focal plane of the objective lens. R is the distance to the optical axis; θ the corresponding scattering angle and Λ the corresponding reciprocal spatial frequency. R_a is the aperture redius and θ_0 the aperture angle. R_s is the screen radius and θ_1 the screen angle. The numerical relationship is valid for a focal length f equaling 2.5 mm and an electron wave length λ =0.05 Å, which corresponds to an accelerating voltage of approx. 60 kV.

the specimen; however, this is not the case because the wave model is no longer valid at that dimension. The reason for this is that the spatial extension of the electron wave (coherence length) is not larger than, e.g., 30 Å. A considerable electron intensity, but little information regarding the specimen, goes through the axial region. Most of the significant information, but little electron intensity, goes through the "far from axis" region. It is only this intensity that is kept in dark-field image formation. In bright-field the nonsignificant intensity is spread over the image, covering the significant information.

Interference Contrast

In crossing the specimen, the electron wave is not only diffracted by an angle Θ but its phase is changed by $\Delta\varphi$ as compared to the undiffracted beam. In optical microscopy, Zernike (1942) succeeded in using this phase information to form an image in his "phase contrast" microscope. In electron microscopy Boersch (1947) first studied the possibility of using this phase shift to get an increase of image contrast. During the last ten years, a number of studies have been made to use this phase information, although not much success has been achieved until recently (Thon and Willasch, 1970; Hanszen, 1969). However, the phase of the diffracted wave plane plays a highly important role in image formation in bright-field (Thon, 1965; 1966).

Interference at the level of the image, between diffracted and undiffracted beams, produces a nonlinear intensity term. This nonlinear intensity is explained as follows. Consider a specimen formed by a virus on a supporting film in a linear system. The image is the sum of the image of the virus plus the image of the supporting film. In a nonlinear system, the complete image will have, in addition, a composite term: supporting film-virus. The nonlinear term often produces a strong contrast on the image. This contrast is generally called "phase contrast." The relation between "phase contrast" image structure and object is generally difficult. In axial electron microscopes this relation is highly complicated because a supplementary phase shift is introduced by the optical system (Scherzer, 1949; Thon, 1965 and 1966). It is preferred, therefore, to call this contrast "interference contrast" and use the word "phase contrast" when it is directly related to the specimen structure.

Thon (1966) had shown that in bright-field, under normal working conditions, the interference contrast produces a filtering effect on reciprocal space frequencies Λ. This effect is dependent upon focusing.

In dark-field, the image formation seems to be linear for the dimensions used in biological observations. As the undiffracted beam is eliminated from the image, it cannot participate in the interference phenomenon. Furthermore, in practice no interference between diffracted beams has been shown to play a significant role. However, at the dimensions used in biological observation, the linearity of image formation has to be considered with care. Spurious structures could appear, as shown by Hanszen (1969), on a light-optical model.

The linearity of image formation in dark-field means, in other words, that elements of the image correspond directly to elements of the object and vice-versa. This is not the case in bright-field for dimensions smaller than, e.g., 30 Å.

Figure 3.10 shows the image and its optical transform taken from a carbon film in both bright- and dark-field. These micrographs were obtained at different focus levels. In bright-field it shows that only some reciprocal space frequencies Λ are contained in the micrograph; the image is strongly dependent upon the focusing. In dark-field, the optical transform is almost independent of focusing, and therefore only the sharpness of the image will vary in a focal series. For more details, the reader is referred to Thon (1966; 1968).

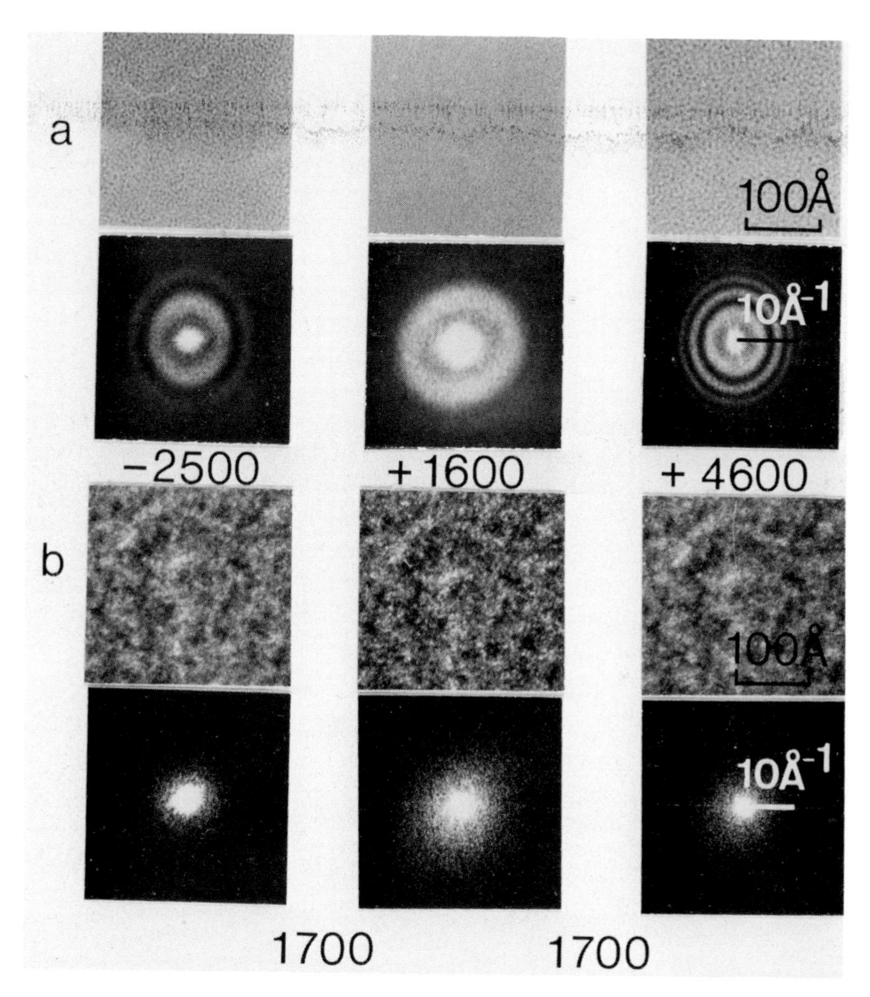

Fig. 3.10 Direct image and optical transform of a focus series of a thin carbon film in bright-field (a) and dark-field by conical illumination (b). The distance from the object plane to the object is given in Ångströms, in bright-field. In dark-field the distance of the object plane between two successive micrographs is also given in Ångströms. The effect of interference contrast and its change with focus is obvious in bright-field. There is nothing similar in dark-field. Acceleration voltage: 80 kV; focal length: $f = 2 \cdot 2$ mm; objective aperture diameter: 25 μ; objective aperture angle: $\theta_0 = 5.7 \times 10^{-3}$ radian. Bright-field: illumination angle: $\alpha_1 = 2.4 \times 10^{-4}$ radian; astigmatism: 450 Å. Dark-field: illumination angle: $\alpha_0 = 7 \times 10^{-3}$ radian; $\alpha_1 = 5 \times 10^{-4}$ radian.

Chromatic Aberration

Inelastically scattered electrons that may reach the image have lost part of their energy in crossing the specimen. In the lens, they behave differently than other electrons, thereby altering the image quality. The thicker the specimen is, the larger will be the number of inelastic electrons produced, which, in turn, will result in a greater loss of energy. The workers who have studied differential cross section have generally considered elastic scattering separate from inelastic scattering. Marton and Schiff (1941), Lenz (1954), and Burge and Smith (1962) have shown that most of the inelastically scattered electrons are found within a very small angle. Those that are scattered elastically are spread within a larger angle. Consequently, the value of the angle Θ_1 in a dark-field optical system is essential to the determination of the number of inelastically scattered electrons appearing on the image formation.

The role of chromatic aberration in dark-field image quality is not, however, completely clear. Boersch (1936b) believed that its effect is important, and a majority of the workers agree with this concept. On the other hand, Ramberg and Hillier (1948) have shown, on a theoretical basis, that the chromatic aberration would not seriously harm the image quality in dark-field and that the "bad" quality obtained in dark-field is due only to spherical aberration. Koike and Kamiya (1970) had the same idea. At present, although high resolution has been achieved only for thin specimens, it is believed that chromatic aberration is one of the most important limitations to the potential applications of dark-field electron microscopy.

OBSERVATION IN DARK-FIELD

The Types of Dark-field

Dark-field image formation can be achieved with one of several methods. The main ones are represented in Fig. 3.11 and are discussed below:

1. Dark-field by axial illumination has been used by several workers (Dupouy *et al.*, 1966a and b; Dupouy, 1967; Thon, 1968; Johnson and Parsons, 1969). The construction of the axial objective aperture is difficult. One method is to weld a thin thread through the aperture (Dupouy *et al.*, 1966a and b). Johnson and Parsons (1969), by melting the central part of this thread, produced a drop that was used as a screen. A thinner thread can thus be used to lessen dissymmetry. Thon (1968) used an aperture prepared by the more elaborate method of Möllenstedt *et al.* (1968), which was a galvanic treatment and a chemical etching of silver

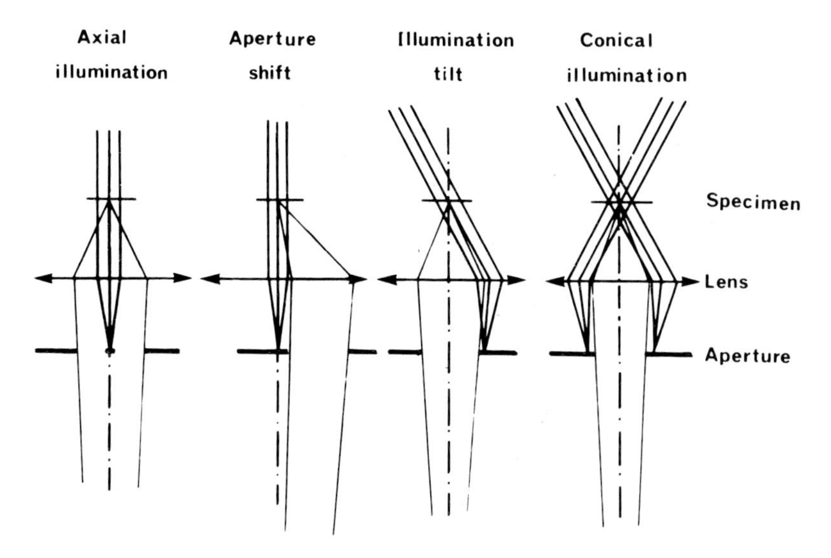

Fig. 3.11 Different types of dark-field image formation; for explanation see text.

foil partially protected by a thin layer of contaminant deposited in the electron microscope.

2. Dark-field by displacement of the objective aperture is the easiest to obtain. Using this method, the paraxial region of the objective is not used. As the spherical aberration increases with the fourth power of the distance to the optical axis, an image of good quality cannot be obtained under these conditions.

3. The method by beam tilt uses the objective lens in its paraxial region. The intensity on the screen is lower than by axial or conical illumination and the dissymmetry of the illumination may cause difficult problems of astigmatism and electrostatic charges. Several microscopes are constructed with a beam-tilt, thereby easily obtaining this type of dark-field (Hale and McLean, 1964). It is with this method that Ottensmeyer (1969) obtained his remarkable micrographs of biomicromolecules.

4. Dark-field by conical illumination has several advantages: it gives good illumination; it lessens dissymmetry; only the bridges carrying the central screen of the condenser aperture modify the cylindrical symmetry of the system. The practical realization of this type of dark-field is relatively simple. The condenser aperture generally has a dimension in the order of 1 mm.

The following discussion is limited to the axial and conical illuminations, which seem to be the most interesting. The characteristics

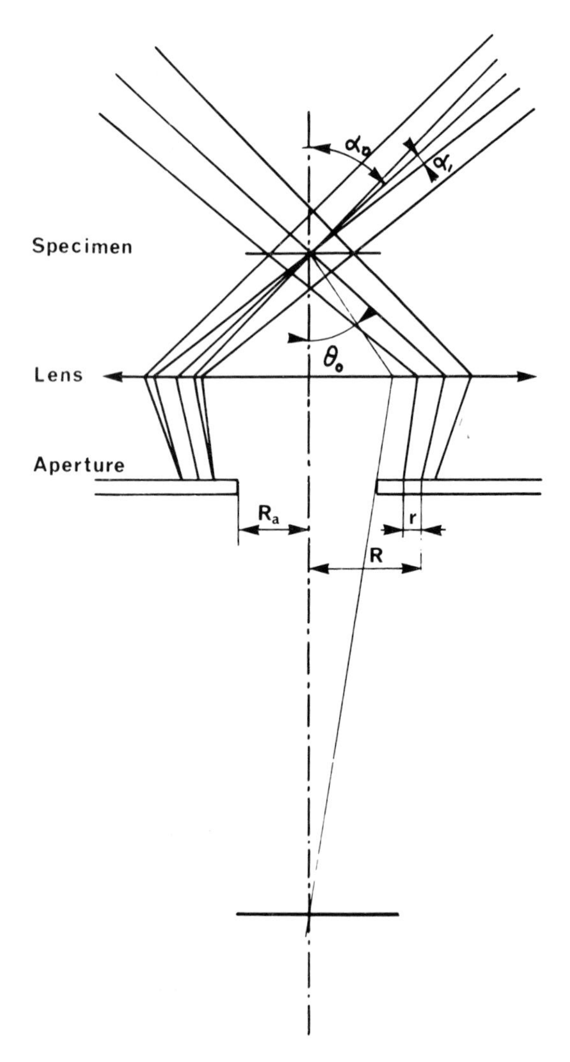

Fig. 3.12 Characteristics of the dark-field by conical illumination. α_0: half angle of the illumination cone; α_1: angle by which the ring of the condenser aperture is seen from the specimen; α_0: objective aperture angle; R_a: objective aperture radius; R: radius of the image of the condenser aperture; *r*: half-width of the image of the ring of the condenser aperture.

of the dark-field by axial illumination have been illustrated in Fig. 3.1. Figure 3.12 defines those of dark-field by conical illumination. If R_a is the radius of the objective aperture that corresponds to an angle θ_0, r is the half-width of the image of the ring, and R is the mean radius of this image, then to achieve dark-field, one should have

$$R > R_a + r$$

α_0 is the half-opening angle of the cone. The width of the ring is seen from the specimen by the angle $2\alpha_1$.

Illumination

The lack of intensity on the image is one of the most serious defects of dark-field and it is made worse by the absence of interference contrast. No Fresnel fringes are present to help a precise focusing of the image and thus it is necessary to focus on the smallest structures, that is, one must use a high electronic magnification. Consequently, a relatively high electron intensity is needed.

From the curves in Figs. 3.5a and b, it is possible to calculate the electron intensity on the image. For example, at 75 kV in dark-field by axial illumination, if Θ_1 is 10^{-3} rad and Θ_0 is 5×10^{-3} rad, then 5.6% of the total intensity in bright-field will reach the image of a 50 Å thick graphite film. A 20 Å layer of biological material will contribute to the image in dark-field less than ½% of the intensity obtained in bright-field, if one takes into account the damage to the material by the electron beam. This lack of intensity forced Ottensmeyer (1969) to make focus series; focusing on the screen was not accurate enough. Dubochet *et al.* (1970a and b), using dark-field by conical illumination, produced pictures at an electronical magnification of 20,000 to 60,000 times by exposing FRD film of Typon, whose sensitivity is comparable to the Kodak Lantern Slides from 1 to 4 sec. This method was also used by Stoeckenius (1970) to obtain sufficient illumination for micrographing at a magnification of 200,000 times.

Two methods are available to increase illumination, both of which have certain disadvantages. The first method is to increase the intensity of the beam on the object. By axial illumination this can be achieved by using the second condenser; it controls at the same time the size of the illuminated spot on the object, that is, the useful field of observation. By conical illumination the second condenser has to form the cross-over of the beam on the specimen. The intensity of the beam can then be controlled either by the bias of the Wehnelt cylinder or by changing the current of the first condenser. Unfortunately, an increase of beam intensity results in increased destruction of the specimen.

Decreasing angle Θ_1 is another possibility to increase the illumination of the image. There are two reasons why this second method is unsatisfactory. First, it increases the proportion of inelastically scattered electrons reaching the image and therefore produces a lower image quality. Second, according to the wave theory, the region near the undeviated beam contains little information relative to the dimension of ~10 Å. These are the dimensions at which dark-field methods are significant.

Resolution

In order to understand the problem of resolving power of a microscope, it is necessary to know what information is contained in the beam and what exactly will be transmitted to the image. Figure 3.13 shows this transfer in the case of both axial and conical dark-field illumination in the hypothetical case of linear image formation, i.e., in the case where interference contrast is negligible.

Consider the situation wherein Θ_1 is zero, in which case the illumination aperture angle α_1 is also zero. Then, by axial illumination, all the information picked up by the beam on the specimen will be transmitted to the image for every reciprocal spatial frequency Λ bigger than the

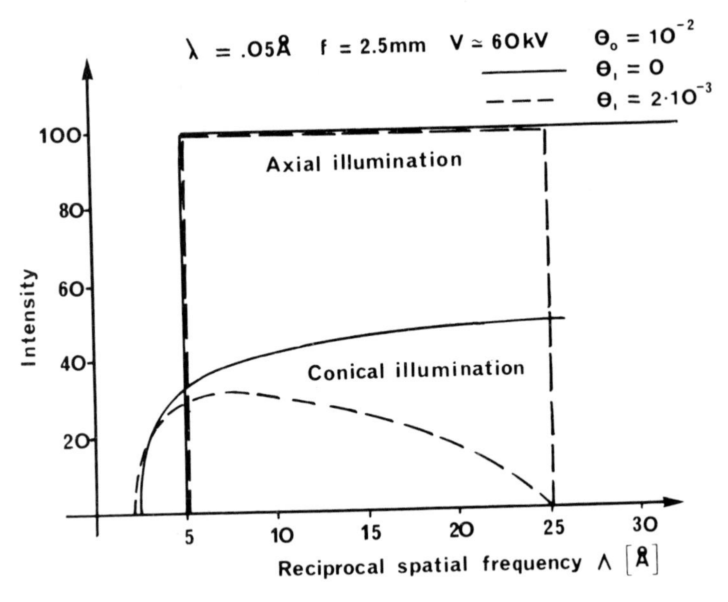

Fig. 3.13 Information transfer through the objective aperture in dark-field by axial and conical illumination.* Wave length: $\lambda = 0.05$ Å; focal length: f = 2.5 mm; objective aperture angle: $\theta_0 = 10^{-2}$.—case where $\theta_1 = 0$.—case where $\theta_1 = 2.10^{-3}$.

*in the hypothetical case of linear image formation, i.e., in the case where interference contrast is negligible.

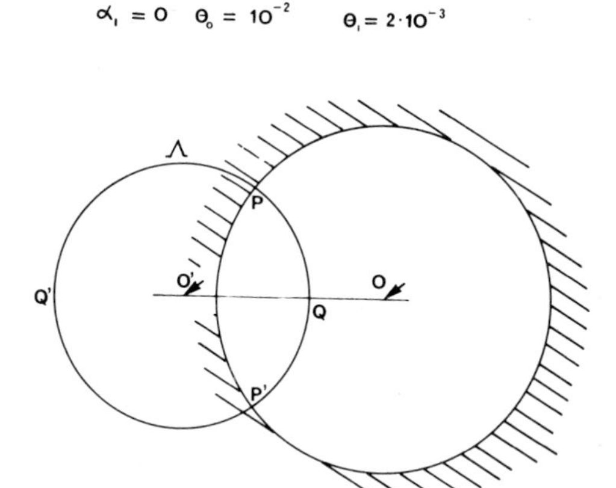

Fig. 3.14 Dark-field by tilted illumination. The objective aperture is centered on the optical axis O. The unscattered beam touches the aperture plane at O′. The intensity scattered by the reciprocal spatial frequency Λ will be on the circle QPQ′P′. Only the intensity of the arc PQP′ will participate in the image formation. PQ′P′ is lost in the aperture. This drawing is adequate only if the aperture angle θ_0 is small. The figure would be valid for: $\alpha_1 = 0$; $\theta_0 = 10^{-2}$; $\theta_1 = 2.10^{-3}$.

one corresponding to the objective aperture Θ_0. By using conical illumination in this ideal case where Θ_1 is zero, an important part of the information content of the beam is cut off by the aperture. This fact is made clear by considering dark-field by beam tilt, which is represented in Fig. 3.14. The method by conical illumination may be regarded as a multitude of beams deviated in every direction by an angle α_0. In this case, part of the beam diffracted up to an angle $\Theta = 2\Theta_0$ will appear on the image. It follows that the resolving power in dark-field by conical illumination may be twice as good as that by axial illumination for the same objective aperture angle Θ_0. The case where Θ_1 is not zero is represented by the dotted curve in Fig. 3.13.

The practical effect of suppressing high reciprocal spatial frequencies on the image is not known and may not be important. On the other hand, the difference in the information transfer by the two methods will more than likely have an effect on the image. Today, there is no comparative study at high resolution that clarifies this difference. It is important to note that a complete information transfer in the wider possible region of reciprocal spatial frequency is not suitable to provide a good image contrast (Lenz and Wilska, 1966 and 1967). Presently, it is not possible to determine which of the two methods of dark-field image formation (axial or conical illumination) is the best one for amorphous specimens.

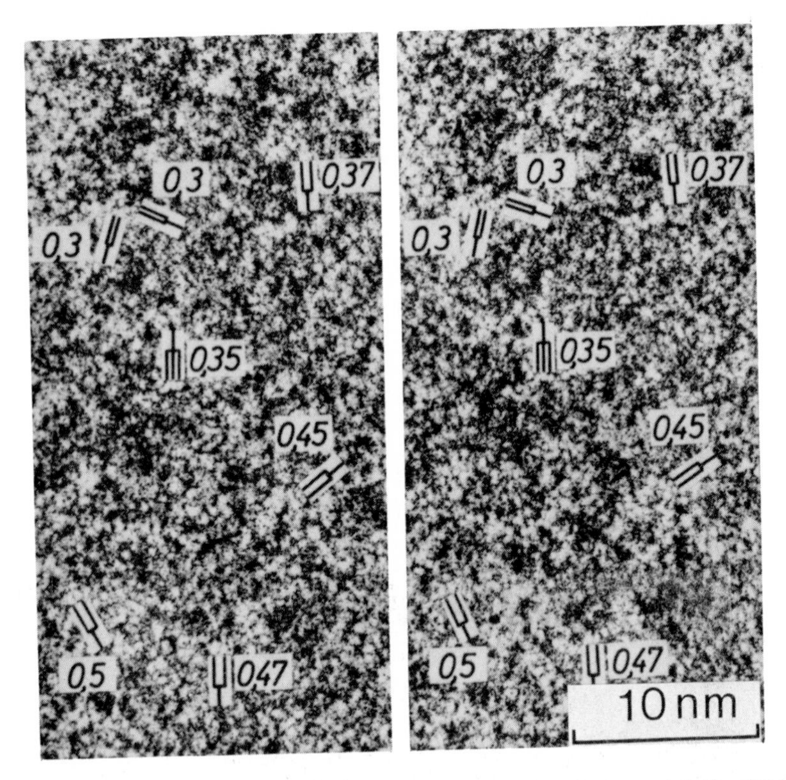

Fig. 3.15 Resolution test on thin carbon film. Dark-field by axial illumination. $\theta_1 = 1.4 \times 10^{-3}$; $\theta_0 = 2.2 \times 10^{-2}$; $\alpha_1 = 3.5 \times 10^{-4}$; $\lambda = 0.037$ Å; (from Thon, 1968). All dimensions are given in nanometer (1 nm = 10 Å).

Thon (1966) obtained the first high-resolution picture of an amorphous specimen. This micrograph is reproduced in Fig. 3.15. It is a thin carbon film whose granular structure is visible. The interpretation of these granularities is not clear.* Nevertheless, this type of specimen is adequate for resolution tests.

Dupouy *et al.* (1969), Ottensmeyer (1969), and Dubochet *et al.* (1970a and b) have obtained high-resolution micrographs of biological specimens. Their success in obtaining high-resolution pictures seems to be due to two factors. The first of which is the use of a small objective aperture angle. In bright-field the weak intensity scattered at wide angle is not sufficiently important to completely destroy the image quality, even though it undergoes spherical aberration and reaches the image when formed without objective aperture. Low resolution is thus achiev-

*Cf. Discussion following the presentation of the paper by Thon (1968) at the 4th European Regional Conference on Electron Microscopy in Rome.

able. In dark-field, on the other hand, the intensity scattered at wide angle takes a much more important role in the image intensity. The general aspect of the image is, however, much more affected by spherical aberration. It is thus necessary that the objective aperture angle should not be larger than the optimum angle Θ_{opt}, which would be (Glaser, 1943)

$$\Theta_{opt} = 1.13 \cdot \sqrt[4]{\frac{\lambda}{C_0}}$$

C_o is the spherical aberration coefficient.

A second factor that has made possible the attainment of an image with good resolution is the use of very thin specimens.

Depth of Focus

Depth of focus p is generally determined by

$$p \simeq \frac{\delta}{\Theta_0}$$

where δ is the resolving power of the microscope. This distance normally is in the order of 500 Å. In fact, a resolution of a few angströms is generally not needed in bright-field, and a rather good image can be obtained when the image is more or less out of focus. This is again due to the predominant role of the paraxial intensity. However, this is no longer true in dark-field by axial illumination and even less so by conical illumination. The depth of focus of the formula has to be respected.

PRACTICE OF DARK-FIELD

Adaptation of Electron Microscope for Dark-field Work

A consideration of the problem relative to the adaptation of an electron microscope to dark-field by axial and conical illumination is worthwhile. Since the information on axial illumination is lacking, conical illumination is here discussed.

Annular Aperture

At present, the annular aperture for the condenser can be obtained from electron microscope manufacturers. The dimensions of this aperture depend upon the microscope and objective aperture used. For the latter, one should choose a diameter slightly smaller than the optimum diameter

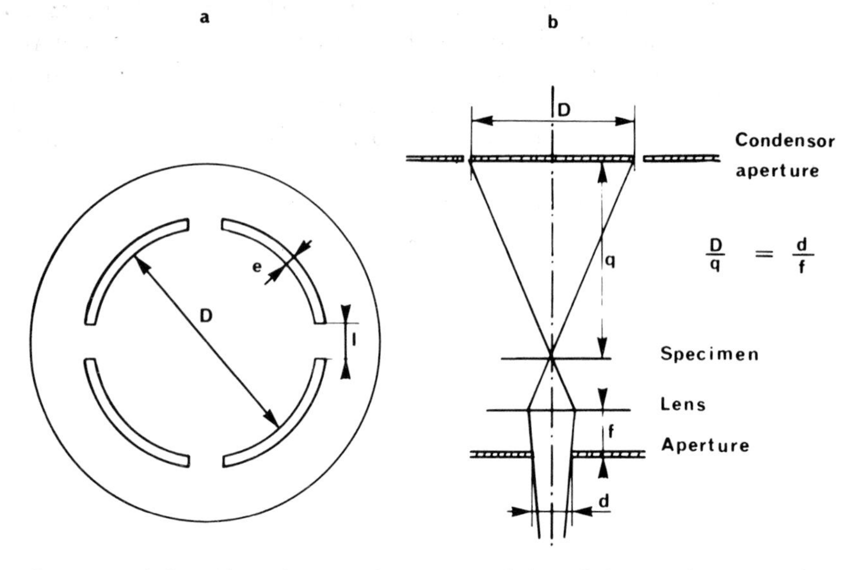

Fig. 3.16 a) Definition of the major characteristics of the condenser aperture. b) Determination of *D* in first approximation. *D* is the internal diameter of the aperture; *e* is the width of the ring; *l* is the width of the bridges; *q* the distance from the condenser aperture plane to the specimen; *f* is the focal length; *d* the diameter of the objective aperture.

$d_{opt.}$, which is deduced from the Glaser formula by

$$d_{opt} \simeq 2\Theta_{opt} \cdot f$$

The most important characteristic of the annular aperture is its internal diameter D (Fig. 3.16a), which can be calculated at a first approximation by considering the objective lens as a thin lens. Referring to Fig. 3.16b we have

$$\frac{D}{q} = \frac{d}{f}$$

where d is the diameter of the objective aperture, q is the distance from the condenser aperture to the specimen, and f is the focal length of the objective lens. For a thick lens, where the specimen and the objective aperture are in a magnetic field, this relation is incorrect. It can, however, be used to make a test aperture, which could simply be two small holes with a distance of D. By comparing the image of the distance D in the objective aperture plane with the objective aperture itself, it is possible to measure the correct value for D. In practice, several apertures of similar diameter are used. The objective of the study determines the selection of the diameter. The width of the ring e may be chosen so that the total surface of the ring is the same as that of the condenser aperture normally used in bright-field. This precaution is taken in order to avoid too much beam destruction on the specimen.

Some characteristic dimensions are:

$$D = 1.72 \text{ mm} \qquad e = 0.04 \text{ mm} \qquad l = 0.2 \text{ mm}$$

This aperture is suitable for the electron microscope AEI 6B, which has a focal length of 2.2 mm and an objective aperture of 25 μm. Such an aperture is easily built by a photographical process described in "Application Data for Kodak Photo-Sensitive Resist."

Almost every electron microscope may be adapted for conical illumination. It is sometimes necessary to remove some aperture that limits the angle of the illumination beam. If there is a coupling between the second condenser lens and the magnification selector, it should be disconnected. It does not seem very important if the image focal plane and the objective aperture plane are not coincident, as long as their distance is small compared to the focal length.

Regulation of the Microscope in Dark-field

In order to obtain good working conditions in dark-field, the electron microscope has to be aligned in bright-field. With a thin specimen and without objective aperture, the annular condenser aperture is inserted. Its centering is achieved when, varying the current in the second condenser lens, the illuminated circle on the specimen is varied concentrically. The condenser stigmator has to be adjusted in order to obtain the smallest possible spot when the cross-over of the beam is placed on the specimen. As the illumination of the optical system is not constructed to work at such a large angle, the spot at the cross-over is much larger in dark-field than in bright-field. Its dimensions define the maximum visible field. The objective aperture is centered so that a dark-field image covers the whole field. A poor centering will produce asymmetric effects on the image (Dupouy *et al.*, 1968). Once the system is aligned, it is often useful to use electromagnetic beam-shift devices to keep the illumination centered. The centering of the objective aperture may be critical (in the order of 1 μm) if the condenser aperture diameter is well adapted to it.

The astigmatism is very difficult to correct in dark-field due to lack of interference contrast. Astigmatism may be avoided by using a very clean aperture and by correcting the residual astigmatism in bright-field. As the objective aperture is strongly heated by the beam, it suffers little contamination and thus seldom needs cleaning.

Specimen Preparation

The specimen thickness is probably the most limiting factor for the application of dark-field in biology. Up till now, the only good results have been obtained on macromolecules prepared from suspensions. Unstained and unshadowed DNA of T4 bacteriophage is shown in Fig. 3.17 and a molecule of tRNA is shown in Fig. 3.18. One should note the granular aspect of the carbon supporting film. Satisfactory results have not yet been obtained with viruses such as TMV or T4 bacteriophage. The same is true for thin sections. Dubochet *et al.* (1970a and b) have indicated that the mass thickness should not exceed 300 Å g/cm³ for accelerating voltage in the order of 100 kV. This represents less than 200 Å of unstained biological material deposited on a 50 Å thick carbon film.

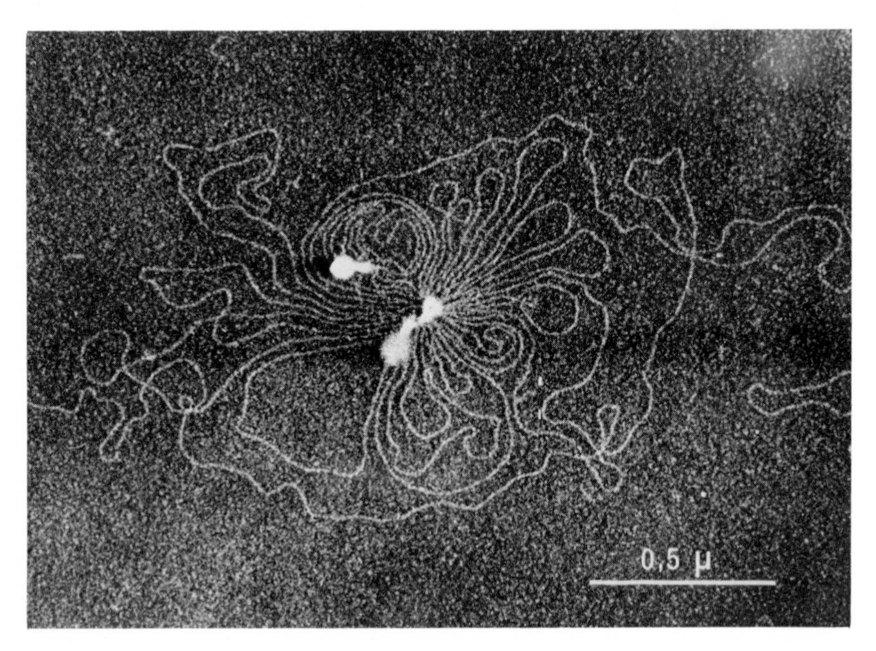

Fig. 3.17 T4 bacteriophage with its DNA spread by the Kleinschmidt method. No staining and no metal shadowing. Dark-field by conical illumination. Acceleration voltage: 100 kV. (From Dupouy *et al.*, 1969).

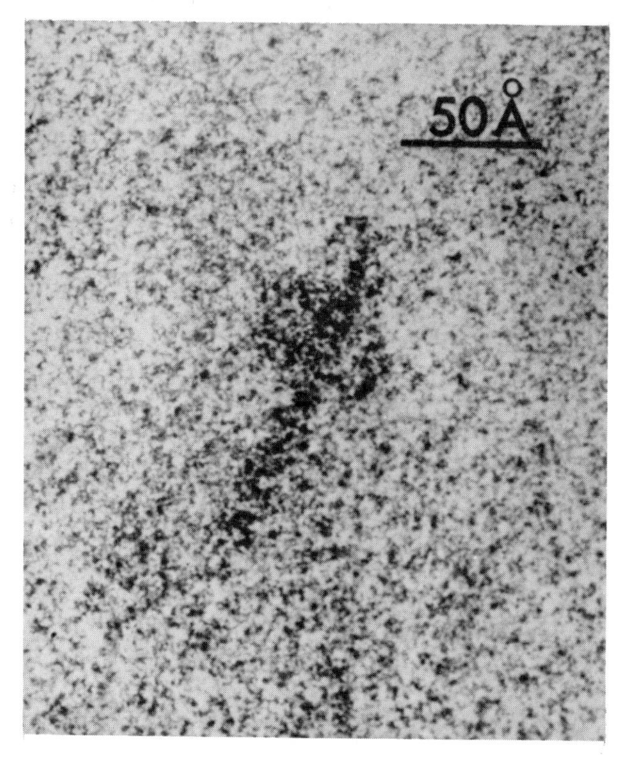

Fig. 3.18 *E. coli* *t*RNA prepared from a suspension in water and deposited on a thin carbon film. Dark-field by tilted illumination; electron microscope: Siemens Elimiskope 1. (From Ottensmeyer, 1969).

Supporting Film

The importance of the thickness of the supporting film cannot be overemphasized, as it determines the contrast of the specimen. The supporting film should not have any visible structure. At present, there is no method to make such a film, although encouraging results have recently been presented by several workers; for example, Müeller *et al.* (1970) have employed aluminum films.

Presently, carbon is the most commonly used material for films. To make the carbon film thin enough, it is better to place it on a perforated film, which can easily be prepared by the method of Fukami and Adachi (1965). The final carbon sheet is evaporated on a plastic film. The minimum thickness of the carbon film, that may be used is in the order of 30 Å. The geometry of Fig. 3.19 describes the method to control this thickness. It is obtained when a layer on the china chip P, which is

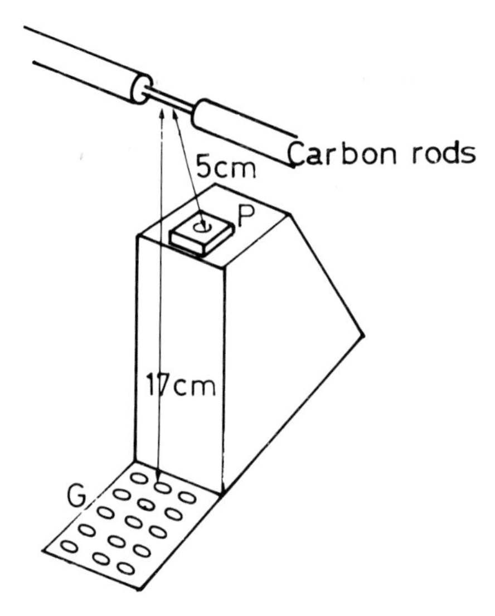

Fig. 3.19 Geometry used for controlling deposition of very thin carbon film by evaporation. *G*: grids with plastic film; *P*: China chip with an oil drop.

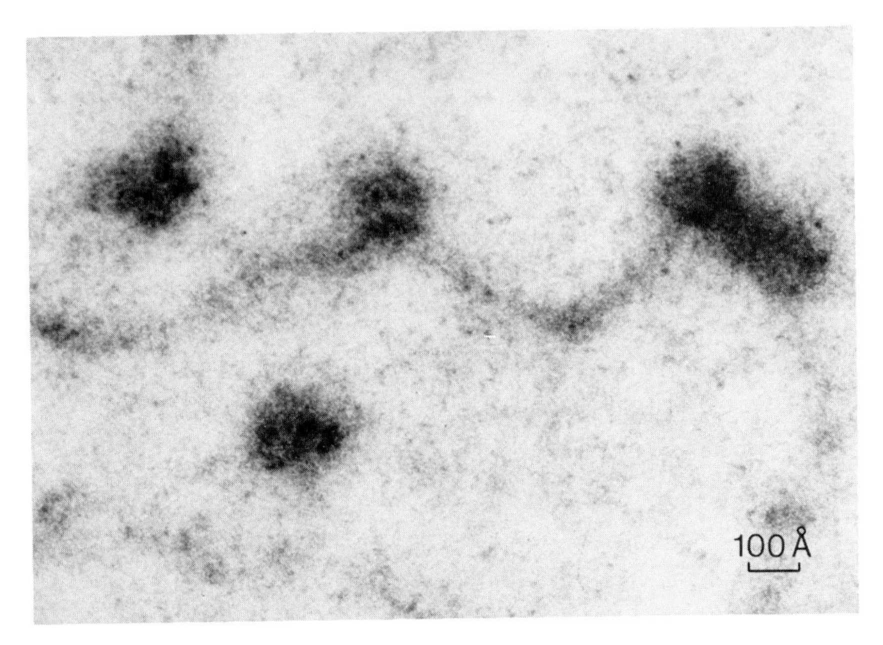

Fig. 3.20 DNA dependent RNA polymerase attached to DNA. This specimen is prepared according to Dubochet *et al.* (1971). It is stained for 10 sec with an aqueous solution of 1% uranyl acetate.

twelve times thicker than the layer on the grids G, produces a slightly brown color. The dissolution of the plastic film may be obtained by placing the grids on a few sheets of blotting paper impregnated with amyl acetate puriss. The slightest impurity is enough to ruin the film. The methods of preparing various types of films have been presented by Hayat (1970).

Transfer of the Specimen onto the Film

Figure 3.4 is an illustration of the necessity to have very clean preparation without any nonvolatile residues in dark-field. Although the film seems to be clean in bright-field, it is practically useless for dark-field. Generally, a cleaning after deposition of the material is not sufficient.

Dubochet *et al.* (1970a and b; 1971) have developed a method for selectively depositing the macromolecules of the suspension onto a carbon film negatively charged with glow-discharges in an atmosphere of amyl amine. Figure 3.20 illustrates this method. There are molecules of DNA-dependent RNA polymerase fixed on DNA. This preparation has been stained by an aqueous solution of uranyl acetate.

Staining and Shadowing

Negative and positive stains are important, not only for producing adequate contrast for bright-field observation, but also for stabilizing the specimen against the damaging effect of the electron beam. Preparations with positive or negative staining are too dense to be used in dark-field; therefore, only a very slight staining (Fig. 3.20) can be used. In this case, no supporting effect is contributed by the stain. Materials with low-scattering coefficients may eventually be used for this purpose.

The technique of metal shadowing may become very useful in dark-field. The thickness of the layer that gives enough contrast is much smaller than that required in bright-field. The metal has less tendency to form microcrystallite, thus limiting resolution. This is deduced from preliminary experiments conducted on this subject. Figure 3.21 is an example of *E. coli* DNA preparation shadowed with Pt-Ir along with polystyrene spheres. The metal layer is almost invisible in bright-field.

LIMITS AND POSSIBILITIES OF DEVELOPMENT OF DARK-FIELD

Between the theoretical resolving power of the presently available electron microscopes and the smallest resolved biological structure, dimensions differ by a factor of about one power of ten. The major reasons

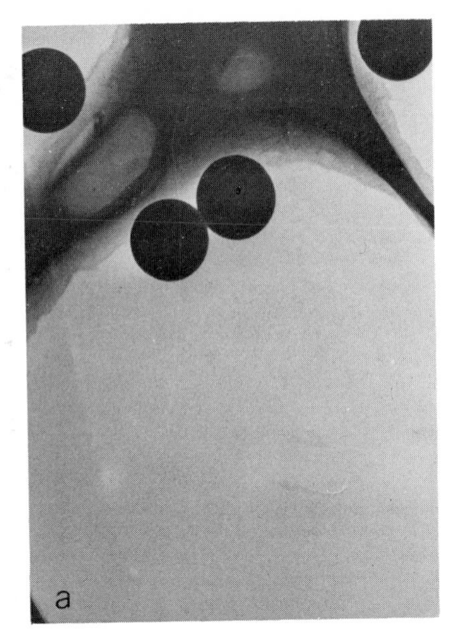

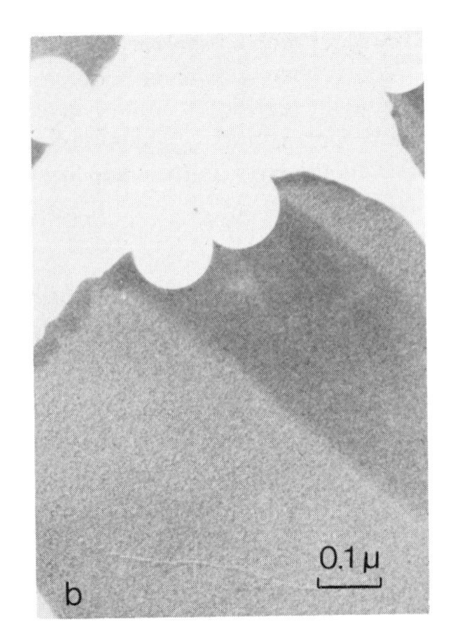

Fig. 3.21 Preparation of *E. coli* DNA with polystyrene spheres of 0.126 μ-m diameter. The preparation is shadowed at an angle of 8° with Pt-Ir. The average thickness of the metal is about 2 Å. Figure 3.21a: bright-field micrograph. Figure 3.21b: dark-field micrograph.

for this gap is the lack of contrast and the staining artifacts. By using dark-field these difficulties can be overcome in a satisfactory manner. New and more fundamental limitations are, however, presented, which were not realized in the past due to lack of contrast. As stated earlier, they are: (1) the thickness requirement of the specimen; and (2) destruction of the specimen by the electron beam.

The use of a very high-voltage electron microscope allows us to use thicker specimens (Cosslett, 1962). In this connection, the curves of Fig. 3.5b are very significant. A five-fold increase of mass thickness seems to be possible when acceleration voltage is increased from 100 to 1000 kV. Dupouy *et al.* (1966b) have illustrated this fact in dark-field.

No paper showing high-resolution dark-field pictures obtained with electron microscopes at very high accelerating voltage has been published. It is, however, probable that high resolution will be obtained very soon in thicker specimens.

The second limitation, that is, the destruction of the specimen by the beam, is more fundamental. As was shown by Thach and Thach (1970), biological materials under the beam of conventional electron microscopes may lose up to 50% of their hydrogen and phosphorous content. Such a transformation cannot be without morphological effects. This has been shown on biological crystals by Glaeser *et al.* (1970). Once more the use of a very high-voltage electron microscope would probably be a method to limit destruction. A few results obtained at present seem to show a decrease in destruction with increase of accelerating voltage for the same density of current. However, one must take into account the fact that at high-acceleration voltage, the amount of information picked up by the beam from the specimen is relatively small. At present, there is too little experimental information available on this subject to draw any definite conclusions.

ACKNOWLEDGMENTS

It is a pleasure to express my thanks to those who helped me in preparing this chapter, and especially, to Dr. Hayat, for giving me the opportunity to contribute to this volume, to Drs. Thon and Ottensmeyer, who provided me with micrographs, and to Dr. Johnson, who supplied an illustration along with very valuable comments and corrections. I am also thankful to Professor Perrier, not only for the micrograph and drawing that he furnished, but even more importantly, for the discussions we had together, which proved very helpful. Professor Kellenberger has followed this work from the beginning, and without his help, the completion of this chapter would have been difficult. I thank Miss Jobin for photographic work and Miss Melnechuk for her patience in typing and correcting this chapter.

References

Anderson, T. F. (1942). "Advances in Colloid Science." (Kramer, E. O. ed.), 1, 353. Academic Press, London and New York.

v. Ardenne, M. (1938). Das Elektronen-Rastermikroskop. Zeitschr. Techn. *Physik,* **11**, 407.

———. (1940a). Uber ein Universal-Elektronenmikroskop für Hellfeld- und Dunkelfeld- und Stereobild-Betrieb. Z. *Phys.,* **115**, 339.

———, (1940b). Ergebnisse einen neuen Elektronen-Uber mikroskop-Anlage. *Naturwiss.,* **28**, 113.

———. (1940c). "Elektronen Ubermikroskopie," p. 304. Verlag Julius Springer, Berlin.

———. (1943). Zur Deutung ubermikroskopischer Dunkelfeldbilder. Z. *Phys.,* **44**, 442.

Boersch, H. (1936a). Uber das primäre und sekundäre Bild im Elektronenmikroskop. I. Eingriffe in das Beugungsbild und ihr Einfluss auf die Abbildung. *Ann. Physik,* (5) **26**, 631.

———. (1936b). Uber das primäre und sekundäre Bild im Elektronenmikroskop. II. Stuckturuntersuchrung mittels Elektronenbeugung. *Ann. Physik* (5), **27**, 75.

———. (1937). Elektronenoptische Abbildung von Dampfstrahlen nach der Dunkelfeldmethode. *Z. Phys.*, **107**, 493.

———. (1943a). Randbeugung von Elektronen. *Z. Phys.*, **44**, 32.

———. (1943b). Nebenbilder in elektronmikroskopischen Abbildungen. *Z. Phys.*, **121**, 746.

———. (1947). Uber die Kontraste von Atomen in Elektronenmikroskop. *Z. Naturforsch.*, **2a**, 615.

Born, M., and Wolf, E. (1959). "Principles of Optics." Pergamon Press. New York.

v. Borries, B., and Ruska, E. (1935). Das Elektronenmikroskop und seine Anwendungen. *Z. VDI.*, **79**, 519.

———. (1938). Uber die Bildenstehung im Ubermikroskop. *Z. Techn. Phys.*, **19**, 402.

———. (1940). Der Einfluss der Strahlspannung auf das Ubermikroskopische Bild. *Z. Phys.*, **116**, 249.

Burge, R. E., and Smith, G. H. (1962). A new calculation of electron scattering cross sections and a theoretical discussion of image contrast in the electron microscope. *Proc. Phys. Soc.*, **79**, 673.

Cosslett, V. E. (1962). High voltage electron microscopy. *J. Roy. Microsc. Soc.*, **81**, 1.

Crewe, A. V., Wall, J., and Welter, L. M. (1968). A high resolution scanning transmission electron microscope. *J. Appl. Phys.*, **39**, 5861.

———, and Wall, J. (1970a). A scanning microscope with 5 Å resolution. *J. Mol. Biol.*, **48**, 375.

———, Wall, J., and Langmore, J. (1970b). Visibility of single atoms. *Science* **168**, 1138.

De, M. L. (1962). On the minimum mass thickness of carbonaceous matter in electron microscopy. *Exptl. Cell Res.*, **27**, 181.

Dubochet, J., Ducommun, M., and Kellenberger, E. (1970a). Specimen preparation method for dark-field electron microscopy of bio-macromolecules. *Experientia*, **26**, 692.

———, Ducommun, M., and Kellenberger, E. (1970b). A new method for the observation of bio-macromolecules in dark-field electron microscopy. *Proc. 7th Inter. Cong. Electron Micros.*, Grenoble **1**, 601.

———, Ducommun, M., Zollinger, M., and Kellenberger, E. (1971). A new preparation method for dark-field electron microscopy of bio-macromolecules. *J. Ultrastruct. Res.*, **35**, 147.

Dupouy, G. (1967). Contrast improvement in electron microscopic images of amorphous objects. *J. Electron Micros.*, **16**, 5.

———, Perrier, F., and Verdier, P. (1966a). Sur une methode permettent d'améliorer le contraste des images en microscopie électronique. *C. R. Acad. Sci.*, Paris, **262B**, 1063.

———, Perrier, F., and Verdier, P. (1966b). Amélioration du contraste des images d'objets amorphes minces en microscopie électronique. *J. Microscopie,* **5**, 655.

———, Perrier, F., and Verdier, P. (1968). Étude du contraste des objets amorphes en microscopie électronique. *Proc. 4th Europ. Reg. Conf. Electron Micros., Rome,* **1**, 155.

———, Perrier, F., Enjalbert, L., Lapchine, L., and Verdier, P. (1969). Accroissement du contraste des images d'objets amorphes en microscopie électronique. *C. R. Acad. Sci.,* Paris, **268B**, 1341.

Faget, J., Fagot, M., and Fert, Ch. (1960). Microscopie électronique en éclairage cohérent. Microscopie interférentielle, contraste de défocatisation, strioscopie et contraste de phase. *Proc. Europ. Reg. Conf. Electron Micros.,* Delft, **1**, 18.

Fukami, A., and Adachi, K. (1965). A new method of preparation of a self-perforated micro plastic grid and its application. *J. Electron Micros.,* **14**, 112.

Glaser, W. (1943). Bildentstehung und Auflösungsvermögen des Elektronenmikroskops vom Standpunkt der Wellenmechanik. Z. *Physik,* **121**, 647.

Glaeser, R. M., Budinger, T. F., Aebersold, P. M., and Thomas, G. (1970). Radiation damage in biological specimens. *Proc. 7th Inter. Cong. Electron Micros.,* Grenoble, **1**, 463.

Hale, K. F., and McLean, D. (1964). A new high resolution dark-field electron microscope technique. *Nature* **201**, 696.

Hall, C. E. (1948a). Dark-field electron microscopy. I. Studies of crystalline substance in dark-field. *J. Appl. Phys.,* **19**, 198.

———. (1948b). Dark-field electron microscopy. II. Studies of colloidal carbon. *J. Appl. Phys.,* **19**, 271.

———. (1951). Scattering phenomena in electron microscope image formation. *J. Appl. Phys.,* **22**, 655.

———. (1966). "Introduction to Electron Microscopy." Chap. 8-9, McGraw-Hill Book Company, New York.

Hanszen, K. J. (1969). Problems of image interpretation in electron microscopy with linear and non-linear transfer. Z. *angew. Phys.,* **27**, 125.

———, Morgenstern, B., and Rosenbruch, K. J. (1964). Aussagen der optischen Ubertragungstheorie über Auflösung und Kontrast im elektronenmikroskopischen Bild. Z. *angew. Phys.,* **16**, 477.

———, and Morgenstern, B. (1965). Die Phasenkontrast und Amplitudenkontrast—Ubertragung des elektronenmikroskopischen Objektivs. Z. *angew. Phys.,* **19**, 215.

Hayat, M. A. (1970). "Principles and Techniques of Electron Microscopy: Biological Applications." Vol. 1, Van Nostrand Reinhold Company, New York and London.

Hillier, J., and Baker, R. F. (1942). The observation of crystalline reflections in electron microscope images. *Phys. Rev.,* **61**, 722.

Johnson, H. M., and Parsons, D. F. (1969). Enhanced contrast in electron microscopy of unstained biological material. I. Strioscopy (dark-field microscopy). *J. Microscopy,* **90**, 199.

Kleinschmidt, A. K. (1968). Monolayer techniques in EM of nucleic acid molecules. *Methods in Enzymology,* **12,** 361.

———, Lang, D., Jacherts, D., and Zahn, R. K. (1962). Darstellung und Längenmessungen der gesamten DNS-inhaltes von T_2 phage. *Biochim. Biophys. Acta,* **61,** 857.

Koike, H., and Kamiya, Y. (1970). Contrast of atom in dark-field microscopy. *Proc. 7th Inter. Cong. Electron Micros.,* Grenoble, **1,** 27.

Langer, R., and Hoppe, W. (1967). Die Erhöhung von Auflösung und Kontrast im Elektronenmikroskop mit Zonenkorrekturplatten. III. Abbildung von Atomen bei nahezu inkohärenter Belechtung. *Optik,* **25,** 507.

Lenz, F. (1954). Zur Streuung mittleschneller Elektronen in kleinste Winkel. *Z. Naturforsch.,* **9a,** 185.

———. (1965). Kann man Biprisma-Interferenzstreifen zur Messung der elektronenmikroskopischen Anflösungsvermögen verwenden. *Optik,* **22,** 270.

———, and Wilska, A. P. (1966/67). Electron optical systems with annular apertures and with corrected spherical aberration. *Optik,* **24,** 383.

Levy, G. B. (1944). Dark-field illumination in electron microscopy. *J. Appl. Phys.,* **15,** 623.

Locquin, M. (1956). Nouvelle methode d'observation en microscopie électronique. Le contraste de phase et le contraste interchromatique, *C. R. Acad. Sci., Paris,* **242,** 1713.

Magnan, C. (1961). "Traité de Microscopie Electronique." Hermann, éd., Paris.

Marton, L., and Schiff, L. I. (1941). Determination of object thickness in electron microscopy. *J. Appl. Phys.,* **12,** 759.

Möllenstedt, G., Speidel, R., Hoppe, W., Langer, R., Katerbau, K. H., and Thon, F. (1968). Electron microscopical imaging using zonal correction plates. *Proc. 4th Europ. Reg. Conf. Electron Micros.,* Rome, **1,** 125.

Müller, M., Koller, T., and Moor, H. (1970). Preparation and use of aluminium films for high resolution electron microscopy of macromolecules. *Proc. 7th Inter. Cong. Electron Micros.,* Grenoble, **1,** 633.

Ottensmeyer, F. P. (1969). Macromolecular fine structure by dark-field electron microscopy. *Biophys. J.,* **9,** 1144.

Pashley, D. W., and Presland, B. A. (1959). The observation of antiphase boundaries during the transition from CuAu I to CuAu II. *J. Inst. Met.,* **87,** 419.

Ramberg, E. G., and Hillier, J. (1948). Chromatic aberration and resolving power in electron microscopy. *J. Appl. Phys.,* **19,** 678.

Scherzer, O. (1949). The theoretical resolution limit of the electron microscope. *J. Appl. Phys.,* **20,** 20.

Schiff, L. I. (1942a). Atomic images with the electron microscope. *Phys. Rev.,* **61,** 391.

———. (1942b). Ultimate resolving power of the electron microscope. *Phys. Rev.,* **61,** 721.

Stoeckenius, W. (1970). Dark-field microscopy using hollow cone illumination. *Proc. 7th Inter. Cong. Electron Micros.,* Grenoble, **1,** 599.

Thach, R. E., and Thach, S. S. (1970). Damage to biological samples caused by the electron beam. *Proc. 7th Inter. Cong. Electron Micros.*, Grenoble, **1**, 645.

Thon, F. (1965). Elektronenmikroskopische Untersuchungen an dünnen Kohlefolien. Z. *Naturforsch.*, **20a**, 154.

———. (1966). Zur Defokussierungsabhängigkeit der Phasenkontrastes bei der elektronenmikroskopischen Abbildung. Z. *Naturforsch.*, **21a**, 127.

———. (1968). Hochauflösende Elektronenmikroskopische Abbildung Amorpher Objekte Mittels Zweistrahlinterferenzen. *Proc. 4th Europ. Reg. Conf. Electron Micros.*, Rome, **1**, 127.

———, and Willasch, D. (1970). Hochunflösungs-Elektronenmikroskopie mit SpezialaperturebIenden und Phasenplatten. *Proc. 7th Inter. Cong. Electron Micros.*, Grenoble, **1**, 3.

Valentine, R. C., Bahr, G. F., and Zeitler, E. H. (1965). Characteristics of Emulsions for Electron Microscopy. *In:* "Quantitative Electron Microscopy." Lab. Invest., **14**, 596.

Yada, K., and Hibi, T. (1967). Fine lattice fringes resolved by the bright and dark-field axial illuminations. *J. Appl. Phys.*, **6**, 1007.

Zernike, F. (1942). Phase contrast, a new method for the microscopic observation of transparent objects. Parts I and II. *Physica,* **9**, 686.

4 In-Focus Phase Contrast Electron Microscopy

*H. M. Johnson**

Electron Optics Laboratory, Dept. of Biophysics
Roswell Park Memorial Institute
Buffalo, N.Y.

*Present Address—Atomic Energy of Canada Limited, Chalk River, Ontario, Canada.

INTRODUCTION

The imaging of biological specimens in the electron microscope has required contrasting procedures, such as heavy metal shadowing, and positive and negative staining, which have indeed brought morphologic advances. However, the elucidation of structure at the molecular level requires new techniques that extract greater image information from the electron beam and permit examination of specimens in their natural state. Electron optical methods are available for contrast enhancement, such as phase contrast, interference contrast (Fert *et al.*, 1962), dark-field imaging (Dubochet, this volume), and electronic manipulation of elastic and inelastic scattering components (Crewe *et al.*, 1970a). This work describes an investigation of phase contrast, evaluating it as a practical contrasting technique for biological images and pointing out current misinterpretations in contrasting mechanisms.

Phase contrast provides very satisfactory contrast for unstained biological specimens in light microscopy (Zernike, 1942a and b; 1955); furthermore, it provides the contrasting technique for the observation of living material at the resolution limit of the light microscope. Because electron microscopical specimens can be considered as phase objects, it has been natural to attempt to apply phase contrasting methods to the electron microscope. Since light phase microscopy equipment has been simpler and cheaper than interference equipment and this is also expected to be the case in electron microscopy, it makes phase the logical first choice of investigation.

Phase contrast is the optical technique of interfering scattered and nonscattered wave components after adjusting the phase of one component relative to the other. Scattering occurs in the specimen plane, phase adjustment in the back-focal plane, and interference in the image plane of the objective lens. Phase adjustment is necessary to achieve destructive (positive phase contrast) or constructive (negative phase contrast) interference between the two components. A phase plate in the objective lens back-focal plane increases the optical part of one component relative to the other, and simultaneously attenuates the nonscattered wave to equal the amplitude of the scattered wave for maximum contrast. The advantage of optimizing the interference between scattered and nonscattered components is illustrated in Fig. 4.1. Contrast has been plotted against relative amplitudes of two interfering waves that arose as scattered and nonscattered components from a hypothetical

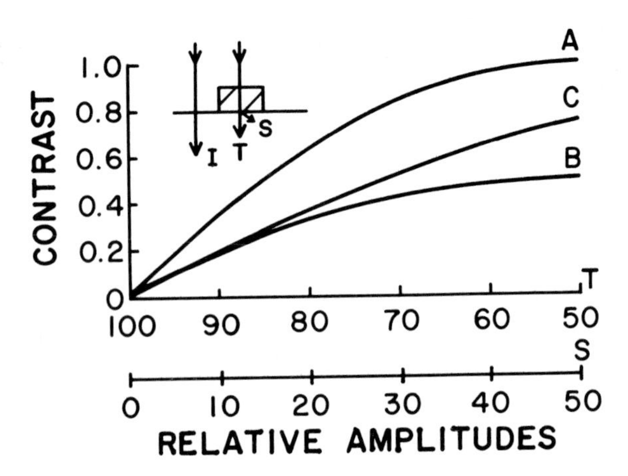

Fig. 4.1 Contrast calculation (equation 4.5) comparing the intensity over the image of a hypothetical scattering particle with that of free space. The particle varied in scattering power from zero to fifty percent of the incident amplitude. Curves A and B were calculated assuming transmitted ($\tilde{T}$) and scattered ($\tilde{S}$) amplitudes interfered according to the superposition theorem (Jenkins and White, 1957) with relative phase differences of π and $\pi/2$ rad, respectively. Calculations for curve C assumed the particle image was illuminated by intensity $|T|^2$ only, the scattering being totally removed by an objective aperture. Maximum contrast was obtained by phase contrast (A) with $\tilde{T}$ and $\tilde{S}$ of equal amplitudes. Even at unequalized amplitudes, if all scattering is coherent, phase contrast exceeds other contrast mechanisms.

specimen, having total phase differences of π rad and $\pi/2$ rad, respectively. A significant advantage in contrast is obtained for destructive interference even when the amplitude ratio is large. To achieve the necessary phase adjustment, a practical phase plate is required.

Zernike (1955) has described his discovery of phase contrast and its application to the light microscopic observation of unstained biological specimens, such as epithelial cells, otherwise invisible. Because absorption in normal electron microscopic specimens is negligible, all such specimens can be considered phase objects. As a result various workers have attempted to imitate the in-focus light optical microscope in electron microscopy. Boersch (1947) first considered phase effects in electron microscopy and proposed an electron optical phase microscope for enhancement of single atom contrast. Experimentally, phase effects were discussed and demonstrated in object space by Hillier and Ramberg (1947), and Ramberg (1949), using Fresnel fringe patterns in the shadow of a transparent edge.

In-focus phase contrast electron microscopy was initially attempted by Agar *et al.* (1949). Phase plates were constructed from collodion films pierced centrally with an etched point. Metal surface replicas were used as specimens. These efforts were hampered by instrumental difficulties, such as low illumination, difficulties in centering the phase plate, and contamination of the phase shifting film. No results were presented. Because of practical difficulties associated with phase plate construction, the ability of phase enhancement was considered limited to object size below 400 Å.

Locquin (1954; 1956; 1957) recognized problems with thin film phase plates and experimented with electrostatic fields to phase shift the nonscattered beam in the back-focal plane. An electrostatic field was generated by the charging of a pointed insulator carried on an objective aperture. The phase shift associated with a charged point such as this was demonstrated by Faget and Fert (1960), with the deflection of electron beam interference fringes at the tip of a single crystal of graphite. Locquin demonstrated enhanced contrast of the internal structure of mycomycetis fibers with this method. The images obtained by Locquin were of low magnification and resemble dark-field images obtained by the central dark ground (Dupouy *et al.*, 1966) or strioscopic (Johnson and Parsons, 1969) technique. The use of such a probe did not permit quantitation of the magnitude of the phase shifts introduced by the field nor isolation of the explicit components experiencing the phase shift.

Recently, Unwin (1970) demonstrated images of reconstituted, negatively stained tobacco mosaic virus (TMV) particles by a similar technique to that of Locquin. Unwin's phase shift was obtained by the charging of a spider's web filament, carried on the objective aperture and positioned to intercept the central beam. He has obtained good contrast of the TMV subunits. The use of such a fiber is a pragmatic operation, permitting no quantitation of conditions appropriate to a specific specimen nor control of the phase shift applied to specific components in the back-focal plane.

Kanaya *et al.* (1958) investigated phase contrast at focus using holey film phase plates. The contrast of membranes in a thin section and of molybdenum trioxide crystals was shown to have been enhanced by visual comparison of photographs and by densitometer tracings. While the phase plates were considered to have provided a $\pi/2$ rad retardation of scattered components, actual quantitation was not provided. These workers made no attempt to assess the relative contrast enhancement of a variety of phase plate films nor to discuss, for example, the contrast enhancement due to secondary scattering by the phase plate film. These experiments were recently resumed (Tochigi *et al.*, 1970) using positive and negative thin film phase plates. However, the reported results were little different from those of the earlier Japanese reports. Quantitation of the phase shift was not provided nor was the relationship between phase plate and specimen discussed.

Fert *et al* (1962) demonstrated that thin films introduced phase shifts into electron beams without destroying the beam coherence, by electron beam interferometry. Subsequently, they modified an electron microscope with a slit condenser, a slit phase plate, and an extra intermediate lens (Faget *et al.*, 1962) for in-focus phase contrast microscopy. Comparative electron micrographs of a collodion film with and without a positive type phase plate ($\pi/2$ rad retardation) demonstrated an enhancement of contrast (Fert *et al.*, 1962). These workers attempted to control the thickness of the phase plate and to control the back-focal plane region undergoing the phase shift. The phase plate was matched to the condenser aperture, aligned parallel to the condenser aperture, and protected from contamination by heating.

However, they seem to have neglected the wide angle scattering components in their consideration, giving attention to the scattering from the particle size alone. This led to the assumption that phase effects were limited to small objects for which the shape transform scattering could be separated from the nonscattered beam. Presumably, these considerations resulted from the practical problems of phase shifting with thin films close to the nonscattered beam. The neglect of consideration of the wide angle scattering was possibly a result of the severe spherical aberration operating on this region of the back-focal plane. In all of these studies, the relationship between scattering and the thickness and composition of the specimen was also neglected. Phase contrast enhancement requires that attention be given to the whole of the scattering distribution. It is not limited to operation on only the low angle scattering.

It has generally been assumed in electron microscopy that almost all of the phase contrast effect could be achieved by slight defocusing of the objective lens (Haine, 1961). This was predicted by Scherzer (1949), who obtained an integrated wavefront distortion due to spherical aberration and defocus of the objective lens out to the limiting angle of the objective aperture. A similar, but not identical, equation was derived by Heidenreich (1964) expressing the phase shift due to these same lens defects for a particular scattering angle. Experimentally, the work of Thon (1966), using carbon film granulation, and Heidenreich (1967), employing graphite and model biological objects, have been based on this concept.

Electron microscopists frequently observe the change of contrast with objective lens defocus. It has usually been assumed that this effect was due to defocus phase contrast. Similarly, the contrast variation of the same specimen viewed at focus in microscopes with different degrees of spherical aberration has been attributed to spherical-aberration-induced phase effects. A different interpretation of these phenomena is described in this report.

A survey indicates, however, that few proven phase contrast images are available in the electron microscopical literature despite numerous claims. By the usual definition, phase contrast is an interference between scattered and nonscattered components in the image plane, and only this type of interference will be considered here. In order to determine whether a comparison between an electron microscope image and a light optical phase contrast image is possible, it must be known whether the electron interference phenomenon giving rise to the image contrast occurred in object or in image space. It is conceivable to have an image consisting simultaneously of interference phenomena in object and in image space. In addition to phase effects, artifacts due to Fresnel diffraction and the displacement of bright- and dark-field images complicate the defocus image. Quantitation of image intensities and contrast is required to demonstrate the presence of a true phase effect and to assess its usefulness for biological electron microscopy.

THEORY OF PHASE CONTRAST

Phase contrast was developed as a light optical technique and was taken over into electron microscopy because of the analogy between light and electron optics. Despite the analogy between particle and wave motion discovered by Hamilton (see Born and Wolf, 1965) one must not expect the analogy between electrons and light to hold too well, considering the difference between the scattering processes of electrons and light by the specimen. Nevertheless, a discussion of the light optical instrument is provided first because the electron optical instrumentation has paralleled that of light optics.

Light Optical Phase Microscopy

The light optical phase microscope is schematically represented in Fig. 4.2. Illumination, collimated by the condenser, is incident on the specimen. Scattering arises in the interaction of the incident beam with the specimen and scattered and nonscattered components enter the objective lens. The nonscattered wave is transformed into a spherical wave focused on the back-focal plane, which then illuminates the whole of the image

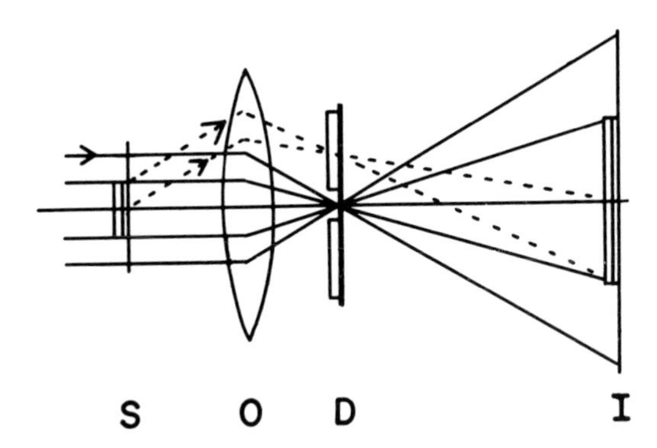

Fig. 4.2 Schematic diagram of the instrumentation of a phase contrast microscope. Parallel illumination from the condenser is scattered and transmitted by the specimen, *S*, enters the objective lens, *O*, and is focused in the image plane, *I*. In the diffraction plane (D, the back-focal plane), a phase plate is installed to increase the optical path of the scattered relative to the nonscattered waves. Amplitude equalization can be accomplished by making the central portion of the phase plate an absorber.

plane. The scattered waves enter the lens, illuminate a portion of the back-focal plane, and are focused onto the image plane. In the back-focal plane of the objective lens, there is the greatest, although not complete, separation of scattered and nonscattered waves. A phase plate in the back-focal plane selectively retards the phase of either the scattered or the nonscattered wave. Thin film phase plates also attenuate the waves that traverse them by scattering. This scattering can be removed by an intermediate aperture. In the case of operation on the nonscattered wave, this serves to equalize the amplitude of scattered and nonscattered components interfering in the image plane. Amplitude reduction of the nonscattered component has been a necessary feature of the light optical phase microscope (Francon, 1961).

The phase and amplitude relationship between scattered and nonscattered waves in the light optical phase microscope has been described by a vector diagram (Zernike, 1942a and b; 1955) representing a model rather than a proof of the relationship for a specific specimen. In such a diagram, the phase of the component, identified with the scattered waves, is retarded by $\pi/2$ rad relative to the phase of the nonscattered component. However, the use of such a model does not adequately represent the scattering from a specimen. A vector model is employed in Fig. 4.3 to represent the possible phase shift manipulations and the change in contrast each brings about.

Figure 4.3 represents phase microscopy with a thin, transparent phase object. Light, incident on a scattering region of the specimen

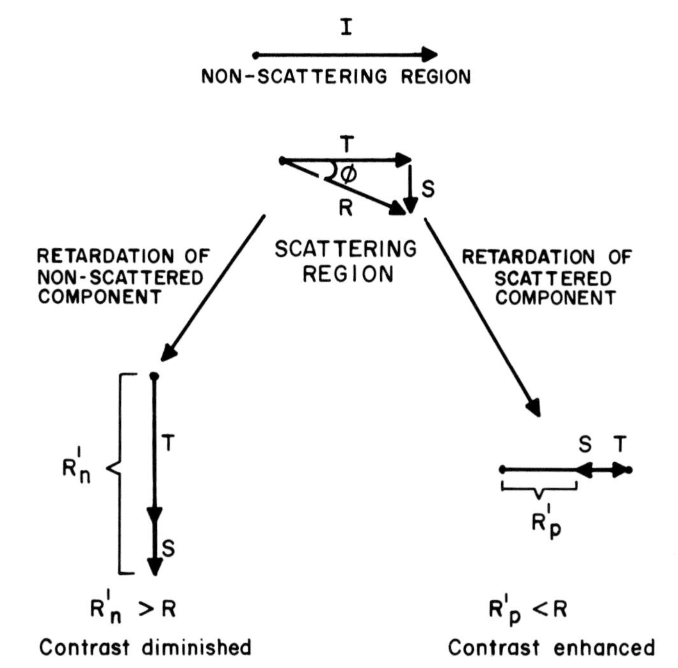

Fig. 4.3 Vector model of wave components in a microscope. The incident wave $\tilde{I}$ was scattered ($\tilde{S}$) and transmitted ($\tilde{T}$) by the specimen assumed to be in phase quadrature. The superposition of $\tilde{S}$ and $\tilde{T}$ in the image plane produced a resultant amplitude $\tilde{R}$ over the specimen image. Normally, $|R|^2 \simeq |I|^2$, contrast is very low. Negative phase contrast (retardation of nonscattered component by $\pi/2$ rad) diminished the contrast while positive phase contrast (retardation of scattered component by $\pi/2$ rad) increased image contrast.

plane, is scattered and transmitted by the phase object. In the image plane the resultant R has an intensity $|R|^2$ only slightly less than $|I|^2$, the intensity over the adjacent nonscattering area. If a phase plate operates on the nonscattered (transmitted) component, T, retarding it by $\pi/2$ rad, T and S are in phase. The resultant R_N is greater than R and contrast is reduced. Phase retarding the scattered component S by $\pi/2$ rad shifts it into anti-phase with T. In the image plane destructive interference occurs with resultant R_P' less than R. Contrast is enhanced by the positive phase manipulation, and the scattering object appears darker than the adjacent nonscattering area.

In Fig. 4.3, no attempt was made to represent equalization of the amplitudes during the phase manipulation. Optimum contrast occurs when $|T|$ and $|S|$ are equal.

Osterberg (Bennet *et al.*, 1946; 1951) employed the Luneberg diffraction integral to examine phase microscopy with amplitude and phase gratings. It was calculated that contrast of a phase grating would be enhanced using a phase plate giving a $\pi/2$ rad retardation to the scattered components. Experimentally, a phase retardation of $\pi/2$ rad of the scattered components gave maximum contrast for a variety of phase-type biological specimens, which were not highly refractile. In light microscopy a phase plate giving a $\pi/2$ phase retardation to the scattered components has become standard. Such a phase plate is not optimum for every specimen, but has been a practical compromise for the general biological specimens routinely examined.

Electron Optical Phase Microscopy

Biological specimens for electron microscopy are thin and the incident beam energy sufficiently great so that no absorption occurs. Every such specimen appears to be a phase object. However, there are complications in electron microscopy not present in optical microscopy. Irradiation of specimens by an electron beam results in scattered and nonscattered components, but the scattering involves a larger amount of inelastic than elastic components. Crick and Misell (1971) have considered the normal electron microscope image to be largely incoherent because of the dominance of the inelastic scattering. This represents a significant difference between light optical and electron optical phase microscopy.

Three inelastic processes have been identified in electron scattering from carbonaceous material (Wittry *et al.*, 1969; Crewe *et al.*, 1970b). The u.v. type scattering involves energy losses of approximately 10 eV.; plasmon excitation scattering, a long-range collective excitation of electrons, involves energy losses of approximately 25 eV.; and excitation of the K-characteristic X radiation involves energy losses of approximately 285 eV. More normally, the dominant loss is approximately 50–75 eV for typical biological specimens where plural scattering has occurred.

Low-resolution inorganic images have been obtained using inelastically scattered electrons only, which were similar to conventional images obtained with combined elastically and inelastically scattered electrons (Kamiya and Uyeda, 1961; Howie, 1963). Henry (1964) obtained weak fringes around carbon black particles on a collodion film in defocused images using only inelastic scattering with 22.8 ± 0.75 eV loss. These fringes would appear to be due to inelastic components from the particle and film, suggesting a degree of coherence in a narrow bandwidth of loss electrons. Colliex and Jouffrey (1970a,b) have also obtained low-resolution images of carbon specimens with specific-loss electrons, although little concerning coherence can be drawn from the images.

Any coherence between scattered components or between a scattered component and the nonscattered wave must take into account the basic phase shift arising in the scattering event. Bonham has suggested (R. A. Bonham, 1970, personal communication) that the appropriate single atom scattering phase shift for inelastic scattering involving losses of optically allowed transitions might be a retardation of $\pi/2$ rad, and that for inelastic scattering involving s-s electron, transitions might be an advance of π rad. It is unclear whether the single atom phase shifts can be utilized in relation to imaging with the inelastic component or whether the imaging arises from the integrated wavefront. According to Bonham (R. A. Bonham, 1970, personal communication) single atom elastic scattering involves zero phase shift, except for a small additional phase shift dependent upon scatter angle and atomic number at a given voltage. This small phase shift, for carbon irradiated by 100 kV electrons, varies from 0.04 rad ($\pi/78$) to 0.21 rad ($\pi/15$) over scatter angles extending from zero to 70 mrad. (Cox and Bonham, 1967a,b). The sign of this phase shift is not clear, but it presumably represents an amount that must be added to that phase shift predicted from consideration of the integrated wavefront (advance of $\pi/2$, Heidenreich and Hamming, 1965).

Apart from the phase shift of the scattered wave itself, arising from the scattering event, it is also required to consider phase shifts due to spherical aberration and defocus of the objective lens. The importance of these phase shifts, however, is dependent upon the degree of coherence achieved in the image plane. It is anticipated from the calculations of Crick and Misell (1971) that it would be impossible to obtain a large coherence effect because of the dominance of inelastic scattering.

In contradistinction to light lenses, spherical aberration is a severe defect in electron lenses. A high-quality light optical lens has been corrected to give a wave distortion in the exit pupil of less than 0.25 wavelengths (Rayleigh quarter wavelength criterion, Born and Wolf, 1965). Electron waves in the outer zones of an unapertured magnetic objective lens may be distorted by hundreds of wavelengths due to spherical aberration. With conventional lenses this can only be improved by increasing the acceleration voltage.

Heidenreich (1964) derived the phase shift, χ, introduced by spherical aberration and defocus of the objective lens for a specific angle of scattering, β, as

$$\chi = \frac{2\pi}{\lambda} C_0\beta^4 - \frac{\Delta f \beta^2}{2} \tag{4.1}$$

where C_o is the objective lens spherical aberration coefficient and Δf is the defocus. The scatter angle, β, is related to wavelength, λ, and object, d, by

$$\beta = \lambda / d \tag{4.2}$$

Thus, in a lens with $C_o = 1.6$ mm, the wavefront aberration on scattering equivalent to $d = 2$ Å is 10π wavelengths at 100 kV.

The spherical aberration term derived by Heidenreich (1964) was four times greater than that derived by Scherzer (1949). The Heidenreich form was derived from a two-dimensional geometry. It has been used in these calculations because of its applicability to a single ray at a specific scatter angle. The Scherzer expression was derived as an integrated wavefront aberration effect out to the limiting angle of the objective aperture, assuming uniform intensity over this scatter angle region.

The net effects of these lens defects (defocus and spherical aberration) on the image have been evaluated by the phase contrast transfer function. This function was introduced into electron microscopy by Hanszen *et al.* (1963) and Komoda (1967) and has been used to evaluate image contrast for various object spacings under specific lens conditions. The expression provided by Komoda (1967) had the form:

$$R(\omega) = -2 \sin \frac{2\pi}{\lambda} W(\omega) \tag{4.3}$$

where $R(\omega)$ related the image plane contrast to spatial frequency ω, and lens aberrations by means of the wave aberration function $W(\omega)$ given by the phase shift of equation (4.1). Eq. (4.3) was programmed for calculation by an IBM 1130 computer using the wave aberration function of Eq. (4.1), at focus by P.E. Engler of this laboratory. The contrast effects on object spacings between 3 and 13 Å are displayed.

Curve C, Fig. 4.4 represents the basic calculation of $R(\omega)$. The function $W(\omega)$ has a basic wavefront distortion of $\lambda/4$, equivalent to the assumption of $\pi/2$ rad phase retardation of scattering components, arising in the scattering event.

Contrast is assumed to be enhanced positively (structure is darker than the surround) for positive $R(\omega)$ and negatively enhanced (structure is brighter than the surround) for negative $R(\omega)$. A 2π rad phase difference exists between curve A and curve E (Fig. 4.4) so that the complete set of curves represents a cylical contrast variation. The effect of $\pi/2$ and π rad phase retardations added respectively to the scattered components (added as a positive quantity to the argument of the sine function in equation (3)) is observed in curves B and A, respectively.

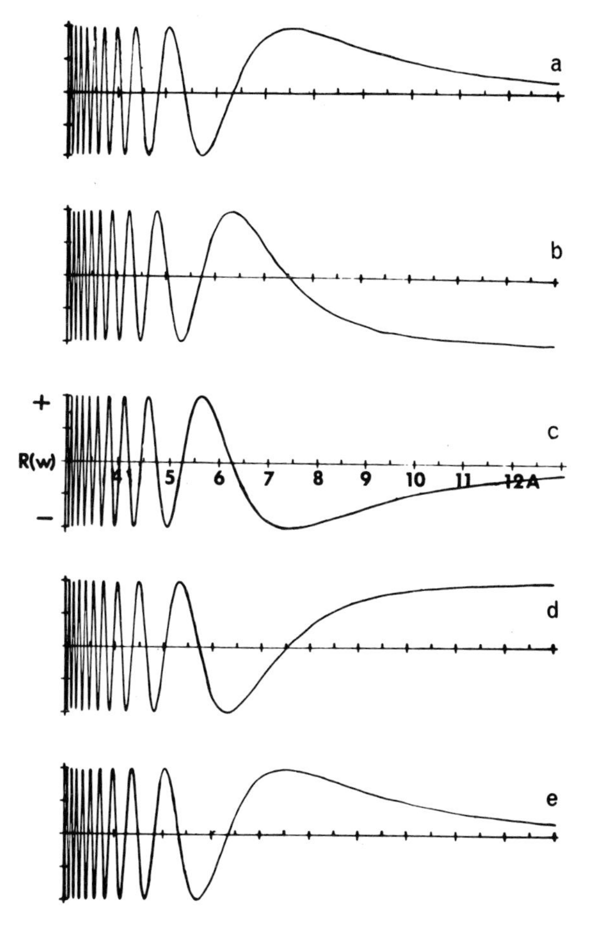

Fig. 4.4 Phase contrast transfer function, R(ω), representing the contrast enhancement (+) or diminution (−) as a function of object size when various phase manipulations are applied. Curve C represents the normal situation, assuming phase quadrature between scattered and transmitted components in the image plane. Positive phase contrast manipulations shown by curves B, A represent $\pi/2$ and π rad applied to the scattered waves, respectively. Negative phase contrast was obtained by retarding the phase of the transmitted components by $\pi/2$ and π rad, respectively, illustrated by curves D and E. These theoretical curves represent only contrast information and give no other imaging characteristics.

Phase retardations (of $\pi/2$ and π rad) added to the nonscattered component are represented by curves D and E of Fig. 4.4, respectively.

The curves of Fig. 4.4 were calculated over a limited object spacing (13 to 3 Å) because of the severe influence of spherical aberration on image contrast (for spacings less than 3 Å). The image contrast for smaller object spacings is seen to fluctuate rapidly. This was interpreted as leading to a loss of control over phase contrast. In the experimental work, objective aperture cut-off was employed at the lower limit shown on these curves to remove this uncontrollable phase effect. The upper limit of object spacing was chosen as the practical limit in the construction of positive thin film phase plates, and is related to the smallest size of phase plate hole that can be centered and to the size of the beam.

The phase of an electron wave can be retarded by an electric or magnetic field that alters the optical path of the wave. A phase contrast microscope requires back-focal plane phase plates capable of phase shifting in a selected spatial frequency band. Four possible phase shifting devices have been considered.

(a) *Evaporated thin films.* The phase of the traversing wave is altered by the inner potential of the film.
(b) *Charged probes.* The electric field created by a charged probe alters the phase of the wave component passing through the field.
(c) *Mini-electrostatic lens.* Generation of a controllable, shaped electrostatic field by a miniature three-element electrostatic lens in the back-focal plane is capable of differential phase shift of all wave components.
(d) *Magnetic field devices.* The magnetic vector potential of a magnetic domain acts to retard and advance the phase, respectively, of two electron beams that straddle the domain on adjacent sides (Aharonov-Bohn effect; Aharonov and Bohn, 1959).

Only the first possibility has been used in this work because of the relative ease of construction and the ability to quantitate the phase shift.

TABLE 4.1 PROPERTIES OF VARIOUS THIN FILM ELEMENTS APPROPRIATE TO THEIR CONSIDERATION FOR ELECTRON WAVE PHASE PLATE CONSTRUCTION

Element	*Density*	*Inner potential*	*(t) $\pi/2$*
Aluminum	$2.70 g \cdot cm^{-3}$[b]	13.0 ± 0.4 V[e]	19.7 nm
Beryllium	1.85[b]	7.8 ± 0.4[d]	26.0
Carbon	1.98 ± 03[c]	7.8 ± 0.6[c]	26.0

[a]Thickness for a $\pi/2$ rad shift was relativistically corrected and calculated according to Eq. 4.4.
[b]Handbook (1964/65).
[c]Buhl (1959).
[d]Jonsson *et al.* (1965).
[e]Johnson and Parsons (1969).

The mean inner potential, V_0 (Heidenreich, 1964) of the phase shifting film can be measured by electron beam interference techniques (e.g., Wahl, 1970). The phase shift, χ, introduced into the wave (wavelength λ and accelerating potential V_a) upon traversing a thickness t of thin film material is (Heidenreich, 1964)

$$\chi = \frac{\pi}{\lambda} \cdot \frac{V_0}{V_a} t \tag{4.4}$$

where V_o is the inner potential of the phase shifting film and λV_a is relativistically corrected. The film density, inner potential, and thickness for a $\pi/2$ phase retardation, for example, is presented in Table 4.1 for several thin film phase plate materials. In the experimental work carbon was chosen because of its low scattering cross section, measured density, and relative ease of handling in phase plate construction.

MEASUREMENT OF IMAGE CONTRAST

Detection Systems

To quantitate an electron optical phase contrast effect, image plane detection systems were constructed capable of the required sensitivity, accuracy, and convenience of operation. Such systems permitted the microscopist to observe intensity data immediately while performing electron optical manipulations.

Photographic plates have been favored for image plane detection in the past. They offer the advantage of a permanent record in a familiar form and the detection of the complete image rather than point-by-point; they are also capable of absolute measurements under carefully controlled conditions, and are nearly the perfect detector over the limited dynamic range (Valentine, 1966). Disadvantages such as limited response range, nonlinearity at optical densities above 2, the inconvenience of controlled development conditions, and the time required to retrieve quantitative image data by densitometry have limited the usefulness of plates.

A recording Faraday cage device (Johnson *et al.*, 1969) and a scintillation detection system (Johnson and Parsons, 1970a; Bradbury, 1969) were constructed for this work. The Faraday cage system consisted of a collection cup connected to a charge measurement device (vibrating reed electrometer). This system proved the most useful in the present work. The scintillation detector system counted light output pulses from a plastic scintillator as a function of electron intensity incident on the microscope side of the scintillator. The light pulses were amplified by a photomultiplier, discriminated in pulse height to remove noise, and counted. Such a system is capable of counting single incident electrons. Other detection systems using a lithium drifted silicon detector (R. M. Glaeser, 1970, personal communication) or television pick-up with video analysis (Southworth, 1966) are also possible and have been explored. However, the bulk of the image quantitation reported in this work has been accomplished with the Faraday cage system.

The complete Faraday cage system for use in the Philips EM300 electron microscope is illustrated in Fig. 4.5. The collection cup was located beneath the aperture and centered in the phosphor-coated base plate. The preamplifier of the electrometer was attached directly to the cup and the main amplifier was located nearby. The electrometer output was displayed on a strip chart recorder. Inherent noise in this system

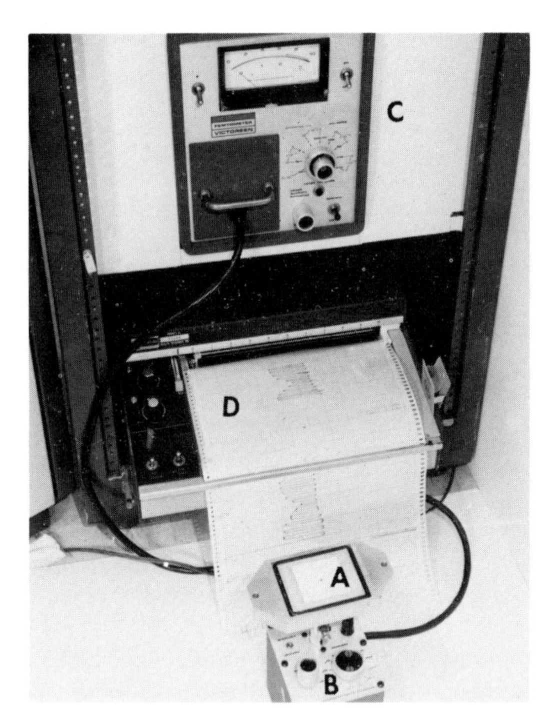

Fig. 4.5 Recording Faraday cage system for the Philips EM300 electron microscope. A cup, set beneath the admittance aperture in the center of a phosphor screen (A), was connected to the preamplifier of a vibrating reed electrometer (B). These components were mounted to a baseplate in the 70 mm. camera position. The output of main electrometer chassis (C) was recorded by the strip chart recorder (D). The recorder displays defocus intensity variations of a polystyrene latex sphere image.

was approximately 2×10^{-15} A and the signal, equivalent to an electron intensity for a 2 sec photographic exposure, was a current of approximately 10^{-13} A. The admittance aperture to the Faraday cage had an area of 3.1 mm^2. Data collection consisted of manually scanning the specimen across the admittance aperture, dwelling for 5 sec at each measurement point. The Faraday cage system is not capable of single electron counting, being limited by its inherent noise. However, in this application, abnormally low image plane intensities were not encountered and the system functioned adequately.

Contrast

Image contrast was determined from intensity data obtained with the Faraday cage system or from optical density data obtained from densitometry of photographic plates over the linear part of the response curve. Contrast was expressed by a function that approximated human visual response at normal light levels (Rushton, 1965).

In the case of the Faraday cage measurements, image contrast, C, was calculated from the expression

$$C = \frac{I_B - I_p}{I_B} \tag{4.5}$$

where I_B was the background intensity and I_p the intensity in the image of a test particle. The contrast obtained from photographic plates was determined from optical density data using the expression

$$C = \frac{OD_B - OD_p}{OD_B} \tag{4.6}$$

where OD_B was the optical density of the background and OD_p was the optical density of a test particle as recorded on photographic plates. It has been shown that these equations yield the same contrast for a given image detail (Johnson and Parsons, 1970b).

CONTRAST EFFECTS ASCRIBED TO PHASE CONTRAST

Because normal electron microscope specimens are phase objects, it has been assumed that phase contrast has been the contrasting mechanism in many electron micrographs. As was discussed earlier, spherical aberration and defocus phase shifts of scattered components can be responsible for such effects. However, the contrast of images is complicated by other optical phenomena that obscure the phase contrast mechanism. Principal among these are Fresnel diffraction and the displacement of dark-field from bright-field images.

Fresnel Fringe Enhancement of Small Particulate Images

Small particulate specimens, less than 20 Å in diameter, exhibit dramatic contrast changes upon defocusing even when imaged with an objective aperture that removes the wide angle scattering components. This contrast variation is usually attributed to phase contrast based on the observations that contrast is inverted on opposite sides of focus and minimum near focus (e.g., Hall and Hines, 1970). However, it will be shown that the dominant contrasting mechanism for such images is due to Fresnel diffraction and this phenomenon obscures any defocus phase contrast effect. In any case, as can be seen from Fig. 4.3, it is not possible to completely reverse the contrast of a particle relative to the background for phase plate systems that do not involve amplitude equalization of scattered and nonscattered components.

The influence of Fresnel fringes has been illustrated by defocus images of two types of specimens, representing the inverse in object density distribution. Type 1 specimens consisted of a dense particle on a less dense substrate (gold particles on a carbon support) and type 2 specimens consisted of low density inclusions in a higher density matrix (holes in a carbon film).

Reisner (1964) related the Fresnel fringe width, y, (the peak to trough distance in the over-focus fringe) to defocus, Δf, and wavelength λ (see also Sommerfield, 1954):

$$y = \sqrt{\Delta f \lambda} \tag{4.7}$$

Thus, in defocusing to achieve phase contrast for particulate images,

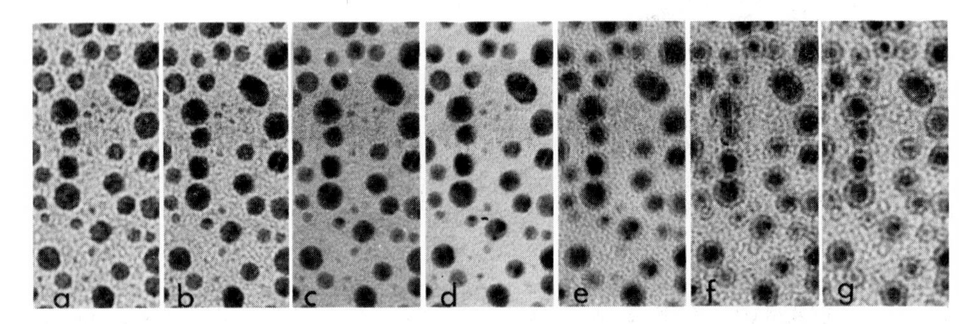

Fig. 4.6 Selected micrographs from a through-focus series of gold particle images obtained with a 10μm. objective aperture in the Philips EM300 at 100 kV. Images displayed represent 1500 Å defocus steps from +4500 (a) to −4500 Å (g). Contrast of small gold clusters was made darker by underfocusing and brighter by overfocusing. Fresnel fringes are apparent in the images and are responsible for the observed contrast changes, rather than a defocus phase contrast mechanism. Magnification: 430,000 diameters (5460–5486).

Fresnel interference fringes are superimposed on the image with a fringe width related to defocus.

To illustrate this phenomenon, images of type 1 and type 2 were recorded in a Philips EM300 using a 10μm objective aperture at 100 kV. Type 1 images consisted of gold atom clusters observed in the shadow of a polystyrene latex sphere after vacuum evaporation of gold at a low angle. A selection of images from a through-focus series (500 Å steps) is illustrated in Fig. 4.6. These images appeared with minimal contrast at focus, got darker on over-focusing and lighter on under-focusing.

Specimens of type 2 consisted of clusters of very small holes formed in collodion films. The plastic film was carbon coated and had a total mass thickness of less than 3 μg.cm^{-2}. Images observed in a through-focus series with this specimen are reproduced in Fig. 4.7, obtained in 2.5μm defocus steps with a 10μm aperture at 100 kV. Qualitatively, these images appeared darker on over-focusing and brighter on under-focusing, opposite in variation to the images of Fig. 4.6, and were nearly invisible at focus.

Contrast variations observed in the images of Figs. 4.6 and 4.7 were identical to those characteristics attributed to a defocus phase contrast mechanism in images of particulate specimens (Hall and Hines, 1970). According to these characteristics, contrast was minimal at focus, reversed on opposite sides of focus, and enhanced at a defocus level related to the particle diameter. An additional characteristic, observed on comparing defocus series of these opposite type of particles, was the opposite sign of contrast between images of these two types of specimens

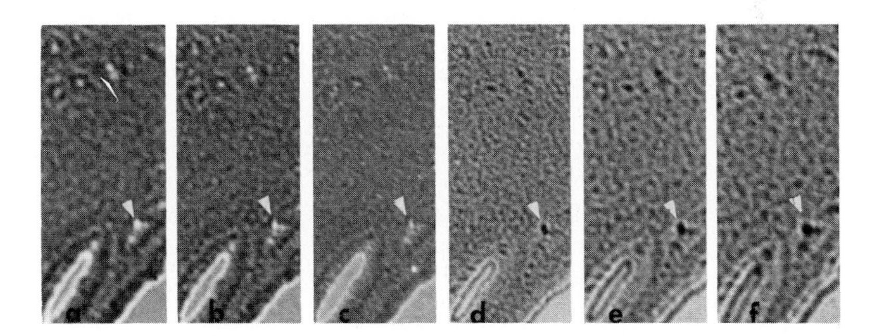

Fig. 4.7 Selected micrographs from a series of underfocus (a, b, c) and overfocus (d, e, f) images of holes in a carbon-collodion film. Contrast of images of holes was made brighter upon underfocusing and darker upon overfocusing, in contradistinction to the images of Fig. 4.6. Fresnel fringes can be identified in the larger holes and were responsible for the contrast effects of defocused hole images (otherwise invisible at focus). Magnification: 250,000 diameters. (6110–6151).

on a given side of focus. These characteristics more properly fit a description related to Fresnel fringes. The fringes are apparent in both series of images (Figs. 4.6 and 4.7). In over-focus images of both types, a dark, then a bright fringe, extends from the denser material into the less dense material.

Fresnel fringes overshadow phase contrast effects in defocus images. This was demonstrated by the inversion of contrast changes on opposite sides of focus for the two classes of specimens. The major influence of Fresnel fringes in defocus images makes difficult any phase contrast observations in defocus images of small objects.

In addition to the object space interference phenomenon of Fresnel fringes, a geometrical effect is present in images that complicates the phase contrast, both in-focus and upon defocusing. This phenomenon is discussed in the next section.

Displacement Image Contrast Enhancement

Striking contrast variations have been observed upon swinging through focus while imaging a biological specimen without an objective aperture. However, upon insertion of an objective aperture the contrast change for a large object, such as an 880 Å polystyrene latex (P.S.L.) sphere, showed little change with defocus. These specimens were too large to have image contrast influenced by Fresnel fringes.

Initially, such a striking contrast change was attributed to phase contrast (Johnson and Parsons, 1967). However, it has since been demonstrated (Johnson and Parsons, 1970b) that this contrast variation was due to the displacement of dark-field images from bright-field images by defocus. At focus, the same mechanism was responsible for the contrast variation of a given specimen imaged in microscopes of different spherical aberration coefficients (Johnson and Parsons, 1970b).

Ramberg (Zworykin *et al.*, 1945) calculated that some of the intensity at the edge of a film, imaged at focus, was distributed into the free space region due to spherical aberration. The displacement of the scattered intensity, a distance y from the edge, was related to the scattering angle and spherical aberration of the objective lens by

$$y = C_0\beta^3 \tag{4.8}$$

Heidenreich (1964) has related the geometrical displacement y, of a dark-field image due to a particular reflection (scatter angle) to the degree of defocus, Δf, by the expression

$$y = \Delta f \beta \tag{4.9}$$

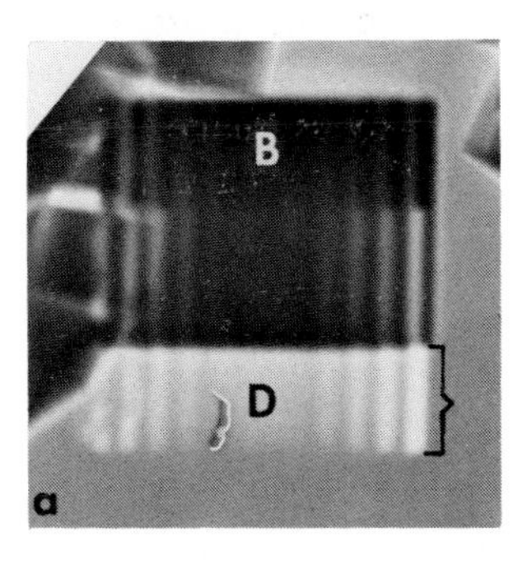

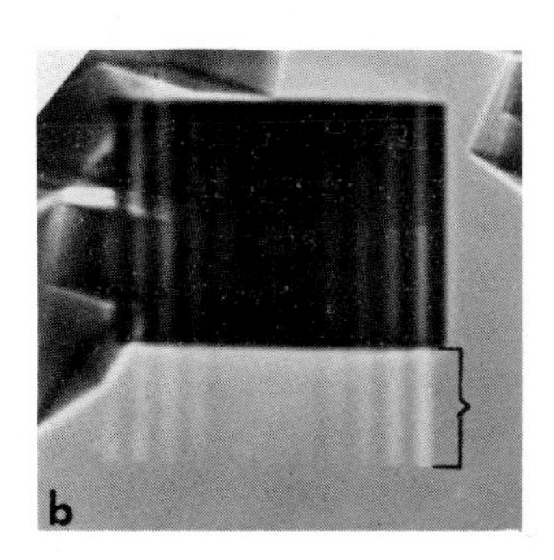

Fig. 4.8 Dark-field image (D) displaced from the bright-field image B of an MgO crystal. The (200) reflection gave rise to the dark-field image. Displacement was accomplished by underfocusing to $+7.5\mu$m. Image (a) was obtained without an objective aperture while (b) was obtained with a 472 Å carbon film phase plate acting on the 200 reflection identified with this crystal. The phase plate displaced the dark-field image by 100 Å (related to object space) and attenuated the image due to phase plate scattering. Magnification: 110,000 diameters. (5397, 5398).

where the magnification factor has been removed. This is illustrated by the displacement of the dark-field image from the corresponding bright-field image for an MgO crystal in Fig. 4.8. Such images have been observed by several workers (Hall, 1966). Heidenreich *et al.* (1968) have used this phenomenon to calibrate objective lens defocus steps. The image displacement with the crystalline image of Fig. 4.8 is distinct because of its one-dimensional shift. For amorphous material the displaced dark-field image expands in two dimensions and is not clearly visible against the bright background.

The simultaneous solution of Eqs. 4.8 and 4.9 yield the relation

$$\Delta f = C_0 \beta^2 \tag{4.10}$$

expressing the defocus level at which complete overlap of dark-field and bright-field images occurs. In this way the defocus displacement is balanced against spherical aberration displacement.

This has been used by Johnson and Parsons (1970b) to explain the contrast minimum in images of 880 Å P.S.L. spheres at an underfocus level of 6μm in the Elmiskop Ia, and at 1.2 μm under-focus in the Philips EM300 (see the recorder tracing in Fig. 4.5). At this defocus level, maximum overlap of dark-field images (attributed to the three carbon-carbon nearest neighbor scattering peaks) with the bright-field image occurred. No phase interaction was observed, only an intensity summation of the various images.

This geometrical effect has been responsible for the misinterpretation of image contrast, such as the change of contrast upon defocusing without an objective aperture and the residual contrast of thin objects at focus as phase contrast. The displacement effect is distinct from any Fresnel fringe effects since it applies also to large objects, which is mainly due to the wide angle scattering and is an amplitude contrast phenomenon.

PHASE CONTRAST EXPERIMENTS

To demonstrate an in-focus phase contrast effect applicable to biological specimens, experiments were planned capable of quantitating the relation between contrast and phase shift for a given specimen. Model biological specimens such as polystyrene latex spheres and carbon films of known thickness were chosen. These specimens exhibited the characteristic carbon-carbon nearest neighbor scatter pattern and the inelastic scattering cross sections of biological material. The use of model specimens avoided problems with dehydration and radiation damage and were capable of mathematical simulation.

For phase contrast to be useful to the biological investigator, it was necessary to establish the object size to which contrast enhancement was applicable. These experiments were undertaken to assess the dependence upon object size and the magnitude of the phase contrast effect.

Phase Contrast Experiments in a Modified Microscope

Initially, in-focus phase contrast experiments were conducted with a four-stage modification of an Elmiskop Ia electron microscope (Johnson and Parsons, 1967; Parsons, 1970). The four-stage modification was necessary because of the lack of coincidence of objective aperture and back-focal planes at 100 kV, and because of insufficient room for phase plates and an anticontamination device in the normal lens. A second objective-intermediate lens assembly was installed beneath the first such section. The second objective pole piece was removed to make room for the phase plate and anticontamination device. Intermediate lens I imaged the back-focal plane of the objective lens onto the phase plate plane. The second intermediate and the projector lenses provided normal magnification. With all lenses maximally energized, a maximum magnification of 10^6 diameters was observed, and a resolution of 5 Å point to point was obtained at 200,000 diameters.

Positive phase plates were constructed from carbon films carried on 100 mesh copper grids. The films were centrally drilled by means of an American Optical Co. prototype laser microscope. Negative phase plates were constructed from uniform carbon films by a contamination spot technique.

Polystyrene latex (P.S.L.) spheres of 880 Å diameter, supported on a carbon-Formvar substrate served as an object in this study. The normal Siemens anticontamination device protected the specimen.

Positive phase contrast experiments using centrally drilled phase plates exhibited an enhanced contrast of the P.S.L. images as determined by the densitometry of photographic plates. However, an accurate measurement of the phase shift of the phase plates was not possible at that time and a quantitative relationship between contrast enhancement and phase shift could not be obtained. This relationship must be demonstrated in order to separate the phase contrast effect from other causes of increased contrast, such as the amplitude contrast effect arising from the extra scattering by the thin film phase plate. A dynamic phase shifting experiment was sought to demonstrate the phase effect.

A dynamic experiment was provided by the contamination build-up over the region of the nonscattered beam focused on a thin carbon film. This self-building phase plate was formed by leaving the phase plate anticontamination device at room temperature and introducing a droplet of low molecular weight hydrocarbon into the phase plate region. The effect is illustrated by the contamination spot images shown in Fig. 4.9. The phase contrast variation of the P.S.L. sphere image relative to the support film is shown in Fig. 4.10 during phase shift of the nonscattered beam. Contrast diminished, then increased repetitively until a dark-field image was obtained with a thick contamination layer. The phase of the nonscattered beam was shifted relative to the whole of the wide angle

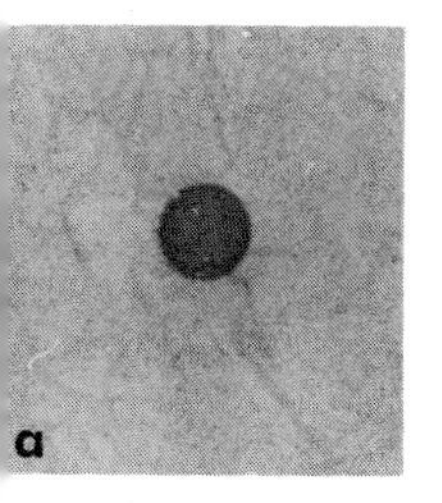

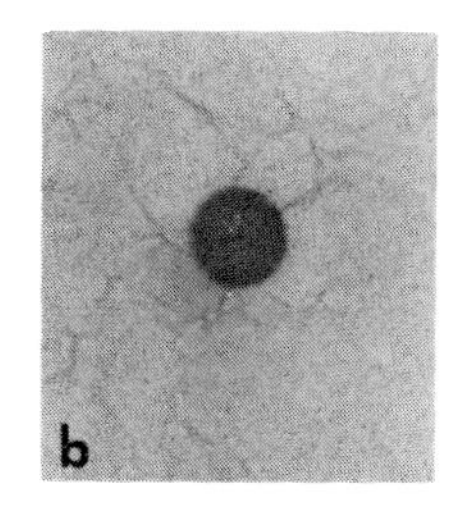

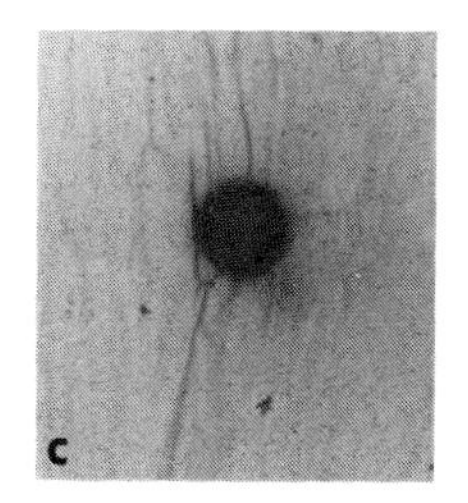

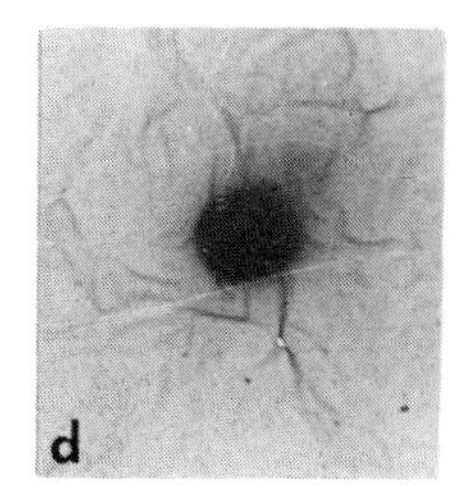

Fig. 4.9 Contamination spot build-up on a uniform carbon film located in the back-focal plane of the objective lens. Contamination times of 5 (a), 10 (b), 10 (c) and 40 (d) min were used to obtain these spots. In negative phase contrast experiments these self-building spots formed a self-retarding negative phase contrast phase plate. The increasing mass thickness of the contamination spot attenuated the nonscattered component and served to equalize the amplitudes (Fig. 4.1) as well as phase shift the nonscattered wave. Magnification: 1880 diameters. (5399–5402).

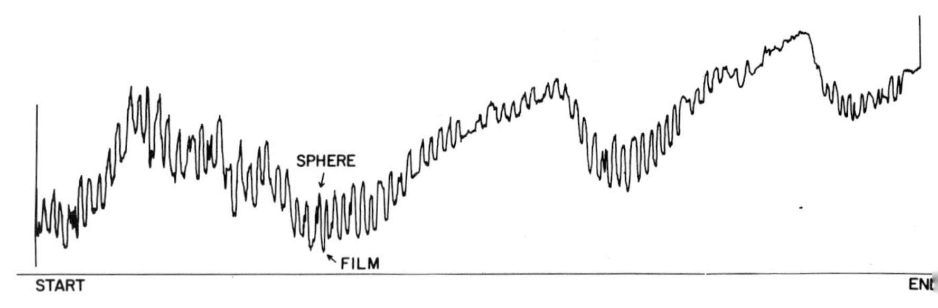

Fig. 4.10 Dynamic phase contrast effect obtained with a self-building phase plate of the negative phase contrast type, and 880 Å polystyrene latex sphere images. Intensity variations obtained from recording micro-Faraday cage detection system during a 30 min build-up of the contamination spot phase plate (Parsons, 1970).

scattering, demonstrating a negative phase contrast effect with large-scale objects.

The positive and negative phase contrast results observed with the four-stage Elmiskop Ia served to demonstrate a phase contrast effect. However, the contrast enhancement did not appear significant and little control over the spherical aberration of the lens was possible. Experiments were transferred to the Philips EM300 for which the spherical aberration of the objective lens was lower, and the aperture and back-focal planes were coincident at 100 kV without further modification.

Phase Contrast Experiments in an Unmodified Microscope

Phase plates were constructed for the Philips EM300 from carbon films carried on 30 μm objective apertures. The films were of measured thickness, determined by the technique of Moretz *et al.* (1968), and were developed for this purpose. The American Optical Co. prototype laser microscope was used to drill central holes in these phase plates, making them of the positive type. Film thicknesses up to 800 Å, equivalent to a phase shift of 4.5 rad (Eq. 4.4) were used in these experiments. One such phase plate is illustrated in the micrograph of Fig. 4.11.

Phase plate films, carried on normal 30 μm objective apertures and centrally drilled with a hole of 6-10 μm, provided a phase shift to spatial frequencies between approximately 1/10 and 1/3.5 $(\text{Å})^{-1}$ at 100 kV. This spatial frequency range was selected on the basis of published curves of the wavefront aberration of the EM300 by Rakels *et al.* (1968). On the basis of the phase contrast transfer function curves of Fig. 4.4, it can be seen that contrast at the higher spatial limit of the phase shifted region begins to oscillate rapidly. Effectively, control of contrast is not

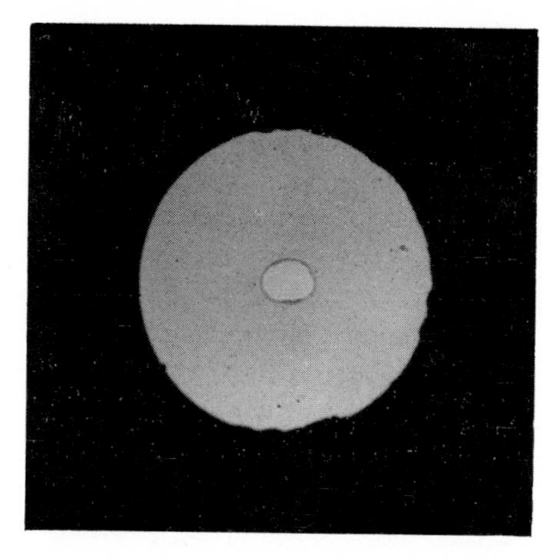

Fig. 4.11 Positive phase contrast phase plate for use in EM300. The 30μm. aperture cut-off scattering at angles greater than 11 mrad. Carbon film phase plate phase shifted scattering components between approximately 3 and 11 mrad. Phase shift was determined from a measurement of film thickness. Central hole was drilled with American Optical Co. laser microscope. Magnification: 1100 diameters. (6322)

possible in this region. In order to assess the phase contrast effect, an element of amplitude contrast was imposed by the aperture, removing those spatial frequencies considered incapable of phase effects.

Specimens consisting of broken carbon films, 65 and 160 Å in thickness, were used in a series of experiments. Contrast of the film was measured relative to free space with an entrance aperture on the Faraday cage equivalent to a diameter of 0.1 μm in object space. Faraday cage measurements were made at focus and at distances greater than 0.5 μm from the film edge.

The results of one phase experiment using carbon film phase plates of 56 to 560 Å are illustrated in Fig. 4.12. The contrast ratio was obtained from a comparison of the contrast measured with the phase plate to that measured at the same image point with a standard 30 μm reference aperture.

This experiment was conducted in the Selected Area mode using a 50 μm intermediate aperture in the image plane of the objective lens and a half-angle of illumination of approximately 10^{-4} rad. Operation in this mode had the advantage of permitting astigmatic correction at high magnification and operation at low magnification without a change in astigmatism. The intermediate aperture removed scattering at angles greater than 0.6 mrad generated by the phase plate film.

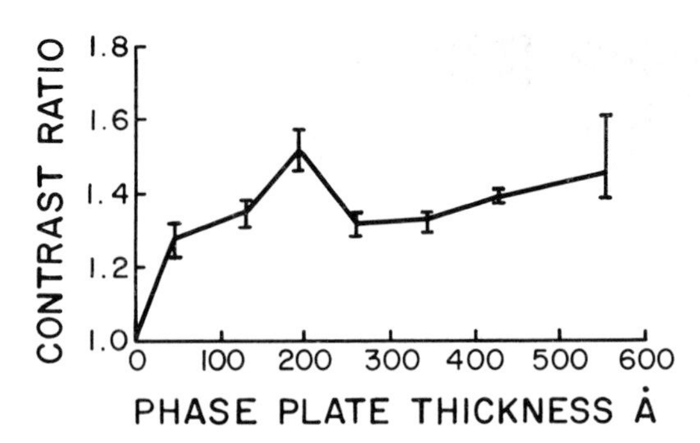

Fig. 4.12 Positive phase contrast results obtained from micro-Faraday cage measurements comparing the contrast of a 1000 Å diameter area of a 160 Å carbon film obtained with a phase plate with that obtained with a normal 30μm reference objective aperture. Phase plate film thickness was varied to vary phase shift of scattering at angles between 3 and 11 mrad. (PC–70–7)

The results of this series were characteristic of all phase contrast experiments of this type. An increase in the contrast ratio was observed with a film thickness of approximately 200 Å, a decrease to a minimum contrast ratio around 300 Å, and a steadily increasing contrast ratio up to the limit of phase plate film thicknesses employed (800 Å). Standard 20 μm and 10 μm apertures, representing purely amplitude contrast, were tested, and contrast ratios of 1.30 and 1.66 respectively, were observed for the same specimen. With the thick phase plate films, the contrast ratio approached that of the standard aperture having a hole diameter equal to that drilled in the phase plate.

In other phase contrast experiments, high-resolution micrographs of carbon film granularity were obtained. Comparative micrographs are shown in Fig. 4.13, obtained with a 472 Å phase shifting carbon film and with a normal objective aperture. These results have indicated an enhancement of granularity and the feasibility of obtaining high-resolution micrographs with carbon film phase plates. Astigmatism, due to charging of the carbon film phase plate, was, therefore, not as serious as would be expected from the microscopists' experience with dirty objective apertures.

Some biological specimens have been imaged with these phase plates in the EM300. Unstained T.M.V. (a sample was kindly made available by Dr. W. O. Bradfute, Wooster, Ohio) was imaged with phase plate film thicknesses of 191 and 472 Å, as well as a standard 30 μm objective aperture. The T.M.V. particles were carried on a 20 Å carbon film suspended on a holey net support film. Contrast of the unstained T.M.V. images, as determined by the densitometry of plates (equation (6)), was

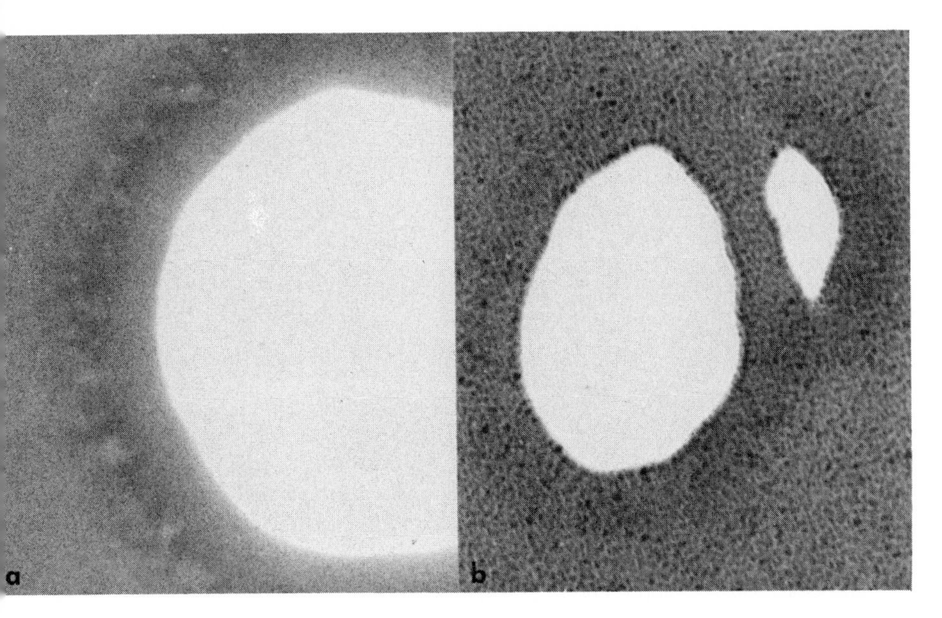

Fig. 4.13 Granularity of a holey carbon film as observed with (a) a normal 30μm objective aperture, $\triangle f = +500$ Å; and with (b) a 512 Å carbon film phase plate on a 30μm aperture, $\triangle f = +300$ Å. Granularity with spacings 8 to 20 Å has been enhanced with the phase plate (B). Magnification: 525,000 diameters. (5366, 5393)

found to be an average of 3% with the standard aperture, 4% with 191 Å thick phase film, and 5% with the 472 Å phase plate. These contrast levels are at the limit of human perception. The micrographs showed the kind of granular appearance illustrated in Fig. 4.13B, which obscured structural detail in the virus particles. Support films remain a problem in this work, as well as in dark-field microscopy (see the chapter by J. Dubochet, this volume). Work in progress continues to overcome these basic problems.

To illustrate the angular distribution of electron scattering from carbon films of the type used as specimens in this study, electron diffraction was performed using films of measured thickness. The resulting pattern for a 127 Å carbon film is illustrated by the densitometer tracing of Fig. 4.14. Marked on this tracing are the limits of scattering angle passed by the intermediate aperture, the angular region phase shifting, and the cut-off angle of a normal 30 μm aperture at 100 kV in the Philips EM300. The curve drawn in this figure to represent the distribution of inelastic scattering is an assumed curve. Present data from energy analysis of carbon film diffraction patterns is too limited to assess the absolute amount and detailed distribution of u.v., plasmon, and K-loss

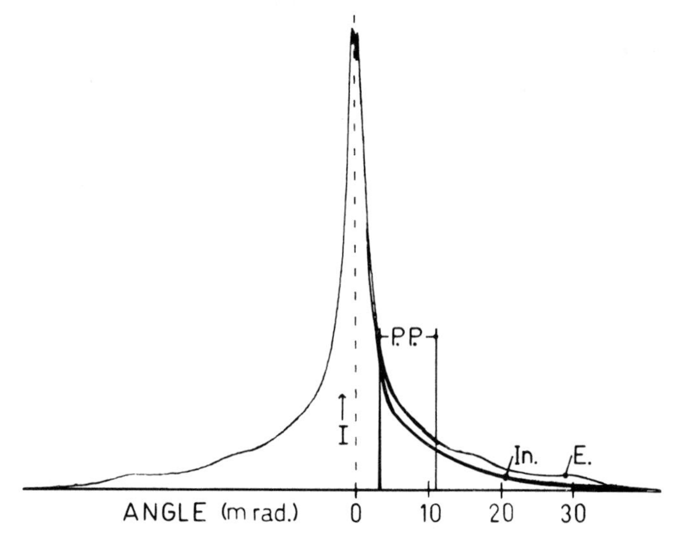

Fig. 4.14 Electron diffraction of 127 Å carbon film at 100 kV illustrating the intensity of the scattering at various angles for the elastic components (E) sitting on top of an assumed level of inelastic scattering (In). Phase shifting experiments were carried out over the limited region PP between 3 and 11 mrad. (6084)

inelastic scattering. However, in other experiments the total scattering into angles greater than 1.8 mrad from a 147 Å carbon film was observed to be 20% of the incident intensity. This determination was made using a 5 μm objective aperture and the internal photometer of the EM300.

The relatively small area under the elastically scattered, nearest neighbor, peaks, and the thermal diffuse scattering (Heidenreich, 1964) is small. It is, therefore, assumed that most of the 20% of scattering measured in this experiment was made up of inelastic components, and this is in agreement with Crick and Misell (1971). Since the inelastic scatter is at best only weakly coherent, it appears that phase contrast effects, even from a thin carbon object, must be dampened by the incoherent inelastic scatter.

Test of Thin Film Phase Plates by Interference Phenomena

The phase shift of thin carbon film phase plates and their ability to preserve the coherence of the electron beams traversing them was made using electron interference phenomenon. An electrostatic biprism (Möllenstedt and Duker, 1956; Wahl, 1970) was constructed for use in a JEM 200 electron microscope. The interference patterns were observed upon application of a positive potential to the biprism wire. Carbon films were interposed into one of the interfering electron beams, and the interference patterns compared in free space and in the shadow of the film.

The image of one such interference pattern in the presence of a 127 Å carbon film is illustratd in Fig. 4.15. In this figure, the densitometer tracings made perpendicular to the interference fringes in the free space

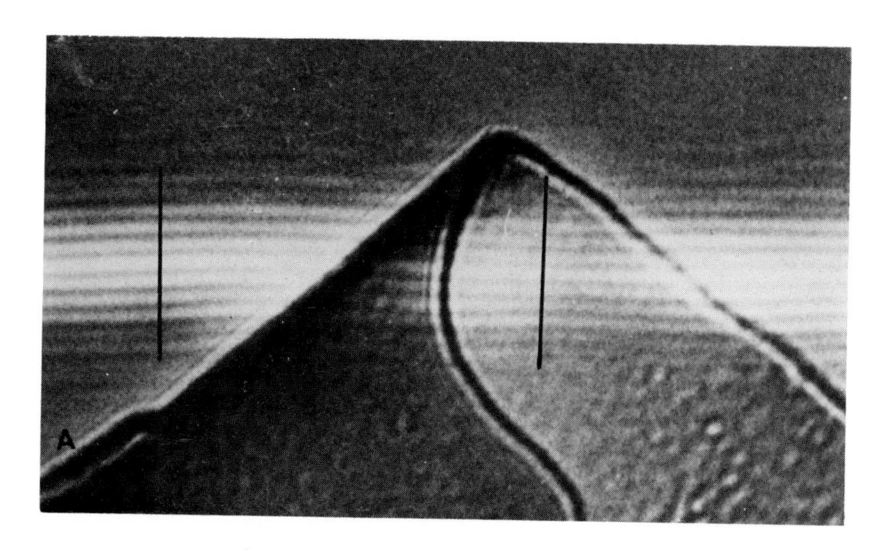

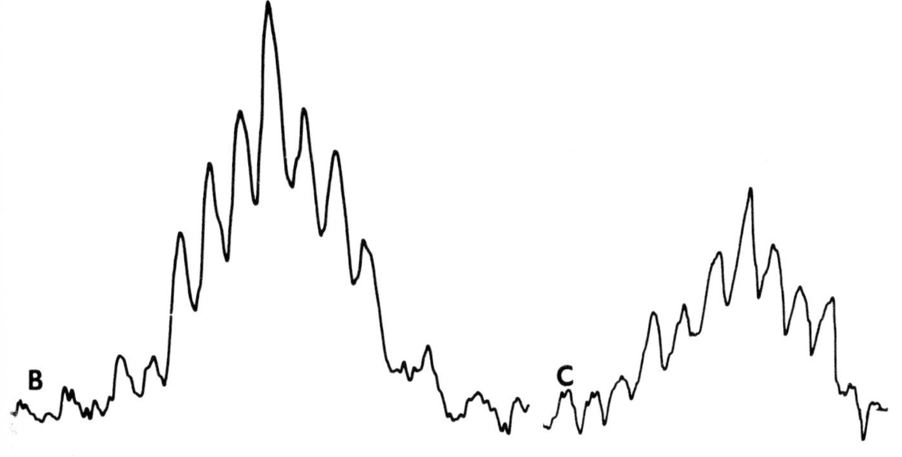

Fig. 4.15 Electron beam interference testing of thin carbon films as phase plates. Phase shift of the 127 Å carbon film was difficult to determine from fringe displacement in film shadow relative to fringes in free space (A). However, densitometric plots comparing the free space (B) and film shadow (C) regions indicate little loss of coherence and some attenuation by the film. Magnification: (A) 48,000 diameters; (B, C) 93,000 diameters. (5730)

region and the film shadow are also shown. It was not possible to obtain an accurate measurement of the phase shift introduced by a given carbon film with this apparatus. However, an appreciation of the ability of such a film to pass the traversing beams without destroying their ability to interfere can be obtained from this analysis. Films of up to 830 Å in thickness were analyzed. While the interference pattern in the shadow of a film of this thickness was not destroyed, the beam traversing the film was severely attenuated by scattering. The scattering, in the case of the 830 Å carbon film, was estimated to have attenuated the amplitude of the wave through the film to 40% (intensity attenuation to 16%) of its initial value.

Coherence has largely been maintained upon passage through these carbon films. However, because of the scattering by the thin films and the likelihood of charging, other methods such as an electrostatic lens or a magnetic domain may be preferable.

A more accurate method for measuring the phase shift of a carbon film (and measuring its inner potential knowing its thickness) was obtained from Fresnel edge interference patterns observed on opposite sides of a transparent carbon film edge (Fig. 4.16). This effect was observed by Hillier and Ramberg (1947) and was suggested as a measuring technique for film thickness by Ramberg (1949). Overfocus Fresnel fringes, observable on both sides of transparent carbon film edges, were obtained in the JEM 200 at 100 kV with an objective lens focal length of approximately 3 cm and overfocused by 0.5 cm with respect to the

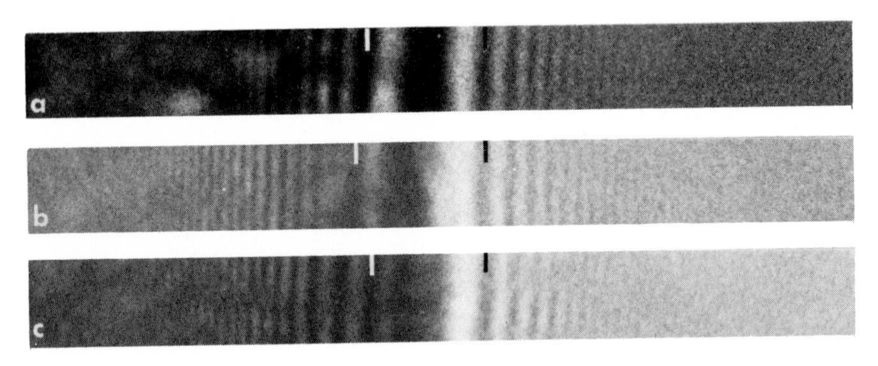

Fig. 4.16 Fresnel interference patterns in free space and in film shadow at the edge of carbon films of thicknesses 127 Å (a), 625 Å (b), and 918 Å (c). The film shadow pattern moved away from the edge then back to the edge as the phase shift of the transmitted wave cycled from 0 to 2π rad. First peak of the free space patterns are marked with a black marker and are in line. First peaks of the film shadow pattern are marked with a white marker. The first 625 Å peak is displaced further from the edge than either of the others. 100 kV, $\Delta f = +4$ mm. Magnification: 25,000 diameters (5758–5798).

plane of the carbon film. The free space pattern was assumed to remain constant. The film shadow pattern was displaced relative to the edge by an amount dependent upon the phase shift given to the beam traversing the film. The position of the shadow pattern relative to the free space patterns was the same for the 127 and 918 Å films, but the shadow pattern with the 625 Å film was displaced farther away from the edge, demonstrating the cylical displacement of the pattern with phase shift. It appears that the measurement of the displacement of the film shadow pattern from the edge is a useful technique for determinations of phase shift and inner potential for films of known thickness. This technique has not yet been fully exploited.

DISCUSSION

Evidence That Phase Contrast Has Been Observed

The investigations discussed here have been directed toward the quantitation of a phase contrast effect in electron microscopy using model biological objects. Demonstrations of negative and positive phase contrast using two different microscopes and different techniques of thin film phase plate construction have been provided. These observations will now be discussed.

Phase Retardation of the Nonscattered Wave

The recorded Faraday cage measurements of Fig. 4.10 have demonstrated image plane contrast variations during the build-up of a negative phase contamination plate. The self-building contamination spot phase plate continuously retarded the phase of the nonscattered wave until the contamination spot increased sufficiently to produce a dark-field image. The contrast variation of the P.S.L. images was obtained by phase shifting the nonscattered wave relative to the wide angle scattering, and not relative to the shape transform scattering of the spheres, since this component was within the region occupied by the nonscattered wave. The observed variation in intensity of the support film was similarly due to the phase shifting of the wide angle scattering from the film. Both spheres and support film, being of carbonaceous material, had similar diffraction patterns (see Fig. 4.15) and similar phase contrast effects. These experiments seem to demonstrate that phase contrast need not be restricted to phase shift of the shape transform scattering, as was apparently assumed by earlier workers (e.g., Faget *et al.*, 1960). These workers apparently considered that phase contrast, at focus, was achieved only with small objects that permitted construction of phase plates of a practical size.

The thickness of the contamination spot giving maximum or minimum contrast could not be precisely determined in these experiments. Furthermore, both elastic and inelastic scattering components were present and a comparison of results with the phase shift to be introduced according to theory cannot be made. However, the contrast was observed to decrease initially, indicating that a retardation of the nonscattered wave was required for negative phase contrast. Since no initial relative phase shift is associated with the nonscattered wave, these data suggest that the average (or effective) phase of the scattered components was retarded relative to the nonscattered wave. Theoretically, on a single atom scattering basis, the scattered wave should have been approximately in phase with the nonscattered wave. If the scattering intensity is sufficiently great, then a coherent scattered wave can be considered as having an advance of $\pi/2$ rad with respect to the nonscattered wave. In either case, a retardation of the nonscattered wave should lead to a resultant decrease in scattered intensity over scattering points and, consequently, to positive phase contrast.

Phase Retardation of a Portion of the Scattered Wave

Positive phase contrast for large scale objects was demonstrated by the results of Fig. 4.12. This series of experiments demonstrated an enhancement of contrast with a carbon film phase plate, approximately 200 Å in thickness (1.2 rad phase retardation), over a limited spatial frequency region. Assuming that a net phase can be assigned to the scattering in this region, the observed enhancement agreed with that indicated by the negative phase contrast experiments. The applied phase shift for optimized contrast was compatible with an effective $\pi/2$ rad phase retardation of these scattered components relative to the nonscattered component, as already concluded from the positive (contamination spot) phase plate experiments.

A contrast minimum was obtained with a carbon phase plate film of approximately 300 Å thickness (retardation of 2.1 rad; expected theoretically: 4.7 rad). However, this contrast minimum was expected to be at a lower level than that of the pure amplitude contrast obtained using the normal 30 μm aperture. This was not observed. The contrast ratio steadily increased as the phase shifting film thickness was increased beyond the experimentally observed minimum of contrast. The explanation of this discrepancy lies in the effect of scattering by the phase plate film discussed in the next section.

Micrographs of carbon film granularity (Fig. 4.13) illustrated the positive contrast enhancement of granularity having dimensions of 8 to 20 Å with a positive, 2.6 rad phase plate. This result is consistent with a $\pi/2$ rad phase retardation of the predominant, coherently scattered component. In the spatial frequency range corresponding to object spacings of 8 to 20 Å, the phase of the scattered components was not significantly complicated by spherical aberration phase shifts. A part of the small angle transform was phase shifted together with part of the wide angle scatter. This result is consistent with the previously described negative phase contrast observations on 880 Å P.S.L. spheres and the positive phase contrast observations on carbon films. They emphasize the presence of a weak coherent scatter component in the angular range of 3-11 mrad (100 kV), having a retardation of $\pi/2$ rad.

Factors Affecting Phase Contrast

The phase plate

Three aspects of the use of thin film phase plates require consideration: scattering, coherence preservation, and charging. Scattering by a thin film phase plate is inevitable. Badde and Reimer (1970) assessed the fractional amplitude transmission, A/A_0, of such a phase plate using the expression

$$\frac{A}{A_0} = \exp\left(-tp\,\frac{(1+\nu)}{(x_a)}\right) \tag{4.11}$$

where t is the thickness for a given phase shift, p is the density of the phase plate material, ν is the ratio of inelastic to elastic scattering cross sections (from Lenz, 1954, and x_a is the Aufhellungsdicke (von Borries, 1949; mean free mass thickness for elastic scattering). However, the comparison of possible phase plate materials by this expression assumes that the Lenz (1954) inelastic calculation, derived from single atom scattering, is appropriate. This is contradicted by the recent work of Crick and Misell (1971), which predicts that the collective (plasmon) losses are more important than the single atom losses. The efficiency of the different materials useful for thin film phase plates will now have to be recalculated.

Effectively, secondary scattering by a thin film phase plate, operating on the scattering components in the back-focal plane, introduces a positional error in the secondarily scattered electrons reaching the image plane (Badde and Reimer, 1970). In the experiments performed in this work, the secondarily scattered electrons, scattered into angles greater than 6×10^{-4} rad were removed by the intermediate aperture and did not contribute to the image. This imposed an element of amplitude contrast, which increased with phase plate thickness. This was consistent with the measurements of the linear increase in scattering intensity for carbon films greater than 100 Å in thickness, demonstrated by Johnson and Parsons (1969). The contrast, imposed by this scattering process, approached the contrast of an objective having a hole of the same diameter as that drilled in the carbon film phase plate. This contrast effect overwhelmed the appearance of further phase contrast effects after the phase plate film thickness reached approximately 300 Å.

Currently, consideration has been given to contamination-constructed phase shifting films capable of differential phase shift across the back-focal plane related to spherical aberration (Thon and Willasch, 1970; see also Müller, 1970 and Fig. 4.10). Such a device offers the possibility of spherical aberration correction as well as a phase plate for phase contrast enhancement when operating on the wide angle scattering components. However, it is apparent from the above discussion that scattering from such film devices is too severe. Other techniques are required for phase shifting without introducing secondary scattering if phase contrast effects are to be exploited in biological imaging.

Ferromagnetic domains and a miniature electrostatic lens provide alternates to phase shifting by means of thin films. According to Wohlleben (1967; 1971, personal communication), insertion of a longitudinally magnetized domain between the nonscattered wave and a given reflection (in a crystalline diffraction pattern) in the back-focal plane introduces a relative phase shift between the two components without either component having to traverse scattered material. Fernández-Morán (1966) demonstrated the image of a ferritin molecule obtained with a ferromagnetic phase shifting device, but no discussion was provided of the geometry of the phase plate nor of the magnitude of the phase shift. An electrostatic mini-lens of the dimensions of a conventional objective aperture has been under design in this laboratory. Such a lens, controlled externally, is considered capable of spherical aberration correction as well as phase contrast enhancement by a properly shaped field. Further developments of infocus phase contrast electron microscopy are dependent upon the development of nonscattered phase shifting devices, as well as reduction of the inelastic contributions to the image.

The results of the biprism interferometry experiments for testing phase shifting films (Fig. 4.15) indicated the preservation of coherence of the component traversing the thin film. The ability for interference was not destroyed for film thicknesses tested up to 830 Å.

A further indication of coherence preservation was provided by the

Fresnel fringes of Fig. 4.16. Many orders of fringes were observed in the shadow of a 928 Å carbon film. This series of fringe patterns also served to establish the correctness of the phase shift relationship with carbon film thickness (Eq. 4.4 and Table 4.1) as described previously.

The fluctuating effect of charging of films less than 100 Å in thickness has been observed by a number of workers (Curtis and Ferrier, 1969; Drahos and Delong, 1965; Dove, 1964). These observations were repeated during the course of the present work with a 50 Å carbon film imaged in the EM300 by the shadow-projection method. However, no effect of phase plate charging has been detected. This may be explained on the basis of the general use of phase plate films thicker than 100 Å, the low intensity of the radiation incident on the phase plate, and the charge neutralizing effect of secondary electrons ejected from the 30 μm objective aperture supporting these phase plates that neutralize the positive charging of the film.

Scattering Components Traversing the Phase Plate

A baseline curve was drawn into Fig. 4.14 to represent the intensity of inelastic scattering in the electron diffraction pattern of a 128 Å carbon film. It is necessary to determine the scattering components actually present in images of this sort in order to assess the role of phase contrast in all biological imaging. For a 147 Å carbon film, 20% of the incident beam was found to have been scattered into an angle greater than 1.8 mrad at 100 kV. However, the angular distribution and energy loss distribution making up these scattered components have not yet been experimentally determined.

Crick and Misell (1971) have calculated the relative distributions of elastic and inelastic scattering from carbon. In the case of a 150 Å carbon film, at 100 kV, they have estimated a total scattering intensity of approximately 23% of the incident beam. This agrees with the experimental measurement given above, considering that a portion of this scattering intensity was found at angles less than 1.8 mrad. The discussion of Crick and Misell (1971) assumed a predominance of plasmon inelastic scattering and weighted the total inelastic scatter intensity four times greater than the elastic scatter within the angular region of the phase plate used in the present phase contrast experiments. Crick and Misell (1971) pointed out that the high level of inelastic scattering for biological specimens makes it unlikely that fine structural detail will be imaged by phase contrast.

The phase contrast effect displayed in Fig. 4.12 was small. This may be explained on the basis of purely elastic scattering of weak intensity, interfering with the nonscattered component. However, in this case the concept of a $\pi/2$ phase advance for a coherent elastic wave is incorrect. On the other hand, a weak phase effect appears possible between specific loss inelastic components, which Crick and Misell (1971) consider dominated by the plasmon scattering. Phase shifts to be associated with the plasmon inelastic scattering are, however, unknown. Further experiments are required to distinguish these possibilities by employing energy analysis of imaging components during phase shifting experiments. Nevertheless, the magnitude of the phase contrast effect (Fig. 4.12) has indicated that phase contrast plays a minor role in conventional electron imaging. This does not mean that in other circumstances, e.g., high-voltage electron microscopy, phase contrast might not prove a useful imaging mode for unstained biological specimens.

Astigmatism and Electronic Instabilities

A comment is required on the role of astigmatism and power supply instabilities in the phase contrast results presented here. Astigmatism was reduced to zero at high magnification, and remained so for the low magnification work, provided the Selected Area imaging mode was used. If the quadrant-to-quadrant astigmatic defocus around a hole image was observed to be 500 Å, it represented a defocus of ±250 Å from the mid-position. The effect of such a defocus can be observed to be small based on the phase contrast transfer function curves of Fig. 4.4.

Operating with the 30 μm objective aperture in the Philips EM300 electron microscope meant a cut-off of spatial frequencies approaching the theoretical limit of the microscope design. The effect of electronic instabilities is uncertain. According to Rakels *et al.* (1968), it was sufficient to maintain the high voltage and objective lens ripples below those levels calculated to affect the resolution limit. However, the calculations by P. E. Engler (Parsons, 1970) indicated that the effect of voltage ripple on the objective lens of the Siemens Elmiskop Ia was to dampen all spatial frequencies slightly and to severely dampen those below 5 Å.

Dark-Field Image Displacement

The contrast changes of P.S.L. spheres with defocus (Johnson and Parsons, 1967; see recorder tracing in Fig. 4.5) have been attributed to an amplitude contrast effect arising from the spherical aberration and defocus displacement of scattered rays (Johnson and Parsons, 1970b). This was illustrated by Fig. 4.8, demonstrating the effect with an MgO crystal. The same effect with amorphous specimens is less obvious than with crystalline specimens because of the two-dimensional displacement of the scattered components in the image plane.

At any particular scattering angle, the inelastically scattered components are subject to the displacements of Eqs. 4.9 and 4.10, as well as the elastic scatter. The prediction of this effect with both scatter components is contained within the discussion of Crick and Misell (1971).

Other Phase Contrast and Interference Effects in Electron Microscopy

The resultant contrast enhancement in present phase contrast experiments was small for large scale objects and less than that obtained with an aperture that removed all of the wide angle scattering. A decision must be made as to the contribution to image plane contrast from coherent electron optical effects and the contribution available from incoherent contrast effects. Perfect coherence has been assumed in the theory of phase contrast. It was implicit in the discussion of phase that interference occurred, and that the main experimental effort required was to determine the conditions of optimized interference (constructive or destructive) for maximum contrast. However, coherence must receive greater consideration, separating observations of coherent effects in object and image space, and recognizing the dominance of inelastic scattering for biological specimens. It has generally been assumed in electron microscopy that because most specimens have the characteristics of phase objects minimal absorption, phase contrast is the dominant imaging mechanism. However, the phase effect has been observed to be weak. This result suggests the need for a re-examination of other observations considered to have arisen from a phase contrast mechanism.

In apparent contradiction to this questioning of image plane coherence, striking interference phenomena, occurring in object space, have been observed in the micrographs of Figs. 4.15 and 4.16. These observations give the impression of a high degree of coherence in the image. The number of Fresnel fringes has been directly related to the half-angle of illumination, a measure of transverse coherence of the source (Hall, 1966). Also, phase effects appear possible in relation to Fresnel fringes. The fringe patterns in Fig. 4.16 demonstrated the effect of phase in the oscillatory displacement of the pattern in the shadow of the film. However, this observation was made in object space and represented the super-position of the projection of the film on the interference pattern at a defocus level. Similarly, the interference phenomenon of Fig. 4.15 represented an object plane effect obtained at a defocus level beneath the film and the biprism assembly. A half-angle of illumination of the order of 10^{-6} rad was required to achieve sufficient lateral coherence for this interference effect. These observations suggest that a higher degree of coherence occurs in object space compared to image space, and that incoherent inelastic scatter does not occur.

With respect to the defocus contrast of small objects, two observations (Figs. 4.6 and 4.7) demonstrated that Fresnel fringes could account for observations usually attributed to phase contrast. It is likely that some other reported phase contrast effects should be related to these observations of Fresnel edge fringes. Johnson (1968) has attributed the defocus contrast of negatively stained myofibrils to phase contrast. In this case, these images were of type 2 specimens. Image contrast appeared negative on underfocusing and positive on overfocusing, consistent with the holey specimen images of Fig. 4.7. It, therefore, appears that they represented a Fresnel fringe effect rather than a phase contrast effect, as suggested.

Hall and Hines (1970) have used defocused images of gold particles on a graphite support to illustrate their phase contrast calculations on clumps of gold atoms. While they suggested the possibility of Fresnel fringe enhancement, they nevertheless related contrast enhancement to defocus levels predicted on the basis of phase contrast. These specimens were of the type 1 discussed previously. The possible agreement between the defocus phase contrast and defocus image contrast is a result of the relatively broad spatial extent of the Fresnel fringe and its effect on the contrast of the particle image over a wide range of defocus. The variation of the fringe width with defocus (Eq. 4.7) indicates that the overlap of fringes on a particle is a function of defocus qualitatively similar to the dependence of the defocus phase shift on particle size for a postulated phase contrast enhancement.

Published data relating the phase contrast enhancement to crystal lattice images (e.g., Yada and Hibi, 1968) is inconclusive. It has generally been assumed (e.g., Heidenreich *et al.*, 1968) that maximum lattice image contrast was obtained at a defocus level satisfying equation 1 for a phase shift of $\chi = \pi/2$ rad. The $\pi/2$ requirement has been carried over from the light microscope analogy. The study by Yada and Hibi (1968) tested the contrast of various lattice images under defocus conditions, but without resolving the uncertainty. They found agreement with equation 10 for the coincidence of bright- and dark-field images, and considered this to be a condition of maximum lattice image contrast. This conclusion implies a phase interaction between the dark- and bright-field images. However, the results of four defocus series of 4.5 Å mica lattice images by Yada and Hibi (1968) illustrate a wide variation in the defocus levels at which the image of greatest contrast was obtained in the respective series. Therefore, there is considerable uncertainty in the degree of agreement with theoretical predictions. It is to be questioned whether agreement with the first overlapped images can yield the optimum conditions for phase contrast; for example, the phase shift given to the dark-field image after underfocusing to meet the overlap requirement of Eq. 4.10 would be approximately 2 π rad according to Eq. 4.1. Similar calculations for lattice spacings of 1 to 5 Å indicate phase shifts of 805 π to 1.3 π introduced by the defocus necessary to achieve overlap. Such large phase shifts for small (1 Å) lattice spacings implies the need for very great phase stability in the scattered waves. It is doubtful that the stability of the objective current and high voltage are sufficient for this.

The best explanation of lattice images appears to be the side-by-side placement (noninterfering) of bright- and dark-field images. In addition, defocus introduces Fresnel fringes that confuse the mechanism for forming such images. Labaw (1960) has discussed defocus changes in lattice image contrast and periodicity related to object space Fresnel interference phenomena. Repeat "Fourier images" (Cowley and Moodie, 1957; Rogers, 1969) formed as a result of Fresnel diffraction occurred in these organic crystals even after radiation damage had caused fading of the Bragg diffraction spots (Labaw, 1960; 1961).

Possibilities of Phase Contrast Imaging in Biological Electron Microscopy

Phase contrast in electron microscopic images of biological specimens at conventional voltages has been observed so far as a weak effect. Examination of the scattering components taking part in image formation and a variety of images attributed to a phase contrast mechanism indicates a greater dominance of the incoherent image formation mechanism than was previously generally assumed. If Crick and Missell (1971) are correct, the dominant scattering component for biological specimens is inelastic, with at best a weak coherence between components in a narrow energy loss bandwith.

High-voltage electron microscopy offers an improvement in phase contrast imaging due to reductions in the two major aberrations affecting coherence at conventional voltages. The effect of spherical aberration is significantly reduced at 1 MV with a conventional lens compared to the use of 100 kV accelerating potential. For a lens with 1.6 mm spherical aberration, the 3 Å reflection has imposed on it a 10 wavelength spherical aberration distortion. To meet the Rayleigh quarter wavelength criterion (Born and Wolf, 1965) at 100 kV, for a 3 Å object spacing, a lens with a spherical aberration coefficient of 40 μm would be required. At 1 MV accelerating potential, the Rayleigh condition would be met by a lens of spherical aberration coefficient equal to 3 mm. The latter represents a practical, attainable figure, while the former condition is unobtainable. The present objective lens of the United States Steel Company high-voltage microscope, for example, has a spherical aberration coefficient of 4 mm at 1 MV (S. Lally, 1971, personal communication). In addition to a reduction in the effects of spherical aberration, chromatic aberration is also lessened at higher accelerating potentials. Exact information on velocity analysis data is not yet available, but qualitative estimates indicate a considerable reduction in inelastic scattering at an acceleration potential of 1 MV, based on observation of electron diffraction results. The potential of in-focus phase contrast now requires a full exploration with the high-voltage electron microscope.

In order to assess the various contrast enhancement techniques, contrast calculations (Table 4.2) have been performed on a model biological specimen. Specimen data were based on Crick and Misell (1971). Amplitude, phase contrast, and dark-field techniques have been compared for a 200 Å carbon step on a 50 Å carbon support film imaged at 100 kV. Calculations have been based on imaging with and without energy filtration, which would remove the inelastic scattering. These

TABLE 4.2A SCATTERING AND TRANSMISSION SPECIMEN DATA*

Component	*50 A carbon intensity*	*Amplitude*	*250 A carbon intensity*	*Amplitude*
Nonscattered (N.S.)	88.4	9.4	53.7	7.3
Inelastically scattered (I.S.)	8.2	2.9	35.0	5.9
Elastically scattered (E.S.)	3.4	1.8	11.3	3.4
Totally scattered (T.S.)	11.6	3.4	46.3	6.8

*Comparison of the contrast of a 200 Å carbon step on a 50 Å carbon support film obtained by various electron optical manipulations. Relative scattering and transmission intensity for 50 and 250 Å of carbon, obtained from Crick and Misell (1971), are presented in part A.

TABLE 4.2B COMPARISON OF CONTRASTING MECHANISMS

Mechanism	*Contrast*
1. Amplitude contrast (T.S. removed)	0.39
2. Phase contrast (T.S. coherent)	0.99
3. Phase contrast (E.S. coherent, I.S. incoherent)	0.22
4. Phase contract (E.S. coherent, I.S. removed)	0.72
5. Phase contrast (E.S. coherent, I.S. removed, amplitudes equalized)	1.0
6. Dark-field (T.S. admitted to image)	−3.0
7. Dark-field (E.S. only admitted, I.S. removed)	−2.3

calculations were optimized; for example, it was assumed in the amplitude (aperture) contrast method that all scattering could be removed by an objective aperture. Nevertheless, these calculations are enlightening in several respects.

Table 4.2 illustrates the significant advantage of dark-field over phase contrast and amplitude contrast techniques for biological specimens. The dark-field contrast was represented as a negative number because of its inversion of contrast. Johnson and Parsons (1969) have shown the contrast advantage of dark-field to be limited to thicknesses not appreciably greater than one mean free scattering path length and most advantageous for specimens less than 100 Å in thickness (80 kV). Others, as discussed by J. Dubochet (in this volume) have obtained very high resolution for biological macromolecules by means of dark field, despite the greater radiation damage of this technique as used and without image intensification.

Phase contrast is seen to be advantageous, according to these calculations, if all of the electron scattering can take part coherently in the imaging mechanism. However, with the level of inelastic scattering calculated for carbonaceous specimens by Crick and Misell (1971), phase contrast will be advantageous for biological imaging only if combined with energy filtration to remove the incoherent components. The use of higher accelerating voltages, involving a reduction in the inelastic-to-elastic scattering cross sections, will increase the effectiveness of phase contrast enhancement of biological images. Work is just commencing on the study of phase contrast in biological imaging at acceleration voltages of 800–1000 kV.

The use of any electron optical contrasting technique requires an awareness of the scattering distribution of the specimen, dehydration and radiation damage, and the possibilities of optical artifacts in the image. As with chemical staining techniques, the use of several optical contrasting mechanisms is required to conclusively demonstrate an image. This appears to require the availability of more sophisticated instrumentation, including energy analysis and high voltage microscopes.

ACKNOWLEDGMENTS

The research discussed here has been directed by Dr. D. F. Parsons, Roswell Park Memorial Institute, to whom the author is grateful for guidance and encouragement. Support was provided by N.S.F. grants GB-5535, GB-8235 and GB-15389. Discussions with Dr. W. J. Claffey and Mr. P. E. Engler have aided the development of this work. The competent technical contributions of Mrs. C. Dean and Mssrs. G. G. Hausner, Jr. and R. Slon are gratefully acknowledged. The American Optical Company has kindly made available their laser microscope for drilling the carbon film phase plates.

References

Agar, A. W., Revell, R. S. M., and Scott, R. A. (1949). A preliminary report on attempts to realize a phase contrast electron microscope. *Proc. Intern. Cong. Electron Micros.*, Delft, 52.

Aharonov, Y., and Bohn, D. (1959). Significance of electromagnetic potentials in quantum theory. *Phys. Rev.*, **115**, 485.

Badde, H. G., and Reimer, L. (1970). Der Einfluss einer streuenden Phasenplatte auf da elektronenmikroskopische Bild. Z. Naturf., **25a**, 760.

Bennett, A. H., Jupnik, H., Osterberg, H., and Richards, O. W. (1946). Phase microscopy. *Trans. Am. Microsc. Soc.*, **65**, 99.

———. (1951). "Phase Microscopy, Principles and Applications." John Wiley and Sons, Inc., New York.

Boersch, H. (1947). Uber die Kontraste von Atomen in Elektronenmikroskop. Z. *Naturforsch.*, **2a**, 615.

Born, M., and Wolf, E. (1965). "Principles of Optics," 2nd ed. Pergamon Press, Oxford.

von Borries, B. (1949). Elektronenstreuung und Bildenstehung im Ubermikroskop. Z. *Naturforsch.*, **4a**, 51.

Bradbury, G. R. (1969). Electronic measurement of electron microscope intensities and energies. *J. Appl. Cryst.*, **2**, 254.

Buhl, R. (1959). Interference microscopy with electron waves. *Zeit. Phys.*, **155**, 395.

Colliex, C., and Jouffrey, B. (1970a). Contribution to the study of the energy losses due to the excitation of deep levels. *C. R. Acad. Sci. B., France*, **270**, 144.

———. (1970b). Filtered images obtained with electrons having undergone energy losses due to excitation of deep levels. *C. R. Acad. Sci. B., France*, **270**, 673.

Cowley, J. M., and Moodie, A. F. (1957). Fourier images. II. The out-of-focus patterns. *Proc. Phys. Soc.*, **B70**, 497.

Cox, H. L. Jr., and Bonham, R. A. (1967a). Elastic electron scattering amplitudes for neutral atoms calculated using the partial wave method at 10, 40, 70, and 100 kV for Z = 1 to Z = 54. *J. Chem. Phys.*, **47**, 2599.

———. (1967b). Table V of Elastic electron scattering amplitudes for neutral atoms calculated using the partial wave method at 10, 40, 70, and 100 kV for Z = 1 to Z = 54. *Document 9539; ADI Auxiliary Publications Project, Photoduplication Service, Library of Congress,* Washington 25, D.C.

Crewe, A. V., Wall, J., and Langmore, J. (1970a). Visibility of single atoms. *Science*, **168**, 1338.

———, Isaacson, M., and Johnson, D. J. (1970b). The energy loss of 20 KeV electrons in biological molecules. *Proc. 28th Ann. Meet. Electron Micros. Am., Houston.* (C. Arseneaux, ed.). Claitor's Book Store, Baton Rouge, p. 262.

Crick, R. A., and Misell, D. L. (1971). A theoretical consideration of some defects in electron optical images. A formulation of the problem for the incoherent case. *J. Phys. D.: Appl. Phys.*, **4**, 1.

Curtis, G. H., and Ferrier, R. P. (1969). The electric charging of electron microscope specimens. *Brit. J. Appl. Phys. (J. Phys. D.). Ser. 2,* **2**, 1035.

Dove, D. B. (1964). Image contrast in thin carbon films observed by shadow electron microscopy. *J. Appl.* **35**, 1652.

Drahos, V., and DeLong, A. (1965). Observation of charges on specimens in a transmission electron microscope. *Czech. J. Phys.*, **B15**, 760.

Dupouy, G., Perrier, F., and Verdier, P. (1966). Amélioration du contraste des images d'objets amorphes minces en microscopie éelectronique. *J. Microscopie*, **5**, 655.

Faget, J., and Fert, C. (1960). Interférometrie électronique et microscopie électronique interférentielle. *Proc. 4th Intern. Cong. Electron Micros. Berlin (1958).* Springer Verlag, Berlin. p. 234.

———, Fagot, M., and Fert, C. (1960). Microscopie électronique en éclairage cohérent. Microscopie interférentielle, contraste de défocalisation, strioscopie et contraste de phase. *Proc. Europ. Reg. Cong. Electron Micros. Delft,* **I**, 18.

———, Fagot, M., Ferre, J., and Fert, C. (1962). Microscopie électronique à contrast de phase. *Proc. 5th Intern. Cong. Electron Micros.* Philadelphia, Academic Press, New York. A-7.

Fernández-Morán, H. (1966). High resolution electron microscopy of biological specimens. *Proc. 6th Intern. Cong. Electron Micros. Kyoto, Maruzen, Tokyo,* **2**, 13.

Fert, C., Faget, J., Fagot, M., and Ferre, J. (1962). Un microscope électronique interférentielle. Description, conditions d'emploi, applications. *J. Microscopie,* **1**, 1.

Francon, M. (1961). "Progress in Microscopy." Pergamon Press, Oxford.

Haine, M. E. (1961). "The Electron Microscope." E. and F. N. Spon, Ltd., London.

Hall, C. E. (1966). "Introduction to Electron Microscopy," 2nd ed. McGraw-Hill Book Company, New York.

———, and Hines, R. L. (1970). Electron microscope contrast of small atom clusters. *Phil. Mag.*, **21**, 1175.

"Handbook of Chemistry and Physics" (1964–65). 45th ed. The Chemical Rubber Co., Cleveland.

Hanszen, K.-j., Morgenstern, B., and Rosenbruch, K. J. (1963). Predictions from optical transfer theory regarding resolution and contrast in the electron microscopical image. *Z. Angew. Phys.*, **16**, 477.

Heidenreich, R. D. (1964). "Fundamentals of Transmission Electron Microscopy." Wiley-Interscience Inc., New York.

———. (1967). Electron phase contrast images of molecular detail. *J. Electronmicroscopy,* **16,** 23.

———, and Hamming, R. W. (1965). Numerical evaluation of electron image phase contrast. *Bell Syst. Tech.,* **44,** 207.

———, Hess, W. M., and Ban, L. L. (1968). A test object and criteria for high resolution electron microscopy. *Appl. Crystall.,* **1,** 1.

Henry, L. (1964). Filtrage magnétique des vitesses en microscopie électronique. Ph.D. Thèse, Faculté des Sciences de l'Université de Paris, Centre d'Orsay. Librairie Masson et Cie., Paris.

Hillier, J., and Ramberg, E. G. (1947). Magnetic electron microscope objective: contour phenomena and the attainment of high resolving power. *J. Appl. Phys.,* **18,** 48.

Howie, A. (1963). Inelastic scattering of electrons by crystals. I. The theory of small-angle inelastic scattering. *Proc. Roy. Soc.,* **A271,** 268.

Jenkins, F. A., and White, H. E. (1957). "Fundamentals of Optics," 3rd ed. McGraw-Hill Book Co., New York.

Johnson, D. J. (1968). Amplitude and phase contrast in electron-microscope images of molecular structures. *J. Roy. Micros. Soc.,* **88,** 39.

Johnson, H. M., and Parsons, D. F. (1967). A four-stage electron microscope for phase contrast and strioscopy. *Proc. 25th EMSA Meet. Chicago.* (C. Arseneaux, ed.). Claitor's Book Store, Baton Rouge. p. 236.

———. (1969). Enhanced contrast in electron microscopy of unstained biological material. I. Strioscopy (dark-field microscopy). *J. Microscopy,* **90,** 199.

———. (1970a). Enhanced contrast in electron microscopy of unstained biological material. II. Defocused contrast of large objects. *J. Microscopy,* **91,** 173.

———. (1970b). Phase contrast in electron microscopy. *Proc. 28th EMSA Meet. Houston.* (C. Arseneaux ed.). Claitor's Book Store, Baton Rouge, p. 48.

———, Hausner, G. G., Jr., and Parsons, D. F. (1969). A recording detection system for quantitation of electron microscope image contrast using a retractable micro-Faraday cage probe. *Rev. Sci. Instr.,* **40,** 1594.

Jonnsson, C., Hoffmann, H., and Möllenstedt, G. (1965). Measurement of the mean inner potential of beryllium by the electron interferometer. *Phys. Kondens. Materie.,* **3,** 193.

Kamiya, Y., and Uyeda, R. (1961). Effect of incoherent waves on the electron microscopic images of crystals. *J. Phys. Soc. Japan,* **16,** 1361.

Kanaya, K., Kawakatsu, H., Ito, K., and Yotsumoto, H. (1958). Experiment on the electron phase microscope. *J. Appl. Phys.,* **29,** 1046.

Komoda, T. (1967). Resolution of phase contrast images in electron microscopy. *Hitachi Rev.,* **49,** 43.

Labaw, L. W. (1960). Electron-microscopic observations of the interference patterns from crystal lattices. *J. Ultrastruct. Res.,* **4,** 92.

———. (1961). Striations on electron micrographs of organic crystals resulting from multiple elastic scattering. *J. Ultrastruct. Res.,* **5,** 409.

Lenz, F. (1954). Zur Streuung mittelschneller elektronen in Kleinste Winkel. *Zeit. Naturfor.*, **9a**, 185.

Locquin, M. (1954). L'influence des modifications pupillaires de l'objectif sur les contrastes en microscopie électronique. *Proc. Intern. Cong. Electron Micros. London. Roy. Micros. Soc. London*, 1956, p. 285.

———. (1956). Le contraste de phase et le contraste interchromatique: Nouvelles méthodes d'observation en microscopie éelectronique. *C. R. Acad. Sci. Paris.*, **242**, 1713.

———. (1957). Contraste de phase et contraste interchromatique. *Proc. 1st. Reg. Conf. Asia. Oceania, Tokyo (1956)*, Electrotechnical Laboratory, Tokyo, p. 163.

Möllenstedt, G., and Duker, H. (1956). Beobachten und Messungen an Biprisma—Interferenzen mit Elektronenwellen. *Zeit. Phys.*, **145**, 377.

Moretz, R. C., Johnson, H. M., and Parsons, D. F. (1968). Thickness estimation of carbon films by electron microscopy of transverse sections and optical density measurements. *J. Appl. Phys.*, **39**, 5421.

Müller, K.-H. (1970). Micro-recording by use of the Elmiskop 101. *7th Intern. Cong. Electron Micros.* Grenoble, France, **1**, 183.

Parsons, D. F. (1970). Problems in High Resolution Electron Microscopy of Biological Materials in their Natural State. *In:* "Some Biological Techniques in Electron Microscopy." (Parsons, D.F. ed.). Academic Press, New York. p. 1.

Rakels, C. J., Tiemeijer, J. C., and Witteveen, K. W. (1968). The Philips electron microscope Em300. *Philips Techn. Rev.*, **29**, 370.

Ramberg, E. G. (1949). Phase contrast in electron microscope images. *J. Appl. Phys.*, **20**, 441.

Reisner, J. H. (1964). Quantitative methods for estimating and improving performance with the electron microscope. *(RCA) Sci. Instr. News.*, **9**, 1.

Rogers, G. L. (1969). Fourier images in electron microscopy and their possible misinterpretation. *J. Microscopy*, **89**, 121.

Rushton, W. A. H. (1965). Visual adaptation. *Proc. Roy. Soc.*, **B162**, 20.

Scherzer, O. (1949). The theoretical resolution limit of the electron microscope. *J. Appl. Phys.*, **20**, 20.

Sommerfeld, A. (1954). "Optics." Academic Press, New York.

Southworth, G. (1966). A new method of television waveform display. J.S.M.P.T.E., **75**, 848.

Thon, F. (1966). On the defocusing dependence of phase contrast in electron microscopical images. *Z. Naturf.*, **21a**, 476.

———, and Willasch, D. (1970). Hochhauflösungs-elektronenmikroskopie mit Spezialaperturblenden und Phasenplatten. *7th Intern. Cong. Electron Micros.* Grenoble, France, **1**, 3.

Tochigi, H., Nakatsuka, H., Fukami, A., and Kanaya, K. (1970). The improvement of the image contrast by using the phase plate in the transmission electron microscope. *7th Intern. Cong. Electron Micros.* Grenoble, France. **1**, 73.

Unwin, P. N. T. (1970). An electrostatic phase plate for the electron microscope. *Deutschen Bunsen-Gesellschaft fur Physikalische Chemie,* **74,** 1137.

Valentine, R. C. (1966). The Response of Photographic Emulsions to Electrons. *In:* "Advances in Optical and Electron Microscopy." (Cosslett, V. E., and Barer, R. eds.). Academic Press, New York. **1,** 180.

Wahl, H. (1970). Electron optics with highly coherent waves. *Deutschen Bunsen-Gesellschaft fur Physikalische Chemie,* **74,** 1142.

Wittry, D. B., Ferrier, R. P., and Cosslett, V. E. (1969). Selected-area electron spectrometry in the transmission electron microscope. *Brit. J. Appl. Phys. (Phys. D. J.) Ser. 2,* **2,** 1767.

Wohlleben, D. (1967). Diffraction effects in Lorentz microscopy. *J. Appl. Phys.,* **38,** 3341.

Yada, K., and Hibi, T. (1968). Factors affecting the contrast of lattice image. I. Focusing of objective. *J. Electron Micros.,* **17,** 97.

Zernike, F. (1942a). Phase contrast, a new method for the microscopic observation of transparent objects. Part I. *Physica,* **9,** 686.

———. (1942b). Phase contrast, a new method for the microscopic observation of transparent objects. Part II. *Physica,* **9,** 974.

———. (1955). How I discovered phase contrast. *Science,* **121,** 345.

Zworykin, V. K., Morton, G. A., Ramberg, E. G., Hillier, J., and Vance, A. W. (1945). "Electron Optics and the Electron Microscope." John Wiley & Sons, Inc., New York.

5 Electron Microscopic Evaluation of Subcellular Fractions Obtained by Ultracentrifugation

Russell L. Deter

Department of Cell Biology
Baylor College of Medicine
Houston, Texas

INTRODUCTION

Since their beginning in the late 1930s and early 1940s, studies of biological specimens with the electron microscope have provided much new information concerning the structural organization of organs, tissues, cells, and subcellular particles. The primary objective of most investigations has been the morphological characterization of these structures in various normal and pathological states. From a functional standpoint, however, an understanding of the physiology of biological systems requires a knowledge of the chemical reactions, mediated for the most part by enzymes, which occur in these structures. Although radioautography and cytochemistry have provided important information about these processes, most of our knowledge has been derived from biochemical studies. Biochemical methods, though relatively specific and quantitative, can generally be applied only to preparations in which cellular constituents are highly dispersed. This is achieved by disrupting tissue architecture and suspending tissue components in large volumes of suspension medium. Because of the inevitable mixing that results, association of functional properties with specific morphological entities is very difficult unless careful morphological study accompanies the biochemical investigations. The necessity of this type of correlative study was recognized by some of the first investigators to use the electron microscope for the study of biological problems (Palade, 1971). It is now a common practice to support biochemical results with electron micrographs whenever particulate fractions of biological tissues are being investigated.

Two problems have plagued this otherwise laudable effort to correlate structure and function at the subcellular level. The first of these concerns the representativeness of the samples being studied. As reviewed on several occasions (de Duve *et al.*, 1962; de Duve, 1964, 1967, 1971; Roodyn, 1965), available evidence indicates that subcellular fractions are not usually homogeneous and contain numerous constituents. The biochemical characterization of such fractions can be quite reliable, however, adequate homogenization and resuspension procedures produce preparations in which constituents are randomly dispersed. With such preparations, each aliquot can be considered a random sample of fraction constituents. In many instances a single aliquot is also a representative sample as variation among replicates has been found to be relatively small.

Achieving sample representativeness in morphological studies is considerably more difficult and has been ignored by most investigators. In view of the morphological complexities of fractions and the infinitesimally small samples usually studied (2,000,000 sections, 0.1 cm $\times$ 0.1 cm $\times$ 5×10^{-8} cm, can be taken from each 0.1 ml of pellet volume!), it is surprising that the relationship of structure to function is not more confused. The degree to which misconceptions have been introduced by morphological studies of nonrepresentative samples will not be fully determined, until comparable investigations taking into account sampling problems, have been carried out. One of the principal purposes of this chapter is to present a method developed by Baudhuin that permits the obtaining of truly representative samples for morphological studies of fractions. In the opinion of this author, the Baudhuin method is the simplest and most versatile of those currently available.

A second problem encountered in comparing morphological results with those obtained by biochemical methods relates to the nature of the morphological data itself. As most biochemical parameters are chemical or physical, the quantitative measurement of these parameters is usually possible. The ability to characterize a process or property quantitatively extends considerably the range of questions that can be asked and permits a much more detailed analysis of the relationships existing between variables. Most morphological studies of fractions do not provide quantitative morphological information, and thus only qualitative or pseudoquantitative conclusions can be drawn. As a result, quantitative biochemical measurements on representative samples are usually compared to qualitative morphological descriptions of preparations whose representativeness is unknown. The grossly different nature of these two kinds of data greatly limits the extent to which they can be reliably correlated and no statistical or other mathematical analysis can be performed. This results in the loss of much valuable information and greatly decreases the efficiency of procedures that are technically difficult and time-consuming. For these reasons, a brief discussion of the quantitative morphological analysis of fractions will be presented. It is hoped that this introduction will stimulate the interest of the reader and encourage further use of a quantitative approach in the morphological study of fractions.

ULTRACENTRIFUGATION

The complex nature of the mixture one obtains by homogenization presents difficult problems for the investigator attempting to correlate structural characteristics with biochemical properties. For this reason, most investigators have either (1) isolated a particular morphological entity and then studied its biochemical characteristics, or (2) evaluated the behavior of certain biochemical properties in systems designed to separate particles spacially and then correlated the results with the behavior, actual or expected, of different particle populations defined morphologically. Ultracentrifugation has been employed most widely for both these purposes although phase distribution, electrophoresis, and differential filtration through membranes of defined pore size have also been used (de Duve, 1971).

Theory

The behavior of particles during centrifugation depends upon interaction between an accelerating force developed by rotor rotation and a resistive force that opposes particle movement through a liquid medium (Clark, 1959). The force of acceleration is given by the following expression:

$$F_a = m_{eff} a_r = m_{eff} \omega^2 x \tag{5.1}$$

where F_a is the accelerating force, m_{eff} the effective mass of the particle, a_r the acceleration along radii, ω the angular velocity and x the distance from the axis of rotation. The resistive force is:

$$F_r = 6\pi r_o n_m \Theta v \tag{5.2}$$

where r_o is the radius of the sphere having a volume equal to that of the particle, η_m the viscosity of the medium, Θ the ratio of the coefficient of friction of the particle to that of the sphere of equivalent volume, and v the particle velocity. This resistive force is known as *Stokes resistance.* At any instant in time when these two forces are equal, the particle will be traveling at a velocity defined in the following manner:

$$F_r = F_a \tag{5.3}$$

$$6 r_o \eta_m \Theta v = m_{eff} \omega^2 x \tag{5.4}$$

$$v = \frac{m_{eff} \omega^2 x}{6 r_o \eta_m \Theta} \tag{5.5}$$

In a fluid medium, the effective mass (m_{eff}) is the mass of the particle (m_p) minus the mass of the fluid displaced by the particle (m_{dis}).

$$m_{eff} = m_p - m_{dis} = \varphi_p(\rho_p - \rho_m) \tag{5.6}$$

In this expression, φ_p is the particle volume, ρ_p is the particle density and ρ_m is the density of the medium. Substitution of Eq. 5.6 into Eq. 5.5 gives the following expression:

$$v = \frac{\varphi_p(\rho_p - \rho_m)\omega^2 x}{6 r_o \eta_m \Theta} \tag{5.7}$$

Transposition of $\omega^2 x$ *to* the left-hand side of this equation gives an expression in which the right-hand side depends only upon the physical properties of the particle and the medium. In a homogeneous medium, particles with different physical properties can be defined by means of this constant, which is known as the sedimentation coefficient (s_p):

$$s_p = \frac{v}{\omega^2 x} = \frac{\varphi_p(\rho_p - \rho_m)}{6\pi r_o \Theta \eta_m} \tag{5.8}$$

Dimensional analysis reveals that the units of the sedimentation coefficient are seconds. Sedimentation coefficients are frequently expressed in Svdberg units where one unit equals 10^{-13} sec. For spherical particles, Eq. 5.8 can be simplified considerably:

$$s_p = \frac{4/3\pi r_o^3(\rho_p - \rho_m)}{6\pi r_o \eta_m} = \frac{2r_o^2(\rho_p - \rho_m)}{9\eta_m} \tag{5.9}$$

A similar expression for nonspehrical particles can be obtained by adding the frictional coefficient ratio, Θ, to the denominator of this expression. Θ has a value between 1.044 and 1.996 for prolate elipsoids, and between 1.042 and 0.782 for oblate elipsoids, with axial ratios between 2 and 20 (de Duve *et al.*, 1959).

As shown by de Duve *et al.* (1959), the location of a particle having a sedimentation coefficient s_i following centrifugation can be determined in the following manner (effect of particle concentration neglected):

$$v_i = s_i \omega^2 x \tag{5.10}$$

$$\left(\frac{dx}{dt}\right)_i = s_i \omega^2 x \tag{5.11}$$

$$\left(\frac{dx}{x}\right)_i = s_i \omega^2 \, dt \tag{5.12}$$

$$\frac{x_{f_i}}{x_{0_i}} \int_{x_{0i}}^{x_{fi}} d(\ln x)_i = s_i \int_0^t \omega^2 \, dt \tag{5.13}$$

$$\ln x_{f_i} - \ln x_{0_i} = s_i W_t \tag{5.14}$$

In these equations, x_{0_i} is the initial position and x_{f_i} the final position of the *i* particle with respect to axis of rotation, *t* the time of centrifugation in seconds and W_t the integral $\int_0^t \omega^2 dt$ with the units of ω being radians squared per second. Equation 5.13 is called the position equation and describes the location of the particle if it is permitted to sediment freely through a homogeneous medium during the course of the centrifugation. This condition is most ideally met in zonal rotors or in rotors having sector-shaped tubes. It may be approximated in swinging bucket rotors with cylindrical tubes if aggregation and rapid sedimentation do not occur when particles collide with the tube walls. In fixed angle rotors, rotor geometry is of such complexity that this equation can give only a very rough estimate of the location of particles after centrifugation.

As seen in Eq. 5.14, the centrifugational force is best described by the integral W_t. The total integral can be subdivided into an acceleration W (W_a), a plateau W (W_p) and a deceleration W (W_d). Instruments (Beckman $\omega^2 t$ Integrator) are now available for direct measurement of these various W's. Centrifugational conditions can also be given in terms of the acceleration of gravity (g) and the time in minutes (min). As indicated by Leighton *et al.* (1968), W is converted to g·min by the following expression:

$$g \cdot \min = (W) \frac{x_{av}}{(981)(60)} \tag{5.15}$$

where x_{av} is the distance from the axis of rotation to the middle of the column of liquid within the tube. The specification of this distance will depend upon the type of rotor and the volume of fluid used. It is not easily defined in fixed angle rotors but is usually considered to be the distance from the rotation axis to a point in the middle of the tube halfway between the miniscus and the bottom of the column of liquid.

Differential Sedimentation in a Homogeneous Medium

The most frequently used method for obtaining particle separation takes advantage of differences in sedimentation velocity that result from differences in physical properties. This procedure is generally known as differential centrifugation and consists of a series of centrifugations in which the centrifugal force is progressively increased. At the end of each centrifugation, the particles remaining in suspension are separated from the pellet by decantation and subjected to further centrifugation. This is continued until only soluble proteins and small molecular weight components remain in the supernatant. The composition of the fractions obtained by this procedure depends upon (1) the physical properties of the particles, (2) the centrifugation conditions (W), (3) rotor geometry, and (4) the decantation procedure. Controllable variables are altered to fit the objectives of the experiment, and marker distributions are used to monitor the technical aspects of the fractionation procedure. The determination of the distribution of markers for the particles of principal interest, along with those for components making major contributions to the protein content of the original suspension (the protein distribution is

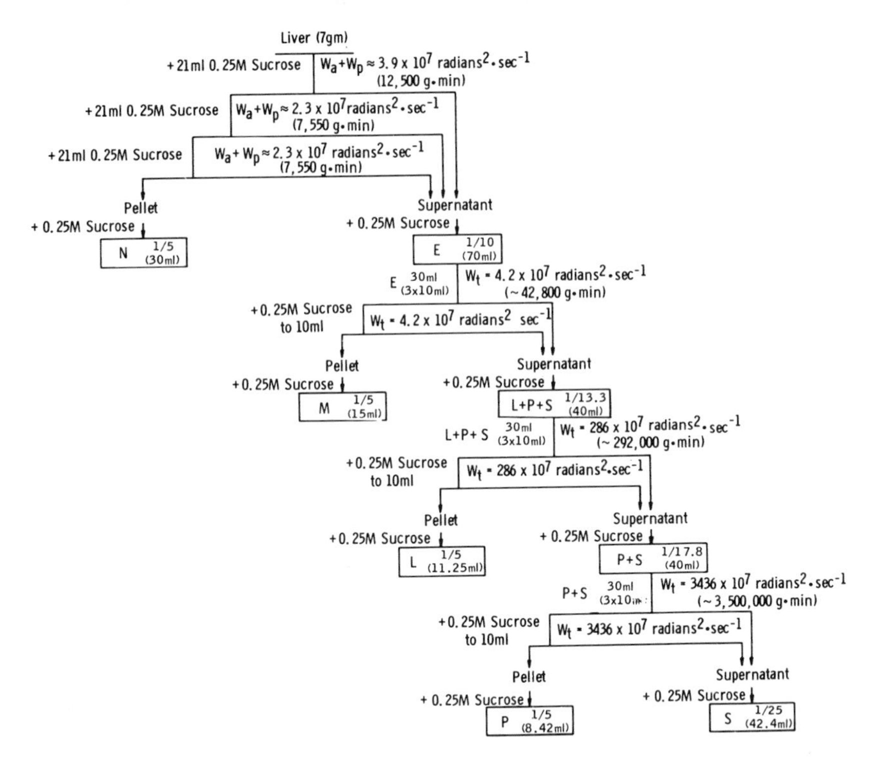

Fig. 5.1 Fractionation procedure for rat liver. E, N, M, L, P, and S signify the cytoplasmic extract, nuclear fraction, mitochondrial fraction, light mitochondrial fraction, microsomal fraction, and soluble fraction respectively. Centrifugation conditions (Type 65 rotor, except for separation of N and E) are given in terms of W, the time-integral of the squared angular velocity (see text). For conversion to g· min, average distances from rotation axis of 19 cm (N, E) and 6 cm(M, L, P, S) were used (Eq. 5.15). At certain steps, the centrifuged volume was subdivided into 10 ml aliquots as indicated by 3 x 10 ml. The final concentration (e.g., 1/10 E-cytoplasmic extract from one gram of liver suspended in 10 ml) and total volume (within the parentheses) of each fraction are given for a fractionation of 7 gm of liver.

frequently determined also), usually provides sufficient information for adequate evaluation.

The fractionation scheme for rat liver outlined in Fig. 5.1 produces distributions similar to those given in Fig. 5.3. A simplified version of this procedure (Fig. 5.2) is presently being used in this author's laboratory. These two procedures are similar to that used by de Duve *et al.* (1955) with some modifications due to the use of the Type 65 rotor. The first step in any fractionation procedure involves making a homogenate. The most common method (de Duve, 1971) employs the Potter-Elveh-

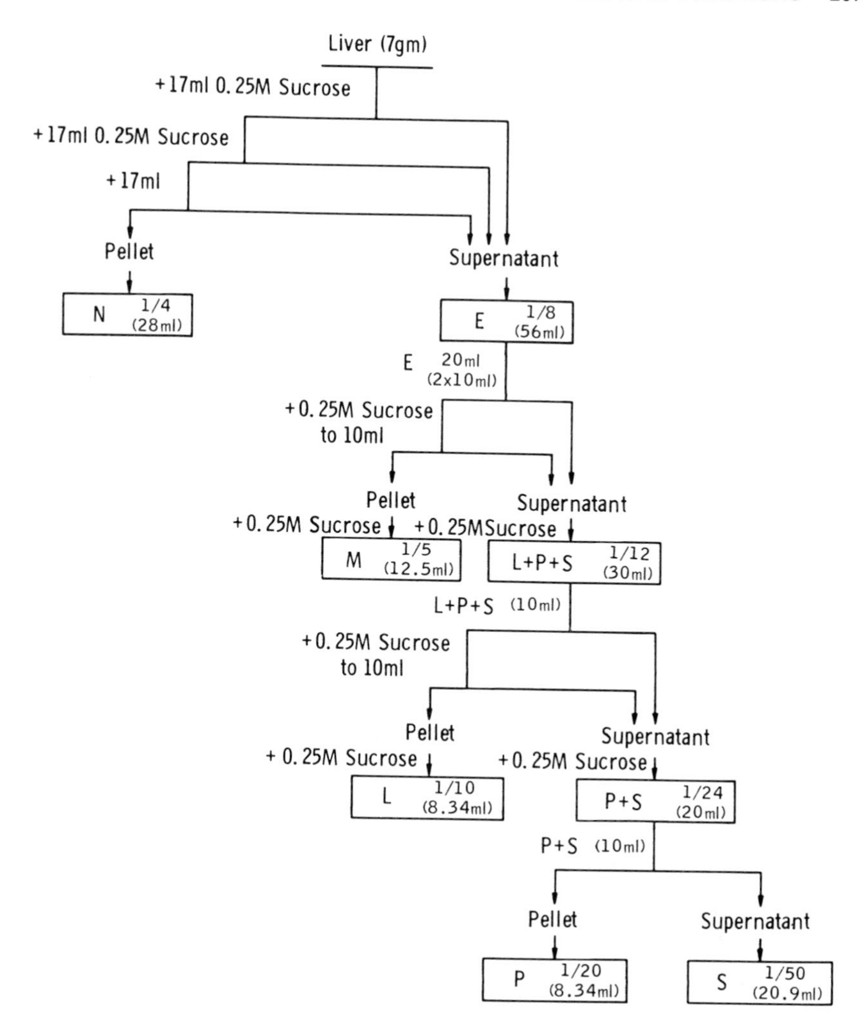

Fig. 5.2 Simplified fractionation procedure for rat liver. Centrifugation conditions are identical to those given in Fig. 5.1. The symbols and numerical values have the same meaning as those given in Fig. 5.1.

jem homogenizer, which consists of a glass tube and a Teflon pestle, the clearance being approximately 0.25 mm. Homogenization is frequently carried out in 0.25 *M* sucrose although other media are used. Simple quarter molar sucrose solutions are near isotonicity and do not contain ions that can cause agglutination of particulate constituents (de Duve and Berthet, 1954). With rat liver, a buffer is not required to maintain a pH near neutrality but this may not be the case with other tissues (de

Duve and Berthet, 1954). It should be pointed out, however, that with some tissues, agglutination occurs in a sucrose medium and improved results are obtained with isotonic salt solutions (Bowers *et al.*, 1967). Three passes of the pestle are sufficient to break up tissues such as liver though more violent methods must be used if large amounts of connective tissue or muscle are present. The homogenization procedure, at least for rat liver, seems to be most effective when the ratio of tissue volume to suspending medium is ~ 1 to 3; the density of rat liver is ~ 1.06 (Weibel *et al.*, 1969).

Preparation of the nuclear fraction has generally been carried out in an International refrigerated centrifuge (2°C). To minimize particle damage, homogenization and separation of the nuclear fraction can be combined. In this procedure the particle suspension is centrifuged after the first pass of the pestle and the supernatant decanted. The pellet is then resuspended in 0.25 *M* sucrose with one pass of the pestle. This is repeated once more. Particles subjected to this procedure are exposed only to the minimum shearing forces needed to release them from the tissue. The preparation of other fractions is also carried out at 2°C in 0.25 *M* sucrose. A rod, made of glass, or teflon-coated metal, is used for resuspension except in the case of the microsomal fraction which requires a motor driven pestle. Resuspension procedures should be kept at a minimum to avoid particle damage and enzyme inactivation, but must be sufficient to give preparations a uniform composition.

The marker distributions shown in Fig. 5.3 are presented in a form where physical significance can be ascribed to each geometric feature of the graph (de Duve, 1967). The ordinate gives the specific activity of the marker in each fraction while the abscissa shows the distribution of protein. The area associated with each fraction represents the percent of total activity found in that fraction. This manner of presentation permits direct visual comparison of data on different markers within the same fraction or on the same marker in different fractions. Comparisons of this kind are quite rapid and frequently reveal relationships that are difficult to recognize when only a simple tabulation of the results is used.

Differential Sedimentation in a Density Gradient

A more sophisticated procedure for the separation of cell particles based on differences in sedimentation velocity, often called *rate sedimentation*, has been utilized by a number of investigators (Kuff *et al.*, 1956; Beaufay *et al.*, 1959; Deter and de Duve, 1967; McCarty *et al.*, 1968). In this procedure, particles are layered over, under, or suspended homogeneously in a continuous density gradient usually made with sucrose. The gradient is

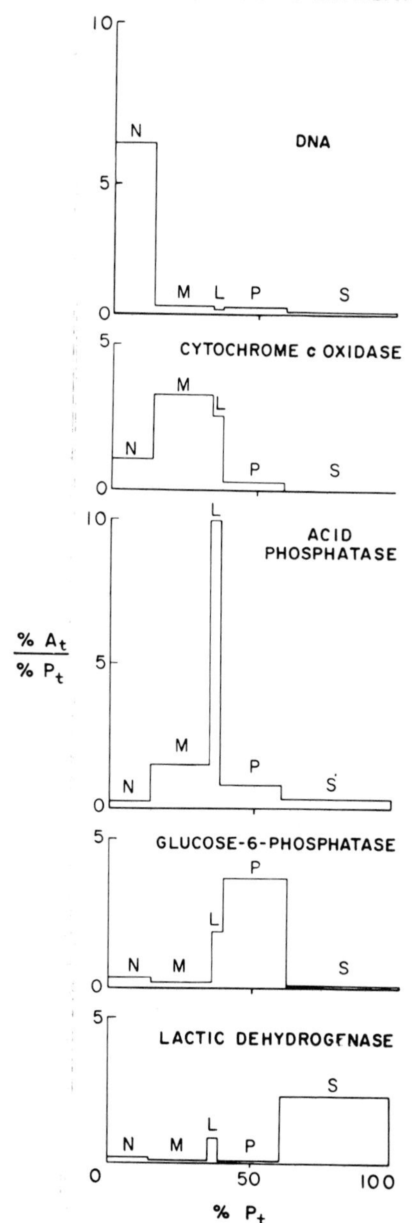

Fig. 5.3 Distribution of markers in fractions obtained by differential centrifugation. Fractions were prepared by the methods described in Fig. 5.1. Enzyme activities were determined using procedures given by de Duve *et al.* (1955) and Deter (unpublished). The Burton method (1956) was used for DNA measurements. The protein content of fractions was determined by the method of Lowery *et al.* (1951). A_t refers to the total enyzme or DNA content of one gram of liver. P_t is the amount of protein per gram of liver.

then centrifuged until partial sedimentation of the particles of interest is obtained (a similar procedure with discontinuous gradients has been used by Baggiolini *et al.*, 1969). The increasing viscosity and density of the sucrose gradient opposes the increase in centrifugal force as the distance from the axis of rotation is increased. Constant velocity sedimentation can be obtained if the proper gradient is chosen (Noll, 1967; McCarty *et al.*, 1968). The choice of the limits of the gradient depends upon the geometry of the rotor, the length of the gradient, and the equilibrium densities of the particles being studied. The presence of a gradient not only produces constant velocity sedimentation but also minimizes thermal and mechanical convections (de Duve *et al.*, 1959).

Figure 5.4 illustrates the results obtained following centrifugation of a suspension containing polyribosomes that was layered over a sucrose gradient. Six of the absorption peaks, starting with the small one on the right, represent populations of polyribosomes that have sedimented out of the original layer. The smallest peak is produced by polysomes containing seven individual ribosomes. The large seventh peak, counting from the right, indicates the location of ribosomes that have not formed polysomes. With increasing centrifugation, these peaks move toward the bottom of the tube, becoming more widely separated (due to the difference in sedimentation velocity) and broader (due to diffusion).

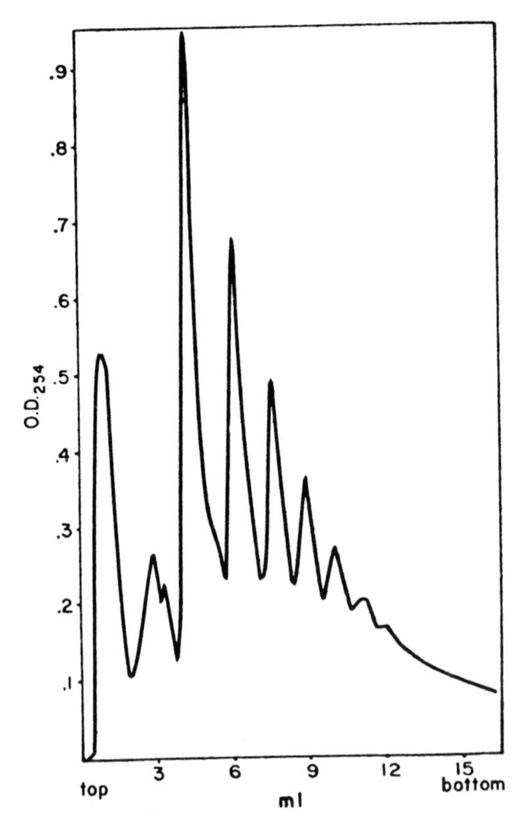

Fig. 5.4 Distribution of polyribosomes following rate sedimentation. Illustrated in this figure are absorption peaks produced by polyribosome populations in the postmicrosomal fraction of rat liver. One hundred microliters of a fraction having an optical density of 2.0 at 260 mm was layered over a continuous convex sucrose density gradient having limits of 1.063 and 1.147. The form of the gradient was that calculated to give constant velocity sedimentation for paricles with a density of 1.41 g/ml. This preparation was centrifuged for three hours at 25,000 rpm in a SW 25.3 swinging bucket rotor. After centrifugation, the gradient was fractionated and the absorption of each fraction measured at 254 mm. From McCarty, K.S. *et al., Anal Biochem, 24*:314, 1968.

Although good spacial separation can be obtained if the sedimentation coefficient distributions of adjacent populations do not overlap significantly, one serious limitation is imposed by placing the particle suspension over the density gradient. When the concentration of particles is high, artifactual localization within the gradient can occur due to drop sedimentation (Beaufay, 1966). This phenomenon, also called the streaming effect (Anderson, 1955), is caused by the movement of small volumes of the original layer into the underlying gradient. It occurs when local particle concentrations are sufficient to produce an overall density that is greater than that of the adjacent region of the gradient. Although methods to circumvent this difficulty have been devised (Beaufay, 1966), the simplest solution is to keep the particle suspension at a high dilution. However, this severely limits the amount of material that can be used in rate sedimentation studies.

Homogeneous distribution of the particle suspension within the gradient prior to centrifugation can eliminate drop sedimentation, provided centrifugation conditions do not cause a local increase in particle concentration (Bauduin, 1971). As illustrated in Fig. 5.5 for a mixture of three classes of particles, an appropriate centrifugation will cause particles with the most rapid sedimentation rate (⬤) to leave a considerable fraction of the gradient closest to the rotation axis and to contribute significantly to the formation of a pellet. Particles with an intermediate sedimentation velocity (●) are cleared from a smaller region of the gradient and they also contribute to the pellet. Particles with a negligible rate of sedimentation (•) do not move and remain homogeneously dis-

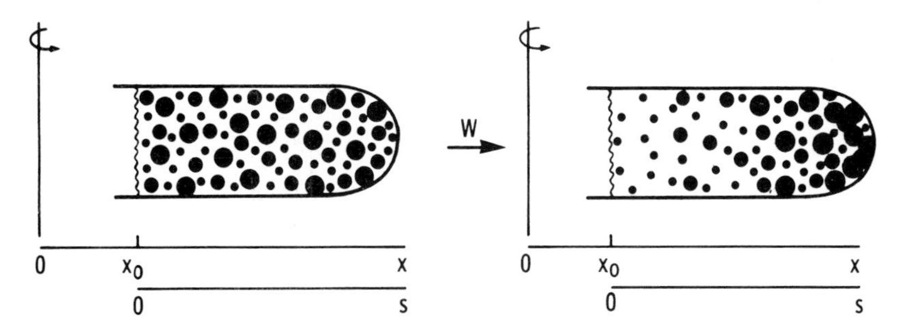

Fig. 5.5 Rate sedimentation procedure (homogeneous particle distribution). In this example, swinging bucket rotor geometry is assumed. The axis of rotation is located on the left of each subfigure. The distance from the axis of rotation is indicated by x, x_0 being the distance to the miniscus. s is the sedimentation coefficient associated with x (see Eq. 5.14). W is the time-integral of the squared angular velocity. The distribution of three particle classes (• ● ⬤) before, on the left, and after, on the right, centrifugation are illustrated. The density of all particles is assumed to be identical.

tributed. If one fractionates a gradient of this type and measures the appropriate marker, cumulative distribution curves reflecting the sedimentation behavior of the particles carrying the marker are obtained. With the appropriate transformations (see Eqs. 5.10–5.14; $v_i = s_i\ \omega^2$ instead of $v_i = s_i\ \omega^2\ x$), the distribution of the marker as a function of the sedimentation coefficient of its host particles can be determined. If the marker concentration is similar in particles of all size classes, its distribution is equivalent to the volume distribution for that particle population. If particle density does not vary significantly between size classes, these distributions can also be considered mass distributions.

Figure 5.6 gives the distributions of marker enzymes for various particles in cytoplasmic extracts from rat liver. The extracts were prepared by differential centrifugation and mixed with the appropriate light and heavy sucrose solutions prior to the formation of the gradient. Gradi-

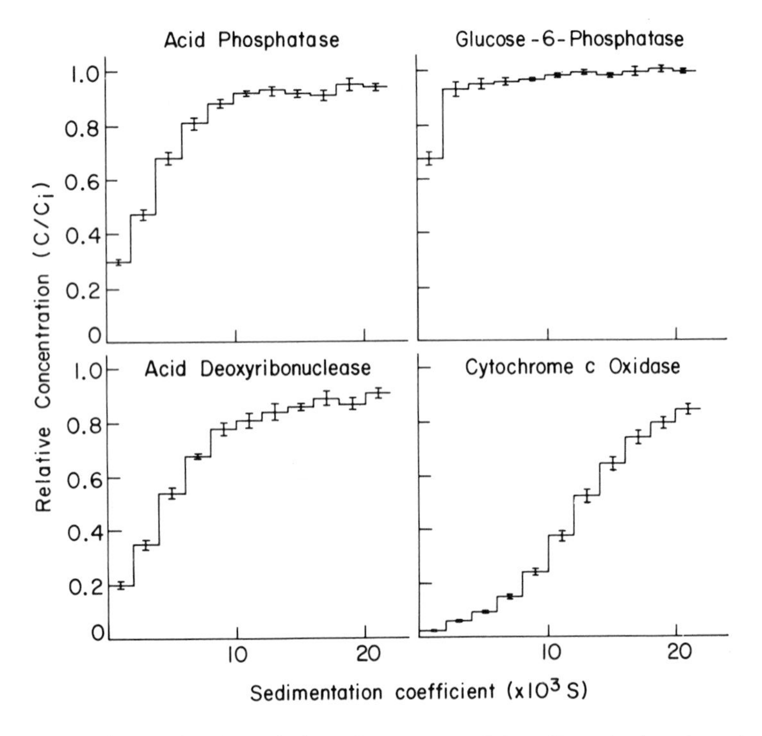

Fig. 5.6 Sedimentation boundaries of enzymes of hepatic cytoplasmic extracts. Boundaries were obtained by centrifuging (W_t = 2.44 × 10^8 radians^2 · sec^{-1}) density gradients (limits: 1.034 and 1.067) containing homogeneously distributed cytoplasmic extracts (1/15 E, final concentration), fractioning the gradients and measuring various enzymes in all fractions. Diagrams represent pooled values (means ± standard error) from seven experiments. C_i and C refer to enzyme concentrations before and after centrifugation, respectively. From Deter, R. L. and de Duve, C., *J. Cell Biol.* *33*: 437, 1967.

ents were made with a gradient maker designed by H. Beaufay, and following centrifugation, were fractionated by means of a tube slicer (de Duve *et al.,* 1959). As illustrated in Fig. 5.6, identification of different particle populations can be made. Information concerning the chemical and/or enzymatic constituents of each particle type is accessible even though the different particles have not been isolated in pure form. With these techniques, evidence suggesting the transfer of enzymes from one particle population to another has been obtained (Deter and de Duve, 1967).

Isopycnic Centrifugation

Particles may be separated not only on the basis of differences in sedimentation velocity but also because of differences in equilibrium density. As shown in Eq. 5.6, sedimentation velocity depends upon a variety of factors such as particle size, shape, and density. If particles are allowed to sediment in a gradient, particle velocity will become zero wherever the density of the medium and the density of the particle are equal. However, the particle density at these points is not necessarily that which would be found in 0.25 M sucrose or some other physiological medium. The density of a particle is a function of (1) the space accessible to the solute forming the gradient, (2) the amount of osmotically active material within the particle, (3) the amount of bound water attached to the matrix, and (4) the density of the dry matrix (Beaufay and Berthet, 1963). Differences in these parameters result in different equilibrium densities for different particles within the same gradient and for the same particle in different gradients. Through selection of the proper medium, spacial separation of particle populations can be achieved.

There are primarily two types of density gradients, the discontinuous and the continuous. The former is a series of layers of constant density, the size and density of each layer being determined by the objectives of the experiment. Particle mixtures are placed at the top, middle, or bottom of such gradients and subjected to centrifugation. Migration of individual components depends upon the nature of the gradient, particle position, particle density, and the magnitude of the centrifugal force field. Particles with sufficient sedimentation or flotation velocity collect at the interface between the two layers that have densities greater and lesser than that of the particle. Although sharp bands are obtained by this procedure, the composition of these bands is often complex. All particles having densities between those of the two adjacent layers will be located at the same interface. Even particles that would be expected to enter one of the layers may be held at the interface if too much material is present. For these reasons, gradients of this type are more useful for obtaining

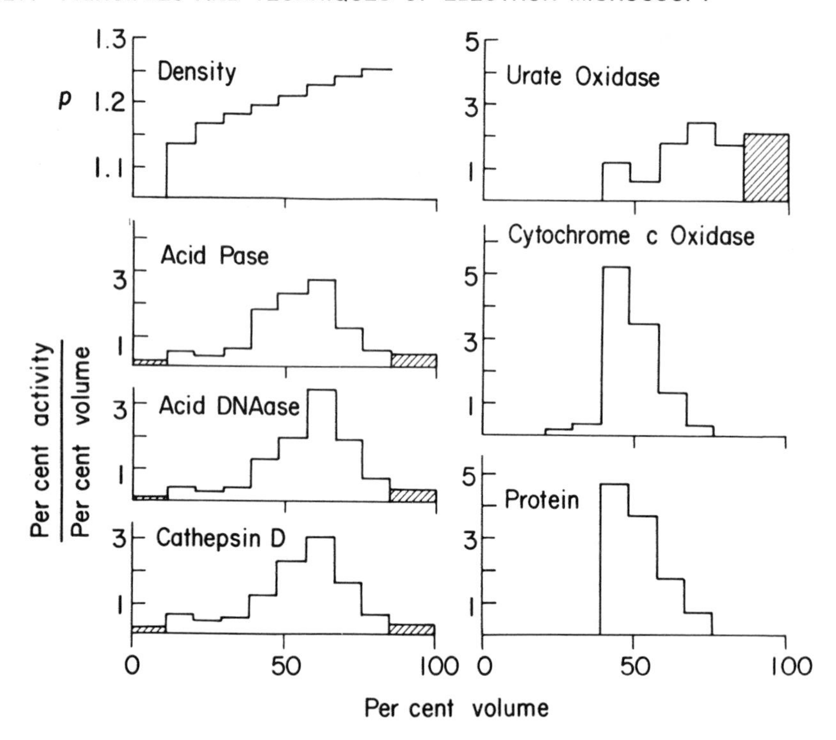

Fig. 5.7 Distribution of markers following isopycnic centrifugation of hepatic M + L fraction (swinging bucket rotor). Enzyme distributions were determined using procedures given by de Duve *et al.* (1955). Protein was measured by the method of Lowery *et al.* (1951) and densities as described by Beaufay (1966). Distributions are presented in terms of the gradient volume recovered, the miniscus of the gradient being at *O* on the percent volume scale. Regions containing the layered M + L fraction (first fraction) and the pellet (last fraction) are shaded in each distribution.

constituents to be used in other investigations than for determining the physical properties of particles.

A more appropriate gradient for analytical isopycnic centrifugation studies is the continuous density gradient shown in Fig. 5.7. In this gradient, the density continuously changes from its minimum to its maximum value as one proceeds from the miniscus toward the bottom of the tube. The particle suspension is placed over or under the gradient or can be homogeneously distributed throughout (Beaufay and Berthet, 1963). Placement at the top or bottom of the gradient will result in migration of particles into the body of the gradient where they will continue to sediment or float until they reach their equilibrium density. Drop sedimentation is not a problem in these gradients provided true equilibrum is achieved. However, if very small particles are present and drop sedimentation occurs, fraction contamination can still be a problem. Slow sedimenting particles carried beyond their equilibrium position may not have sufficient time to reach equilibrium before centrifugation is terminated. Drop flotation is possible with particle suspensions initially located at the bottom of the gradient but the high viscosity of dense sucrose solutions makes this unlikely except for very light particles (Baudhuin, 1971). Homogeneous distribution within the gradient minimizes drop sedimentation or flotation but yields fractions that are contaminated by particles not sedimented under the centrifugation conditions employed.

Separation of particle populations by isopycnic centrifugation is illustrated in Fig. 5.7. An aliquot of a M + L fraction from rat liver (0.6 ml, concentration: 1/1) was layered over a continuous sucrose gradient having limits of 1.17 and 1.26. Centrifugation of this gradient was carried out in a SW-39 rotor ($W_t = 1.52 \times 10^{11}$ rad ^{2}/sec). The centrifugation conditions used are sufficient to allow subcellular particles similar to mitochondria to reach their equilibrium density (Beaufay *et al.*, 1959). The gradient was fractionated using a tube slicing device and the distribution of various markers determined. The data presented in Figure 5.7 indicates that at least three particle populations are present, one marked by cytochrome c oxidase, which contributes most of the protein of the fraction, a second marked by urate oxidase, and a third containing the three lysosomal enzymes. This type of centrifugation procedure can be applied to any subcellular fraction or even cell suspensions, the nature of the gradient used being determined by the types of constituents studied and the research objectives.

Similar experiments can be carried out using zonal rotors (Fig. 5.8) or other swinging bucket rotors. In each case, the centrifugation conditions that allow particles to reach their equilibrium position must be determined. This is accomplished by carrying out a series of centrifugations in which the centrifugal force is progressively increased. When marker distributions remain unaltered with further increase in centrifugal force, the particles have reached their equilibrium density. In selecting centrifugal conditions, a recent observation by Wattiaux *et al.* (1971) should be kept in mind. These investigators have observed that isopycnic centrifugation at very high speeds subject particles to severe hydrostatic pressure and extensive particle damage can occur. This indicates that there may be maximal centrifugal forces which cannot be exceeded in isopycnic centrifugational studies of a given particle population.

ELECTRON MICROSCOPY OF ISOLATED FRACTIONS

With this background on how subcellular fractions are obtained by ultracentrifugation, we can now turn our attention to the problem of evaluat-

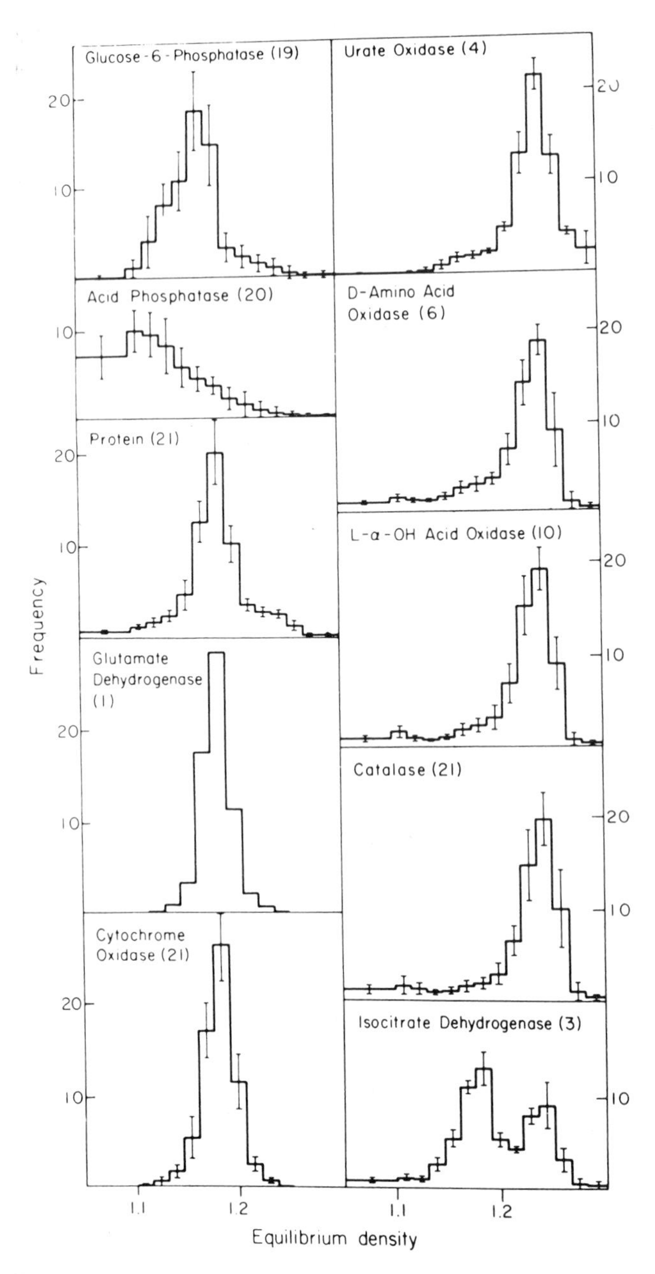

Fig. 5.8 Distributions of markers following isopycnic centrifugation of hepatic λ fraction (zonal rotor). In these investigations, λ fractions (similar to the L fraction described previously) were isolated from the livers of Triton WR-1339 treated rats by differential centrifugation. Eight milliliters of these fractions were injected into

ing fractions with the electron microscope. As indicated previously, the usual objective is to obtain morphological information that can be compared with chemical or physical measurements. As this latter type of data are generally quantitative in nature and obtained from representative samples, the procedures to be described are designed to give morphological data with the same characteristics. Even for qualitative evaluations, these techniques are quite useful since they minimize sampling problems.

Fractions Obtained by Differential Centrifugation

As in other electron microscopic investigations of biological specimens, the initial goal of the preparative procedure is stabilization of the morphology of constituents being studied. Because most permeability barriers are destroyed by homogenization and subcellular particles are very small, fixation of fraction components is a relatively easy matter. As shown in Table 5.1, aliquots of the various fractions are mixed with a glutaraldehyde-phosphate buffer solution, the final glutaraldehyde concentration varying with each fraction (to date, there has been no evidence to indicate that glutaraldehyde concentrations between 0.75% and 1.5% are critical). Fixation is done at room temperature with the time of fixation being as short as 5 min or as long as several hours. Fixed preparations have been kept overnight in the cold before being filtered without excessive particle deterioration (Baudhuin, 1971). Flexibility in the conditions of fixation is of great value as collection and fixation of fractions are usually carried out the same day. The considerable technical problems associated with fractionation work frequently impose limitations on the time available for the preparation of material for morphological studies.

The most critical aspect of the preparation of fractions for electron microscopic study concerns the formation of pellets, which are subsequently sectioned and examined in the microscope. Since fractions are usually heterogeneous and the amount of fraction material examined infinitesimally small (1,570,000 section, 10^{-1} cm $\times$ 10^{-3} cm $\times$ 5 $\times$ 10^{-8} cm, can be obtained from the usual pellet of the M + L fraction, prepared by filtration, whose volume is 7.86 $\times$ 10^{-4} cm^3), it is essential that

---

the rotor of Beaufay (1966) followed by 26 ml of a linear density gradient (limits: 1.15 and 1.27) and a 6 ml cushion (density = 1.32). After centrifugation (35,000 rpm for 37 min) the fluid was removed, fractionated, and the markers' distributions determined. Rotor loading and unloading was carried out while the rotor was running. Mean frequency (de Duve, 1967) values ($\pm$ standard deviation) for each fraction are plotted against fraction density. The number of experiments with each marker is given in parentheses. From Leighton, F. *et al.*, *J. Cell Biol.* *37*:482, 1968.

TABLE 5.1 PRESSURE FILTRATION

Fraction ([1]conc.)	*Fraction Volume*	*[2]Fixative Volume*	*[1]Final Dilution*	*Volume Filtered*	*Protein in Filtered Volume*	*Filter Pore Size*	*[3]Filtration Time*
—	ml	ml	—	ml	μg	μm	min
Nuclear (1/4)	0.1	2.4	1/100	0.5	181	0.1	5
Mitochondrial (1/5)	0.1	2.9	1.150	0.5	124	0.1	3
Light Mitochondrial (1/10)	0.2	1.2	1/70	0.5	133	0.1	13
Microsomal (1/20)	0.1	1.4	1/300	0.5	79	0.01	8

[1]Specified fraction from one gram of liver suspended in the volme indicated (in milliliters).
[2]Fraction and fixative volumes added together to give the fixed particle suspension that is filtered.
[3]Filtration pressure: 1.05 kg per cm^2.

a representative sample be obtained. The sampling procedure must be kept at a minimum, however, as examination of a number of fractions is often required.

At the present time, most investigators use centrifugation to obtain pellets for morphological studies. This procedure, with the exception of a method described by Maunsbach (1966), produces pellets that vary in composition in very complex ways (Baudhuin *et al.*, 1967). When the fixed angle rotor is used, as is usually the case, many phenomena occur (See Pickels, 1943, for a detailed discussion) due to the complex geometry of this type of rotor. The most important of these is the rapid sedimentation, due to aggregation, which occurs when particles traveling along radii collide with the wall of the centrifuge tube. These aggregates, complex mixtures of many constituents, sediment rapidly and accumulate along the periphery of the pellet. Suspension components which do not aggregate on striking the tube wall bounce back into the medium. They will again go to the tube wall where aggregation and rapid sedimentation will occur or they again return to the medium. As a consequence of these phenomena, the pellet formed can vary considerably in composition in almost every spacial direction.

The use of swinging bucket rotors reduces these problems but there is still the possibility of side wall interaction unless sector-shaped tubes are used. Tubes of this type, capable of withstanding high centrifugal forces, are not currently commercially available. Obtaining a truly representative sample from pellets prepared by centrifugation is therefore a formidable task. Most investigators appear to be content with samples whose relationship to average fraction characteristics is unknown at best and significantly distorted at worst.

The problem of preparing pellets from which representative samples can be obtained by simple sampling methods was elegantly solved by Baudhuin *et al.* (1967). The Baudhuin method, which has been extensively used by the author, relies on pressure filtration to produce a pellet of known dimensions. In the apparatus shown in Fig. 5.9, a 13 mm Millipore MF filter (Millipore Corp., Bedford, Mass.) is placed over a support screen and held in place by a vacuum grease coated O-ring. The pore size of the filter is chosen so that the particles of interest will be retained (Table 5.1). A specially made plexiglass filtration chamber (C), whose threads are also coated with vacuum grease, is then attached and the assembled device placed in a holder where it can be held vertical. An aliquot of the fixed particle suspension is layered over the filter using a 1 ml pipette (0.5 ml is ordinarily layered but samples of 1 ml or greater can be used with the chamber originally described by Baudhuin, *et al.* (1967). A layer of buffer solution (e.g., phosphate buffer, 0.05 M, pH

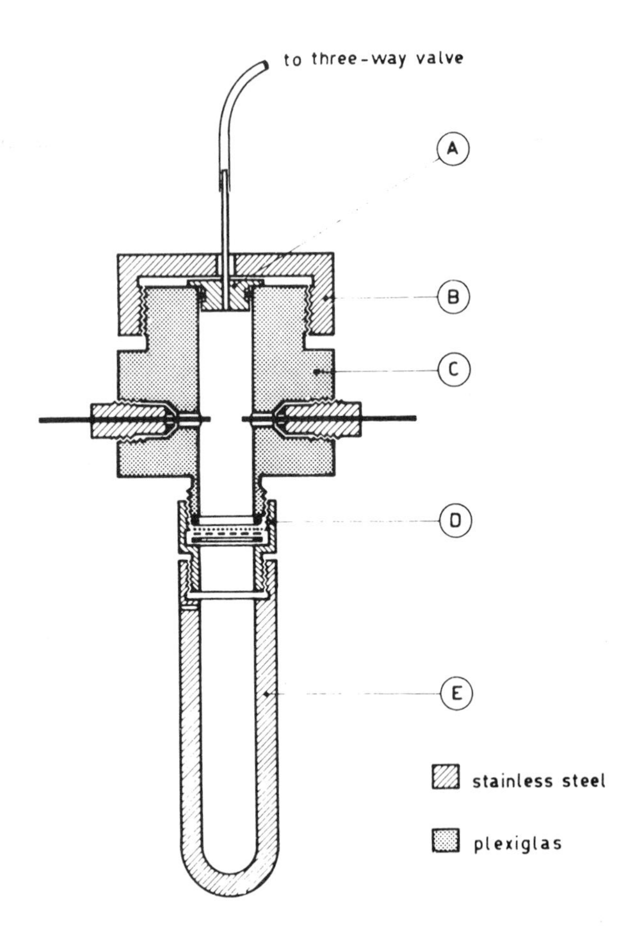

Fig. 5.9 *Filtration Unit of Baudhuin.* Parts D and E are components of the Filterfuge Tube Kit (International Equipment Co., No. 1199), which is no longer commercially available. C is a specially constructed plexiglass chamber with two conductivity electrodes. B is a screw cap holding the plug A in place. The metal tubing passing through A is connected to a cylinder of nitrogen by means of plastic tubing. A three-way gas valve is interposed between the filtration unit and the nitrogen tank. From Baudhuin, P. *et al., J. Cell Biol. 32*:181, 1967.

7.4) having a density less than that of the fixed particle suspension is then carefully added using a 2 ml pipette. The tip of the pipette is placed against the wall of the chamber and the buffer allowed to flow out over the surface of the suspension to minimize mixing. The chamber is filled with buffer until the electrodes shown in Fig. 5.9, or another reference point, are reached. A volume equal to the volume of the sample plus 0.2 ml is then added. The chamber is closed and a plastic tube connected to a tank of nitrogen is attached. In the automated system described by Baudhuin *et al.* (1967), electric valves are placed between the nitrogen tank and the filtration chamber. These valves are opened by a manually operated switch and closed when the level of the buffer goes below the level of the electrodes. This arrangement is very convenient when many samples are being processed but it is not absolutely essential. Filtration can be controlled equally well by manually opening and closing the valve on the nitrogen tank.

As can be seen in Table 5.1, filtration is relatively rapid, the rate depending upon (1) the pressure used, (2) the pore size of the filter, (3) the types of particles in the suspension, and (4) the amount of protein in the sample. The particle composition is determined by the fractionation procedure. Limits on the amount of protein filtered are determined by the requirement that the pellet must have adequate mechanical stability and a thickness that permits photographic recording at a convenient magnification. The pore size of the filter can be varied somewhat but a significant loss of the particles of interest must obviously be avoided. For these reasons, control of the nitrogen pressure is the principal means for regulating the filtration rate.

On termination of the filtration, the filtration chamber is opened and carefully detached from the metal tube. Care is taken not to agitate the remaining liquid so that the pellet is not disturbed. The liquid is removed with an absorbent tissue and the wire screen supporting the filter with its pellet is quickly taken out. Any remaining fluid is eliminated by placing the wire screen and filter on a tissue. The pellet is overlaid with a filter covered with red blood cells (*Wibo et al.*, 1971, have found the use of a red blood cell layer unnecessary in preparing pellets of the microsomal fraction), the red blood cell layer being placed against the upper surface of the pellet. Preparation of this red blood-cell-covered filter requires filtration of a blood-gluteraldehyde mixture (0.1 ml of heparinized mouse blood plus 9.9 ml of 1.5% gluteraldehyde in 0.075 *M* phosphate buffer, pH 7.4) through a 47 mm diameter Millipore-MF filter (1.2 μ pore size) under suction. Disks, 13 mm in diameter, are cut from the large filter with a special cutting tool.

After addition of the second filter, an O-ring is placed on the screen-filter-pellet-red blood cells-filter sandwich and the entire assemblage placed in a clamp. The O-ring compresses the outer edges of the filters against the metal screen blocking the direct flow of liquids over the pellet. This prevents disruption or loss of pellet material in subsequent processing. The sandwich and its clamp are put in a sucrose-phosphate buffer mixture where they are allowed to wash for 5 min. Postfixation in osmium tetroxide, dehydration in a graded series of alcohols, transferring to propylene oxide, and embeddment in Epon are carried out as indicated in Table 5.2. Until the propylene oxide step is reached, the sandwich is kept in its clamp except for passage through the osmium tetroxide solution. A smaller volume of the osmium tetroxide solution can be used if only the sandwich is placed in this solution.

After the last incubation in 100% ethanol, the clamp, O-ring, and screen are removed. This must be done quickly to avoid drying of the pellet. The sandwich is then dropped into a weighing bottle containing propylene oxide where most of the two filters dissolve. What remains is a very thin two-layered structure made up of the pellet and the red blood cell layer which is rather delicate and easily broken. This sandwich is picked out of the propylene oxide with a small watch glass and the excess propylene oxide quickly removed. A mixture of propylene oxide-Epon (Table 5.2) is added and the watch glass covered. After a 30–45 min infiltration, the first mixture is removed with an absorbent tissue and

TABLE 5.2 [1]PREPARATION OF PELLETS FOR ELECTRON MICROSCOPY

[2]*Process*	*Solution*	*Time (min)*
Fixation	1. Glutaraldehyde (1.5%)—phosphate buffer (0.075 *M*, pH 7.4)	5–15
	2. Sucrose (0.15 *M*)—phosphate buffer 0.075 *M*, pH 7.4)	5
	3. OsO_4 (1%)—phosphate buffer (0.075 *M*, pH 7.4)	20
	4. Sucrose (0.15 *M*)—phosphate buffer (0.075 *M*, pH 7.4)	5
Dehydration	1. 25% ethanol	5
	2. 50% ethanol	5
	3. 75% ethanol	5
	4. 95% ethanol	5
	5. 100% ethanol	10
	6. 100% ethanol	10
	7. 100% propylene oxide	overnight
Embedding	1. Propylene oxide (67%)—Epon mixture[3] (33%)	30–45
	2. Propylene oxide (33%)—Epon mixture (67%)	30–45
	3. Epon mixture (100%)	2–3 days at 60°C

[1]This preparative scheme is a modified version of that given by Baudhuin *et al.*, *J. Cell Biol.* *32*:181, 1967.
[2]All procedures carried out at approximately 25°C (room temperature) except final polymerization.
[3]Epon Mixture: 7 ml *A* + 3 ml *B* + 0.2 ml DMP-30.

A	*B*
Epon 812—62 ml	Epon 812—100 ml
Dodecenyl succinic anhydride—100 ml	Nadic Methyl anhydride—89 ml

replaced with a second mixture, more concentrated in Epon. The second mixture is removed after 30–45 min and a small quantity of the embedding medium placed on the watch glass. The pellet is then transferred to a small bottle cap containing a thin layer of the plastic mixture used for the final embedding. This transfer is facilitated if the sandwich on the watch glass is agitated during the infiltration to prevent sticking. Additional embedding material is added to cover the pellet and the cap placed in the oven for at least 48–72 hr at 60°C. For various embedding methods, the reader is referred to Hayat (1970 and 1972).

Pellets of at least the M + L fractions of rat liver, prepared in the manner just described, have two characteristics that greatly simplify the sampling procedure. First, because the lines of force responsible for flow through filter, and thus pellet formation, are perpendicular to the plane of the filter, the random distribution of particles present in the suspension is retained along the pellet diameter (Baudhuin *et al.*, 1967). Wall effects are not observed and the differences in pellet composition as one goes from medial to lateral is only that expected from the Poisson distribution. Secondly, the thinness of the pellet permits its complete examination at 4000 × magnification. Any stratification along a line perpendicular to the filter surface is included in the sample. Because of these characteristics, sections may be taken at any point within the pellet, the number being determined only by the size of the sample needed.

Sectioning is carried out by first cutting a wedge from the embedded sandwich and trimming it so that sections perpendicular to the plane of the filter will contain the entire pellet thickness. The ring of compression at the periphery of the sandwich must be avoided in this process. Ordinarily, the red blood cell layer is some distance from the upper surface of the pellet and may be mistaken for the pellet. This can be avoided by examining thick sections stained with toludine blue (Hayat, 1970 and 1972). Sections (50–80 nm thick, 600 μm wide and up to 1500 μm long), with the long axis of the pellet parallel to that of the section, are collected on uncoated 300 mesh bar grids. With practice, sections can be oriented so that long regions of the pellet fall in the spaces between the grid bars. Staining with uranyl acetate (Frasca and Parks, 1965) and lead citrate (Venable and Coggeshall, 1965) provide adequate contrast for examination in the electron microscope. For staining methods, the reader is referred to Hayat (1970 and 1972).

The electron microscope appearance of sections prepared from pellets of the N, M, L, and P fractions is shown in Fig. 5.10. As shown in Fig. 5.10a, the nuclei of the nuclear fraction are significantly altered and frequently appear in clumps. Because of their small number, a pellet having a uniform layer of nuclei cannot be obtained unless more tissue is filtered

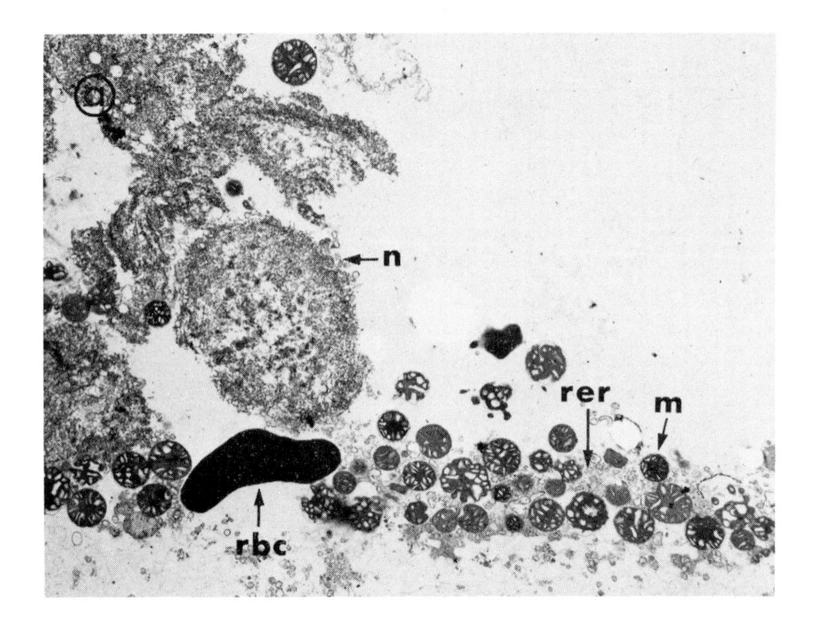

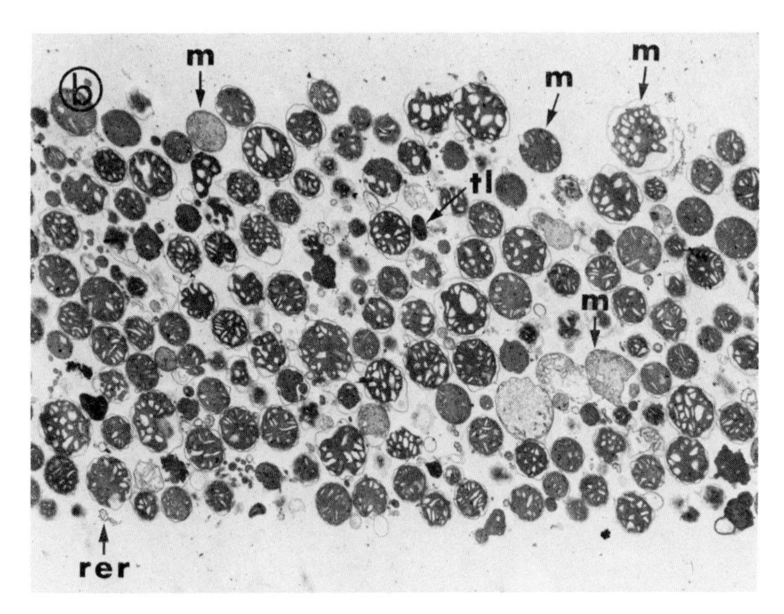

Fig. 5.10 Pellets of fractions obtained by differential centrifugation. In this figure sections of pellets from the nuclear (10a), mitochondrial (10b), light mitochondrial (10c), and microsomal fractions (10d) of rat liver are illustrated. The filtration method of Baudhuin was used to prepare the pellets (see Tables 5.1 and 5.2). Abbreviations: n-nucleus, m-mitochondrion, tl-telolysosome, rer-rough

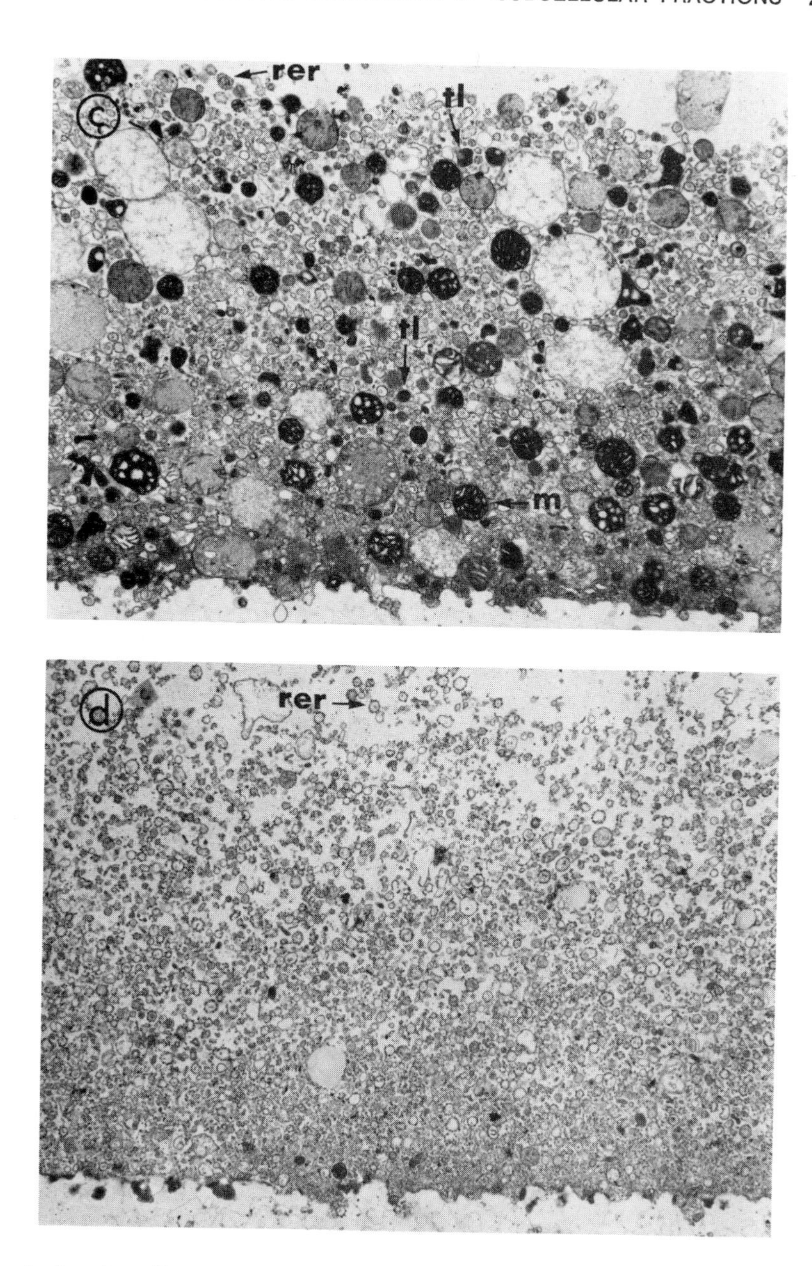

endoplasmic reticulum, rbc-red blood cell. Final magnification: 10a-4900x, 10b-6200x, 10c-6200x, 10d-8400x.

than that given in Table 5.1 (Baudhuin, 1971). The pellets of the M fraction (Fig. 5.10b) are relatively uniform but do contain several subcellular constituents. Mitochondria, primarily in the condensed state, are the main component although lysosomal particles and rough endoplasmic reticulum are also present. In the L fraction (Fig. 5.10c), mitochondria and endoplasmic reticulum predominate but telolysosomes can be easily found. The P fraction (Fig. 5.10d) is composed primarily of small membrane bound vesicles, with and without attached ribosomes, although free ribosomes and fragments of mitochondria and the cell membrane are seen. These morphological observations are consistent with the marker distributions shown in Fig. 5.3. Step-by-step procedures for the isolation and processing of cells and cell organelles for electron microscopy are given by Hayat (1972).

The evaluation of micrographs such as those presented in Fig. 5.10 can, of course, be either qualitative or quantitative. However, three important limitations are imposed by a strictly qualitative study. The first concerns the amount of the pellet that must be examined in order to adequately characterize pellet composition. For a sample to be adequate, it must be obtained in a manner that does not introduce bias and must be large enough to permit estimation of specified parameters with a predetermined precision. In qualitative studies, there is no way to objectively evaluate sample adequacy. With quantitative methods, however, statistical procedures can be used to evaluate the distribution of constituents within the pellets, allowing the design of a sampling system that gives unbiased samples (e.g., Baudhuin *et al.*, 1967; Baudhuin, 1968). Similarly, the sample size needed for parameter estimates can be calculated once an estimate of the variance has been obtained (Geunther, 1965). Satisfaction of these objective criteria of sample adequacy greatly increases the confidence that can be placed in the conclusions derived from any particular sample.

The other limitations of a qualitative evaluation are those discussed previolsly, namely, the limited correlation possible between qualitative and quantitative data and the difficulties associated with applying statistical or other mathematical analyses to qualitative results. In complex situations such as those encountered in the study of fractions, these limitations prevent the extraction of information that the experimental results could otherwise provide. Without quantitation, the full potential of a morphological study of fractions cannot be realized.

Quantitative morphological analysis of fractions is basically an application of the principles of stereology, recently reviewed by Weibel (1969) and presented in another chapter of this publication. Relative and absolute volumes, surface area, length, number, or size distribution are among the parameters that can be determined for any morphologically identifiable particle population. However, if quantitative morphological studies are to be made, several precautions must be taken in the preparation of the electron micrographs. It is essential that a microscope configuration be chosen that minimizes distortion and variation in magnification. Accurate measurement requires that all areas of the micrographs be as free as possible from artifacts that change particle dimensions. In all cases, the exact magnification must be known.

Distortions, frequently found at low magnifications, can be evaluated by determining the line spacing of a grating replica in various parts of a recorded image. The exact magnification of pellet constituents is also obtainable from the grating replica if the replica is photographed under conditions comparable to those used in making negatives of the pellet. For pellets prepared in the manner described here, a magnification of approximately 4000 × is usually the most convenient for M + L fractions. Lower magnifications may be needed for the N fraction and higher ones for the P fraction. It has been found in the author's laboratory that a single pair of calibration negatives, one taken at the beginning of a series of pellet negatives and a second taken at the end of the series, is sufficient for magnification determination. The variation in magnification between these two negatives is 1.38 (±1.35 S.D.)%(Deter, 1971). Final measurements are considerably facilitated if the magnification of all prints is identical. This can be achieved by matching the first calibration negative to a standard print of the replica and then printing all the subsequent negatives under identical enlarger conditions (Deter, 1971).

An essential feature of the preparation of electron micrographs for quantitative morphological analysis is the requirement that no component subjected to measurement be included more than once in the sample. To achieve this, only one section out of those appearing on a grid can be photographed. With most cytoplasmic particles, all sections collected on the same grid will contain profiles of particles seen in the section on which measurements are made. Only by restricting photography to a single section can duplication in the sample be avoided. Ordinarily, sections on different grids do not contain profiles of the same particles. However, with very large structures such as nuclei, with very thin sections, or with a very efficient sectioning process, it may be necessary to consider whether or not the interval between the sections included in the sample is sufficient.

Fractions Obtained by Rate Sedimentation

To date, the Baudhuin method has not been used in morphological studies of the fractions obtained in rate sedimentation experiments. The processing of fractions from gradients used in these investigations should be similar to that described for fractions obtained by differential centrifugation. However, as the protein content and particle composition are not identical in different parts of these gradients, preliminary studies would be necessary to determine appropriate filtration conditions for different fractions.

Although a direct comparison of marker and particle distributions following partial sedimentation has not been made, a somewhat more indirect approach has given gratifying results. In the first of these studies (Deter and de Duve, 1967; Baudhuin and Berthet, 1967), the sedimentation coefficient distribution of cytochrome c oxidase was compared to that of mitochondria, the latter calculated from the mitochondrial size distribution according to Equation 5.9 (Fig. 5.11). A similar comparison (Baudhuin, 1968; Deter, 1971) has been made between the sedimenta-

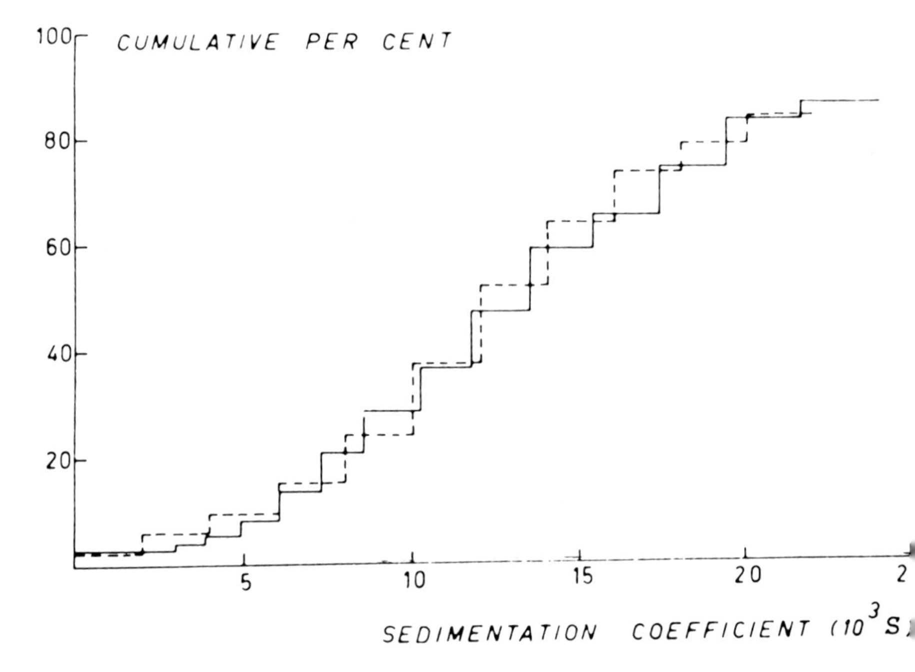

Fig. 5.11 Comparison of the sedimentation coefficient distributions of cytochrome c oxidase containing particles and mitochondria of rat liver. The cumulative sedimentation coefficient distribution of cytochrome c oxidase containing particles (---), obtained from rate sedimentation studies on cytoplasmic extracts (Deter and de Duve, 1967) is an average of seven experiments. A similar distribution for mitochondria (—) was calculated from a pooled size distribution (4 individual distributions), assuming a density of 1.10 for all size classes (see Eq. 5.9). The first class was set equal to 2.4 percent of the cytochrome c oxidase activity of the P + S fraction. From Baudhuin, P. and Berthet, J. J. Cell Biol *35*:631, 1967.

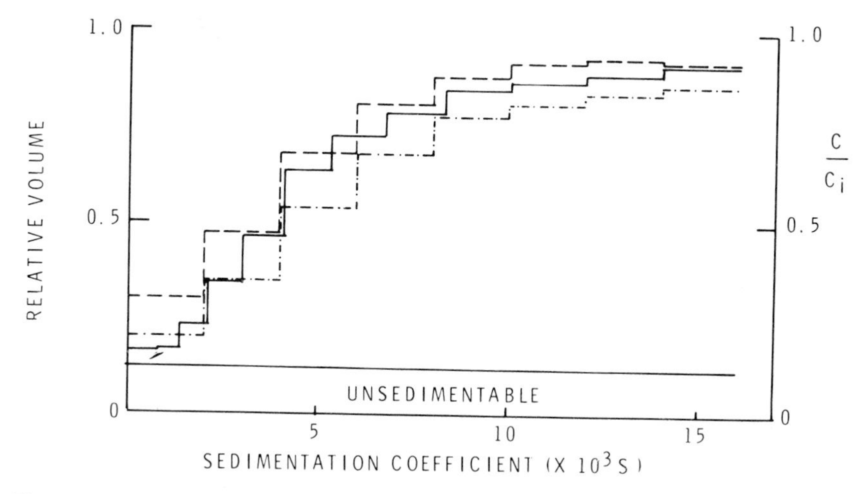

Fig. 5.12 Comparison of the sedimentation coefficient distributions of acid phosphatase containing particles, acid DNAase containing particles and pericanalicular dense bodies of rat liver. The cumulative sedimentation coefficient distributions of acid phosphatase containing particles (———) and acid DNAase containing particles (—·—·—), obtained from rate sedimentation studies of cytoplasmic extracts (Deter and de Duve, 1967) are averages of seven experiments. A similar distribution for pericanalicular dense bodies (—) was calculated from a pooled size distribution (12 individual distributions), assuming a density of 1.10 for all size classes (see Eq. 5.9). The first class was set equal to the average of the mean acid phosphatase activity of the P + S fraction and the mean unsedimentable acid DNAase activity. From Deter, R. L. J. *Cell Biol.* *48*:473, 1971.

tion coefficient distributions of pericanalicular dense bodies, acid phosphatase, and acid DNAase as shown in Fig. 5.12. The surprisingly good correlation of biochemical and morphological results suggests that enzyme concentrations are either relatively constant or subject to the same fluctuation in all size classes and that the preparative procedures used in the electron microscopic studies do not significantly alter particle volume. These are two postulates that are often assumed but rarely subject to experimental verification. Comparisons of this type provide an alternative to cytochemistry for the localization of enzymes or other constituents to specific cellular structures. It is applicable in many situations where appropriate cytochemical methods are not available and is a quantitative procedure.

Fractions Obtained by Isopycnic Centrifugation

Preparation of pellets by filtration from fractions obtained from equilibrium density gradients has been carried out by Leighton *et al.* (1968)

and Wibo *et al.* (1971). The procedures used were similar to those previously described except for the gluteraldehyde fixation. To avoid damage due to osmotic shock, the fractions were first mixed with one fourth their volume of a solution containing 7.5% gluteraldehyde in phosphate buffer. A short time later the fixed fractions were further diluted with a 1.5% gluteraldehyde-phosphate buffer solution and filtered. In cases where the protein content of the fraction is low, albumin can be added before fixation to increase pellet stability (Baudhuin, 1971). Quantitative morphological studies on pellets prepared in this way have been used to evaluate fraction composition and have provided results that correlate well with those obtained with biochemical methods (Table 5.3 and Fig. 5.13).

GENERAL COMMENTS

Preparation of Fractions

Although a complete discussion of the problems associated with the biochemical procedures used in preparing fractions is outside the scope of this chapter, some comments seem warranted. The first of these concerns the process of homogenization. As the purpose of this procedure is to produce a uniform particle suspension representative of the original tissue, it is necessary that certain objectives be realized. Homogenization must effectively rupture cells containing the particles of interest but must do so without causing significant particle damage. If certain cell populations are spared or if loss of certain particle subclasses occurs, subsequent fractionation will be carried out on a sample that is already biased. The homogenization conditions also must minimize absorption of soluble constituents to particles, loss of constituents from particles, or marker activation or inactivation. If any of these phenomena occur, the usefulness of marker distributions in evaluating the fractionation procedure will be compromised.

The study of the distributions of markers is a most important aspect of the use of centrifugation, or any other physical method, for separating and characterizing particle populations. Marker distributions, in addition to providing information about technical procedures, permit the location of specific particle populations and make study of their composition and physical properties possible. To be useful for these purposes, it is essential that the conditions under which markers are assayed give quantitative results. As stressed by de Duve (1967), marker measurements must be made on every fraction and a balance sheet kept to determine whether or not all of the marker activity in the starting material has been re-

TABLE 5.3 COMPOSITION OF FRACTIONS AS DETERMINED BY BIOCHEMICAL AND MORPHOMETRIC METHODS*

Constituent	Fractions							
	9		14		20		21	
	% Total Protein	% Particulate Volume	% Total Protein	% Particulate Volume	% Total Protein	% Particulate Volume	% Total Protein	% Particulate Volume
Endoplasmic Reticulum	3.4	3.5	1.9	2.5	1.9	0.7	0	1.9
Lysosomes	0.8	0.2	3.8	3.6	1.9	3.0	4.9	5.5
Mitochondria	95.1	89.3	39.2	57.2	4.9	0	1.4	2.5
Peroxisomes	0.7	0.6	26.1	24.5	91.3	94.8	82.0	82.7
Unidentified	0	6.4	29.0	12.2	0	1.5	11.7	7.4

*From Leighton *et al., J. Cell Biol. 37:* 482, 1968.

The percent of total protein contributed by each fraction component was determined from measurement of the relative specific activity of a marker enzyme and application of a mathematical procedure known as *linear programing.* The percent of particulate volume was estimated from measurements of the fractional area occupied by profiles of each constituent in sections of pellets prepared from the various fractions.

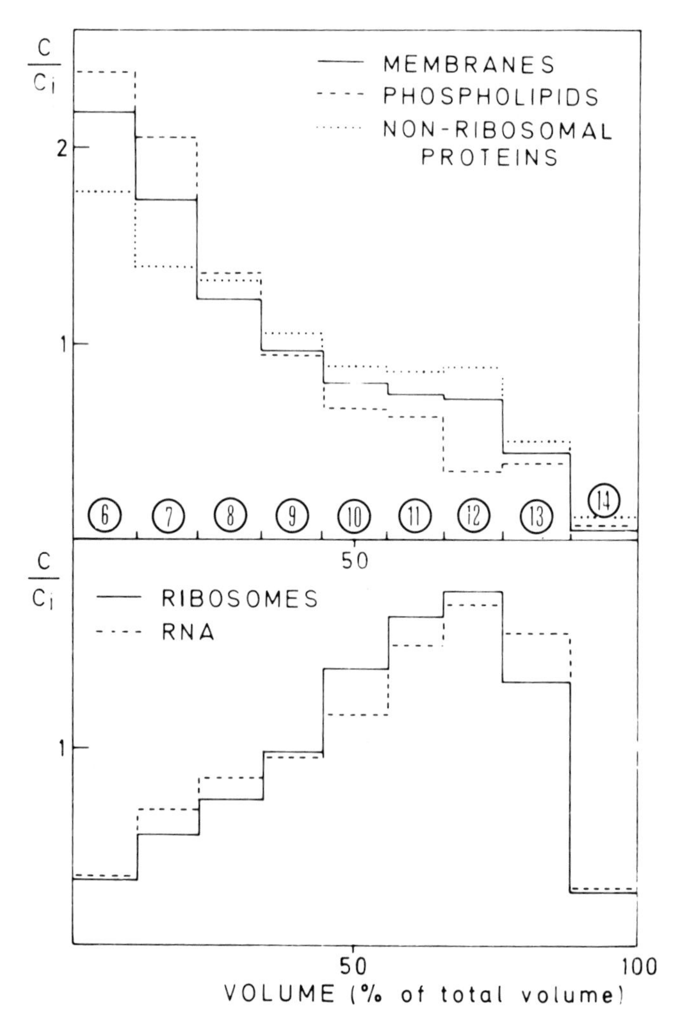

Fig. 5.13 Correlation of morphometric and biochemical characteristics of fractions obtained by isopycnic centrifugation of a microsomal fraction. Isopycnic centrifugation of a microsomal fraction was carried out in the rotor of Beaufay (1966). The gradient limits were 1.10 and 1.24. The circled numbers indicate gradient fractions. All data are normalized, C being the amount per volume of fraction, C_i the corresponding value in the pooled fractions. From Wibo, M. *et al., J. Cell Biol. 51*:52, 1971.

covered. Balance sheets provide a means for detecting phenomena such as marker activation or inhibition and false localization due to technical artifacts. Through such studies, the validity and internal consistency of the centrifugational methods employed in obtaining fractions can be evaluated.

Electron Microscopy

Most of the problems associated with the electron microscopic study of fractions have been presented in the foregoing discussion. There remains only one further point that should be emphasized. In considering the use of morphological methods in the study of fractions, it seems that one is often tempted to apply these methods prematurely. Morphological techniques are laborious and time-consuming, and to be justified from a cost-benefit point of view, there must be a very high probability that they will produce useful information. This author's experience indicates that only a thorough biochemical evaluation of the fractionation procedure and its resulting fractions, *prior* to the application of morphological techniques, provides the foundation needed for achieving this objective. If biochemical criteria cannot be satisfied, it is not likely that morphological studies will yield reliable results.

ACKNOWLEDGMENT

The author would like to express his sincerest appreciation to Dr. Pierre Baudhuin for his assistance in the preparation of this chapter. This work has been supported in part by National Science Foundation Grants GB 8411 and GB 29635.

References

Anderson, N. G. (1955). Studies of isolated cell components. VIII. High resolution gradient differential centrifugation. *Exper. Cell Res.* 9:446.

Baggiolini, M., Hirsch, J. G., and de Duve, C. (1969). Resolution of granules from rabbit heterophil leucocytes into distinct populations by zonal sedimentation. *J. Cell Biol.* 40:529.

Baudhuin, P. (1968). L'Analyse Morphologique Quantitative de Fractions Subcellulaires. Thesis. University of Louvain, Belgium.

———. (1971) personal communication.

———, Evrard, P., and Berthet, J. (1967). Electron microscopic examination of subcellular fractions I. Preparation of representative samples from suspensions of particles. *J. Cell Biol.* 32:181.

———, and Berthet, J. (1967). Electron microscopic examination of subcellular fractions. II. Quantitative analysis of the mitochondrial population isolated from rat liver. *J. Cell Biol.* 35:631.

Beaufay, H. (1966). La Centrifugation en Gradient de Densite. Application à l'Étude des Organites Subcellulaires. Thesis. University of Louvain. Belgium.

———, Bendall, D. S., Baudhuin, P., Wattiaux, R., and de Duve, C. (1959). Tissue fractionation studies. 13. Analysis of mitochondrial fractions from rat liver by density-gradient centrifuging. *Biochem. J.* 73:628.

———, and Berthet, J. (1963). Medium composition and equilibrium density of subcellular particles from rat liver. *Biochem. Soc. Symp.* 23:66.

Bowers, W. E., Finkenstaedt, J. T., and de Duve, C. (1967). Lysosomes in lymphoid tissue. I. The measurement of hydrolytic activities in whole homogenates. *J. Cell Biol.* 32:325.

Burton, K. (1956). A study of the conditions and mechanism of the diphenylamine reaction for the colorimetric estimation of deoxyribonucleic acid. *Biochem. J.* 62:315.

Clark, W. M. (1959). "Topics in Physical Chemistry," p. 48. The Williams and Wilkins Company, Baltimore.

De Duve, C. (1964). Principles of tissue fractionation. *J. Theor. Biol.* 6:33.

———, (1967). General Principles. *In:* "Enzyme Cytology." (Roodyn, D.B. ed.), p. 1. Academic Press Inc., New York.

———. (1971). Tissue fractionation—past and present. *J. Cell Biol.* 50:20D.

———, and Berthet, J. (1954). The use of differential centrifugation in the study of tissue enzymes. *Int. Rev. Cytol.* 3:225.

———, Pressman, B. C., Gianetto, R., Wattiaux, R., and Appelmans, F. (1955). Tissue fractionation studies. 6. Intracellular distribution patterns of enzymes in rat liver tissue. *Biochem. J.* 60:604.

———, Berthet, J., and Beaufay, H. (1959). Gradient centrifugation of cell particles. Theory and application. *Progr. Biophys. Biophys. Chem.* 9:325.

———, Wattiaux, R., and Baudhuin, P. (1962). Distribution of Enzymes between Subcellular Fractions in Animal Tissues. *In:* "Advances in Enzymology," (Nord, F. F. ed.), 24:291. Interscience Publishers, Inc., New York.

Deter, R. L. (1971). Quantitative characterization of dense body, autophagic vacuole and acid phosphatase-bearing particle populations during the early phases of glucagon-induced autophagy in rat liver. *J. Cell Biol.* 48:473.

——— and de Duve, C. (1967). Influence of glucagon, an inducer of cellular autophagy, on some physical properties of rat liver lysosomes. *J. Cell Biol.* 33:437.

Frasca, J. M., and Parks, V. R. (1965). A routine technique for double-staining ultrathin sections using uranyl and lead salts. *J. Cell Biol.* 25:157.

Guenther, W. C. (1965). "Concepts of Statistical Inference," p. 122. McGraw-Hill Book Company, New York.

Hayat, M. A. (1970). Principles And Techniques Of Electron Microscopy: Biological Applications, Vol. 1. Van Nostrand Reinhold Company, New York.

———. (1972). Basic Electron Microscopy Techniques. Van Nostrand Reinhold Company, New York.

Kuff, E. S., Hogeboom, G. H., and Dalton, A. J. (1956). Centrifugal, biochemical and electron microscopic analysis of cytoplasmic particulates in liver homogenates. *J. Biophys. Biochem. Cytol.* 2:33.

Leighton, F., Poole, B., Beaufay, H., Baudhuin, P., Coffey, J. W., Fowler, S., and de Duve, C. (1968). The large-scale separation of peroxisomes, mitochondria, and lysosomes from the livers of rats injected with Triton WR-

1339. Improved isolation procedures, automated analysis, biochemical and morphological properties of fractions. *J. Cell Biol.* *37*:482.

Lowry, O. H., Rosebrough, N. J., Farr, A. L., and Randall, K. J. (1951). Protein measurement with the Folin phenol reagent. *J. Biol. Chem.* *193*:265.

Maunsbach, A. B. (1966). Isolation and purification of acid phosphatase-containing autoflourescent granules from homogenates of rat kidney cortex. *J. Ultrastruct. Res.* *16*:13.

McCarty, K. S., Stafford, D., and Brown, O. (1968). Resolution and fractionation of macromolecules by isokinetic sucrose density gradient sedimentation. *Anal. Biochem.* *24*:314.

Noll, H. (1967). Characterization of macromolecules by constant velocity sedimentation. *Nature* *215*:360.

Palade, G. E. (1971). Albert Claude and the beginnings of biological electron microscopy. *J. Cell Biol.* *50*:5D.

Pickels, E. G. (1943). Sedimentation in the angle centrifuge. *J. Gen. Physiol.* *26*:341.

Roodyn, D. B. (1965). The classification and partial tabulation of enzyme studies on subcellular fractions isolated by differential centrifuging. *Int. Rev. Cytol.* *18*:99.

Venable, J. H., and Coggeshall, R. (1965). A simplified lead citrate stain for use in electron microscopy. *J. Cell Biol.* *25*:407.

Wattiaux, R., Wattiaux-DeConinck, S., and Ronveaux-Dupal, M. F. (1971). Distribution d'enzymes mitochondriaux après centrifugation isopycnique d'une fraction mitochondriale de foi de rat dans un gradient de saccharose: influence de la pression hydrostatique. *Arch. Int. Physiol.* Biochem. 79:214.

Weibel, E. R. (1969). Stereological principles for morphometry in electron microscopic cytology. *Int. Rev. Cytol.* *26*:235.

———, Stäubli, W., Gnagi, H. R., and Hess, T. A. (1969). Correlated morphometric and biochemical studies of the liver cell. I. Morphometric model, stereologic methods and normal morphometric data from rat liver. *J. Cell Biol.* *42*:68.

Wibo, M., Amar-Costesec, A., Berthet, J., and Beaufay, H. (1971). Electron microscope examination of subcellular fractions. III. Quantitative analysis of the microsomal fraction isolated from rat liver. *J. Cell Biol.* *51*:52.

6 Stereological Techniques for Electron Microscopic Morphometry

Ewald R. Weibel
with the collaboration of
Robert P. Bolender
Anatomisches Institut, University of Bern, Bern, Switzerland

1. INTRODUCTION

1.1. The Problem of Measuring Structures on Sections

Cells, tissues, and organs are compact arrays of microscopic structures. In other words, they are three-dimensional objects that are customarily studied on thin sections if the relationship between structures needs to be preserved. An ultrathin section, however, randomly cuts through the solid organelles and produces essentially flat profiles. Through experience, one is usually able to subjectively interpret these profiles in terms of three-dimensional structures, but often this interpretation may be erroneous; for example, a circular profile may be derived from a spherical, ellipsoidal, conical, or cylindrical structure. A single profile, therefore, does not provide enough information to draw a conclusion regarding the three-dimensional shape of the organelle unless additional information is available. Similarly, the size of a profile is not necessarily representative of the size of the structure from which it arose.

Fortunately, quantitative relationships exist between the average dimensions of a large number of organelles and those of their profiles on sections. When certain conditions are satisfied, the aggregate of profiles on the unit area of a section is quantitatively representative of the aggregate of organelles contained in the unit volume, so that measurements obtained from sections can be used to derive structural dimensions by means of simple methods.

These methods are based on mathematical developments that consider the probability of obtaining certain profiles by randomly cutting given structures. This can be demonstrated by a simple case: The sectioning of a sphere with diameter D produces a population of profiles in the form of circular disks of diameter d. If the section passes through the equator, the disk has the diameter $d = D$, but as the plane of section moves away from this central plane, d becomes smaller and smaller, approaching zero as it moves toward the pole (Fig. 6.1). However, not all diameters have the same probability; if d is measured in small increments, Δd, the highest frequency is found in the class

$$[D > d > (D - \Delta d)]$$

and the lowest frequency in the class

$$[\Delta d > d > 0]$$

This finding of geometrical probability, which may not be immediately

TABLE 6.1 LIST OF BASIC SYMBOLS FOR STEREOLOGY

Symbol	*Alternative symbols used by other authors*	*Definition*	*Dimension*
V		Volume of structure or test volume	cm^3
S		Surface of structure	cm^2
M	L	Length of linear feature of structure	cm
N		Number of structures (particles)	cm^0
D		Linear dimension of structure ("caliper diameter")	cm
T		Section thickness	
A		Area on section (planar)	cm^2
L		Length of test line	cm
P		Number of test points	cm^0
B	L	Profile boundary length on section	cm
I	P, N	Intersections of surface trace with test line	cm^0
Q	P, N	Transsections of linear feature with test area	cm^0
A_T		Test area	
L_T		Test line	
P_T		Test point set	
V_{Vi}		Volume density of component *i*	cm^0
S_{Vi}		Surface density of component *i*	cm^{-1}
M_V	L_V	Density of length in volume	cm^{-2}
N_V		Numerical density	cm^{-3}
A_A		Profile density on test area	cm^0
B_A	L_A	Boundary density on test area	cm^{-1}
Q_A	P_A	Density of transsections on test area	cm^{-2}
N_A		Numerical profile density on area	cm^{-2}
L_L		Intercept density on test lines	cm^0
I_L	P_L, N_L	Intersection density on test lines	cm^{-1}
N_L		Intercept number on test lines	cm^{-1}
P_P		Point density	cm^0

Note: This symbolism corresponds to the principle of notation adopted by the International Society for Stereology. The author has introduced four additional letters (*M, B, I, Q*) to eliminate ambiguities that could occur in some formulations if the basic symbols *L* and *P* were used exclusively.

apparent, is graphically demonstrated in Fig. 6.1. It follows from this distribution that the average diameter of the section disks is

$$\bar{d} = \frac{\pi}{4} \cdot D, \tag{6.1}$$

a relationship that allows one to infer D if one can measure $\bar{d}$ on sections.

The above development indicates that, in general, exact relationships exist only between dimensions of structural components and *averages* of many measurements carried out on random profiles of the structures. In

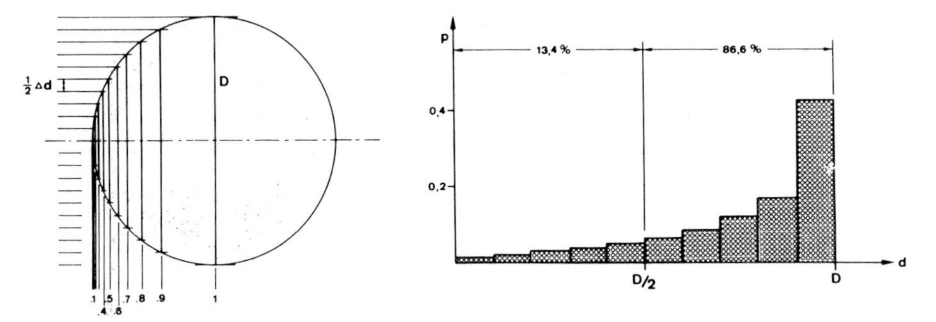

Fig. 6.1 Frequency distribution (relative frequency *p*) of profile diameter *d* obtained by sectioning spheres of diameter *D.*

biology, this does not introduce a serious limitation because the majority of morphometric problems are primarily concerned with obtaining overall data related to a large population of identical or similar structures, rather than with information concerning individual objects. Such "average" data can be obtained by applying the rather simple methods of stereology.

1.2. Fundamental Parameters of Structural Composition

It is appropriate to ask first what are the most informative parameters characterizing the structures such as the organelles of a certain cell type. Evidently, this will depend upon the objective of the study; in general, a good part of the information lies in a few fundamental parameters:

V_V: *volume density,* i.e., the volume of the component related to the containing volume (e.g., mitochondrial volume contained in the unit volume of cytoplasm), also called *relative* or *fractional volume.*

S_V: *surface density,* i.e., the surface area of the component per unit containing volume (e.g., the area of endoplasmic reticulum contained in the unit volume of cytoplasm).

M_V: *line density,* i.e., the length of a filamentous structure per unit containing volume.

N_V: *numerical density,* i.e., the number of particles per unit containing volume.

$\overline{D}$: *mean caliper diameter* of the structures, i.e., the mean distance between parallel planes made to be tangent to the particle over all possible orientations of the particle.

It should be noted that the first three parameters do not require that the structure be made of discrete particles; the shape of the structures is, hence, irrelevant. However, this is not the case for N_V and $\overline{D}$ where both the shape and frequency of particles must be considered.

Many additional parameters can be derived from these basic parameters; some of these will be discussed below.

1.3. The Principle of Delesse

In 1847, the French geologist Delesse proved that the volume density of the various components making up a rock can be estimated on random sections by measuring the relative areas (or areal density A_A) of their profiles. Since this principle is fundamental for all stereological methods, a plausible explanation and then a mathematical derivation of the Delesse principle are given as an introduction to stereological reasoning.

Take a cube of tissue containing a certain component of any shape whose volume density is to be measured, and cut it completely into thin serial sections of constant thickness t. This produces thin slices of the tissue containing slices of the component. On each section the area of the profiles a and the area of the section A (which remains constant) can be measured. Add all the profile areas and multiply by t; this should be equal to the component volume. Then, add all the section areas and multiply by t; this is equal to the volume of the cube. Divide the two to obtain the volume density:

$$V_V = \frac{(\Sigma a) \cdot t}{(\Sigma A) \cdot t} = A_A \qquad (6.2)$$

where the slice thickness, t, cancels and what remains is the ratio of the sum of profile area to the sum of section area, or the areal density A_A.

This can be shown formally on the same model using a mathematical derivation:

Suppose that a cube (Fig. 6.2a) with volume

$$V_T = L^3 \qquad (6.3)$$

contains granules of any shape and size all belonging to a component i that together have a volume

$$V_i = \rho \cdot L^3 \qquad (6.4)$$

ρ being the fraction of the cube volume occupied by i. Consider, now, a thin slice of this cube of thickness dx parallel to the (z, y) plane having a volume

$$dV_T = L^2 \cdot dx \qquad (6.5)$$

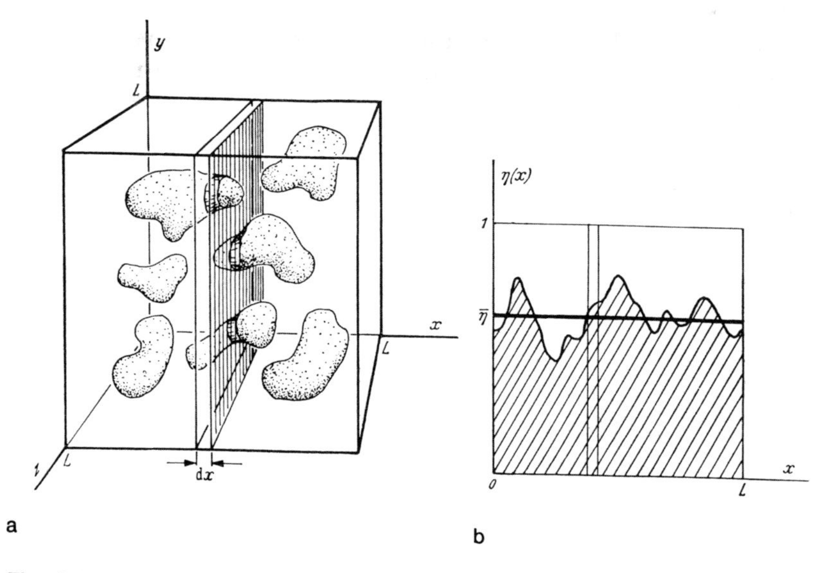

Fig. 6.2 (a) Model for deriving Delesse principle by serial slicing parallel to plane *yz* with slices of thickness $dx \to 0$; (b) Variation of planimetric fraction $\eta(x)$ around mean η as slice moves along *x*-axis. (From Weibel, 1963).

In this slice located at position x a volume

$$dV_i = \eta(x) \cdot dV_T = \eta(x) \cdot L^2 \cdot dx \tag{6.6}$$

will contain slices of the granules, η being the fraction of the slice volume occupied by i . L and dx are constant wherever we place the slice, but $\eta(x)$ will vary with x, as is indicated in Fig. 6.2b. If $dx \to 0$, the total volume of the granules V_i is

$$V_i = \int_0^L dV_i = L^2 \cdot \int_0^L \eta(x)dx = \rho \cdot L^3 \tag{6.7}$$

But,

$$\frac{1}{L}\int_0^L \eta(x)dx = \bar{\eta} \tag{6.8}$$

is the average value of the coefficient $\eta(x)$ between 0 and L, so that it follows from Eqs. (6.7) and 6.8) that

$$\bar{\eta} \cdot L^3 = \rho \cdot L^3 \tag{6.9}$$

or

$$\bar{\eta} = \rho \tag{6.10}$$

It is evident from Eq. (6.4) that

$$\rho = V_i/V_T = V_{Vi} \tag{6.11}$$

is the volume density of the structures in the tissue. Similarly, it follows from Eq. (6.6) that

$$\eta(x)L^2 = A_i(x) \tag{6.12}$$

is the area of all profiles of i on the section at x, so that, with

$$L^2 = A_T$$

as the area of the test field, we find that

$$\eta = \frac{A_i}{A_T} = A_{Ai} \tag{6.13}$$

is the areal or planimetric density of profiles on the sections. Hence, we may rewrite Eq. (6.10) as

$$\overline{A_{Ai}} = V_{Vi} \tag{6.14}$$

It can, therefore, be concluded that the volume density is represented on sections by the areal density of profiles. More exactly: *The areal density of profiles on sections is an unbiased estimate of the volume density of structures.*

This derivation of the Delesse principle should serve as a demonstration that measurements made on sections can indeed be extrapolated to the three-dimensional structures. Identical, but somewhat more complicated considerations can prove that S_V and M_V can also be measured on sections, whereas N_V can only be estimated if certain assumptions are possible. In the following discussion no further proof of the methods will be given; this can be found in the treatises of Weibel (1963; 1972), De Hoff and Rhines (1968), and Underwood (1970).

2. BASIC STEREOLOGICAL METHODS

2.1. Estimation of Volume Density V_V

The principle of Delesse requires the estimation of the relative area of profiles on a section. There are two straightforward approaches:

a) *Planimetry*

Using any planimeter, polar or electronic, trace out the area of all profiles contained in a section, add the areas and divide by the section area to obtain $A_A = V_V$ (Fig. 6.3a).

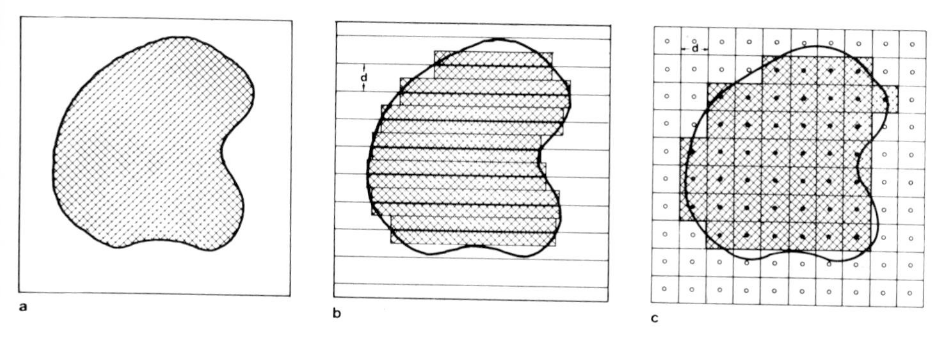

Fig. 6.3 Estimation of the area of a profile by planimetry (a), by lineal analysis (b), and point counting (c).

b) *Cut-and-weigh*

Trace all profiles on heavy paper or use directly an electron micrograph. Weigh the sheet and cut out all profiles, then weigh the combined profiles. The weight ratio between the profiles and the sheet is an estimate of $A_A = V_V$. This method was, in fact, proposed by Delesse.

These two methods are not very convenient in practice because they require too much work to be done on the individual profile. Two alternative methods have, therefore, been devised that are more efficient and yield more reliable results.

c) *Linear integration*

Areas can be estimated by cutting the profile into narrow strips, measuring their length and multiplying with their width (Fig. 6.3b). The validity of this principle can be proven on the same basis as the above proof of the Delesse principle (Weibel, 1963). Rosiwal (1898) has shown that to determine A_A, one can randomly place, project, or draw on the section a test line of length L_T and measure the total length of the line segments L_i included in the profiles. A_A is then directly related to the ratio

$$\frac{L_i}{L_T} = L_L = A_A \qquad (6.15)$$

This method is used today mostly in automatic scanning instruments based either on the scan of the specimen stage of the microscope or on a TV scan.

d) *Point counting*

A common way of measuring areas is to divide them into small squares of unit area $a = d^2$, where d is the distance between the lines, and counting the squares contained in the profile (Fig. 6.3c). The problem of accounting for squares cutting the profile boundary can be simplified by counting only those squares whose center point is within the profile. Alternatively, one can simply count those squares whose upper left-hand corner is in the profile. It can be easily shown that the number of points of a square lattice P_i contained within the profile and related to the total number of test points on the section P_T is an estimate of A_A:

$$\frac{P_i}{P_T} = P_P = A_A \tag{6.16}$$

This principle was independently proposed by Glagoleff (1933) and by Chalkley (1943).

All these principles are entirely independent of the shape of the structures and, consequently, of the shape of their profiles. It is not necessary that the "probes" (section, lines, points) are regularly arranged. In particular, the test lines need not be straight, continuous, or parallel. Test points can be arranged in any manner; square or triangular lattices or random point sets are equally admissible. It is, however, required that the line or point arrangement is independent of the structural arrangement; in other words, the textures of the section and of the test system are stochastically independent.

2.2. Estimation of Surface Density S_V

The surface of structures appears on sections as the boundary trace of the profiles (Fig. 6.4). It can be shown that the length of all the profile boundaries found per unit area of section, B_A, is proportional to S_V where

$$S_V = \frac{4}{\pi} \cdot B_A \tag{6.17}$$

The length of the boundary trace can be directly measured either by fitting a thread to the trace or by using a map measuring wheel. These methods are rather cumbersome, and can be replaced by an efficient point counting method.

It can be shown that B_A is proportional to the number of intersections I_L formed by the boundary trace with a test line system of known length L_T:

$$B_A = \frac{\pi}{2} I_L \tag{6.18}$$

A derivation of this is given by Weibel (1969).

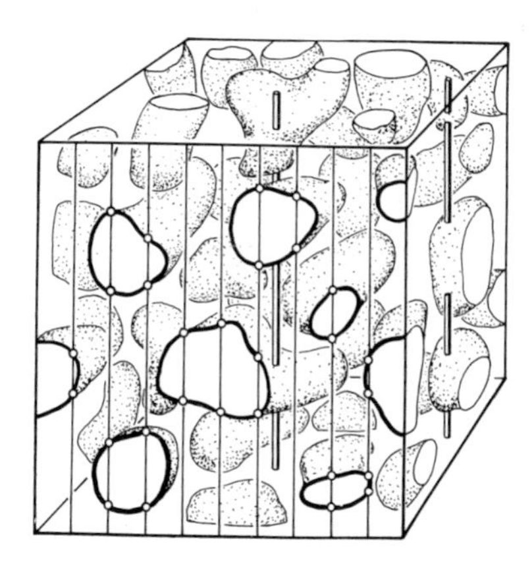

Fig. 6.4 Model for surface estimation by intersection counting; lines on section equal linear probes in tissue. (From Weibel *et al.*, 1966).

By combining Eqs. (6.17) and (6.18), it is found that S_V is directly related to the number of intersections I_i formed with test lines of length L_T:

$$S_V = 2 \cdot \frac{I_i}{L_T} = 2 \cdot I_L \tag{6.19}$$

This principle was first proposed by Tomkeieff in 1945 and, since that time, it has been independently rediscovered at least six times (Underwood, 1968).

It is important to note that the objects to be analyzed with this principle may have any shape; however, there is one requirement: the structure should be isotropic, i.e., the objects should not be oriented in a preferential direction. The reason is that they are tested not with (isotropic) points, but with (anistropic) lines. In Fig. 6.5, a trace showing a preferential horizontal orientation is compared with a horizontal and a vertical test line system. It is evident that the vertical lines form a larger number of intersections. In such a case, at least two orientations of the test lines at right angles to each other must be used.

2.3. Estimation of Length Density M_V

Test lines crossing a boundary trace of a section profile are actually cutting across the surface of the objects. By inversion of Eq. (6.19) it is easily understood that the length of lines or curves in space (filaments, axes of tubules, triple lines between cells, etc.) per unit volume of tissue, M_V, is proportional to the number of times a line crosses the unit area of a section plane, Q_A, (Fig. 6.6) according to:

$$M_V = 2 \cdot Q_A \tag{6.20}$$

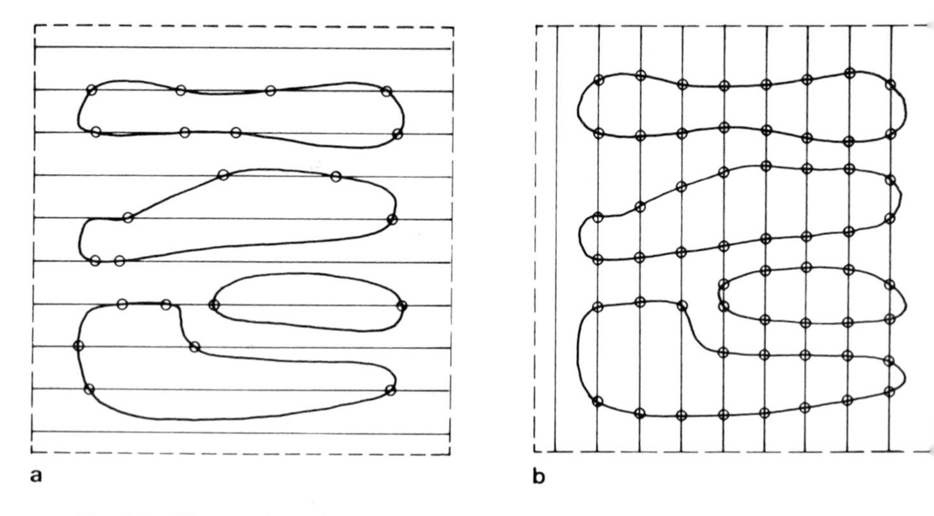

Fig. 6.5 The number of intersections may depend on the orientation of the test lines with respect to the boundary curves.

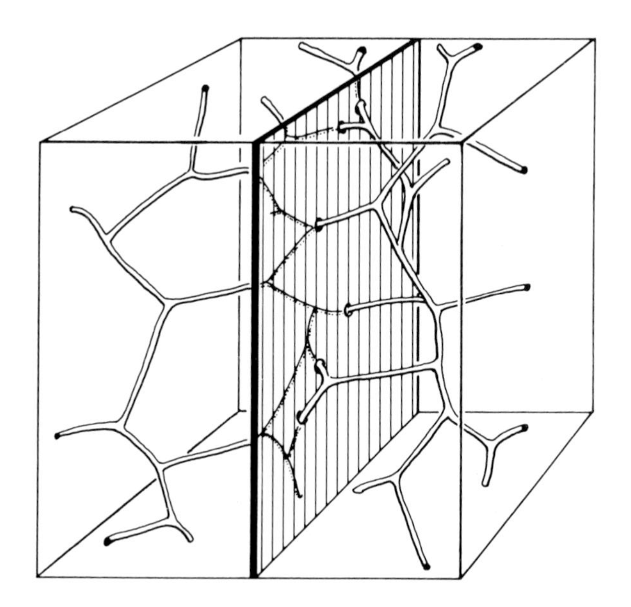

Fig. 6.6 Definition of transections of filamentous structures across section plane.

2.4. Estimation of Numerical Density N_V

The larger the number of objects in the unit volume, N_V, the larger will be the number of profiles per unit area of a random section N_A. This

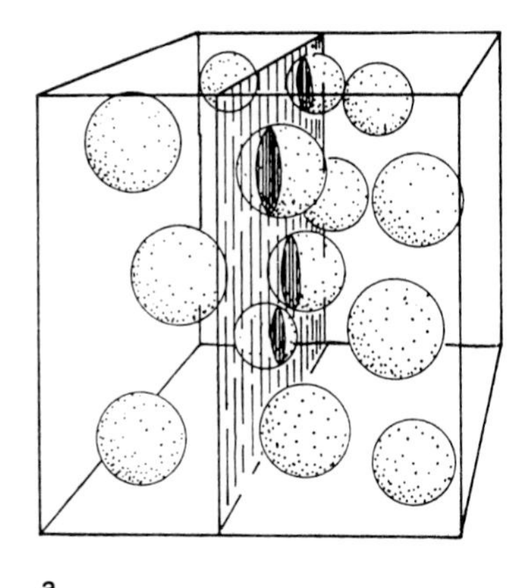

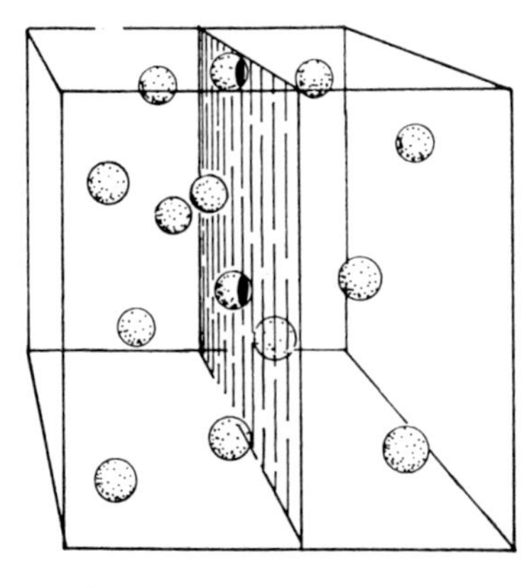

Fig. 6.7 When the same number of particles is contained within a volume, large (a) and small (b) spheres yield different numbers of profiles on sections. (From Weibel, 1963).

intuitively plausible statement is only true if size and shape of the objects are the same. This is evident in Fig. 6.7 where the same number of large and small spheres is cut, yielding profile numbers that are proportional to the sphere diameter.

De Hoff and Rhines (1961) have shown that

$$N_V = N_A/\overline{D} \tag{6.21}$$

an exact method that requires two independent determinations, namely, a count of N_A on sections and an estimation of mean particle diameter $\overline{D}$, an enterprise not easily performed, as shall be shown.

An alternative method has been proposed by Weibel and Gomez (1962), which allows N_V to be determined from N_A and V_V estimated on sections if some assumptions about the shape of the structures can be made:

$$N_V = \left(\frac{K}{\beta}\right) \cdot \frac{N_A{}^{3/2}}{V_V{}^{1/2}} \tag{6.22}$$

The factor K depends on the relative size distribution of the particles. $K = 1$ if all particles have the same size and > 1 if the size is not uniform. When the standard deviation is of the order of 25% of the mean diameter, K is ~ 1.07. For many practical purposes K can, therefore, be disregarded. The dimensionless coefficient β depends on shape. It relates the particle volume, v, to the mean profile area, a, by

$$v = \beta \cdot a^{3/2} \tag{6.23}$$

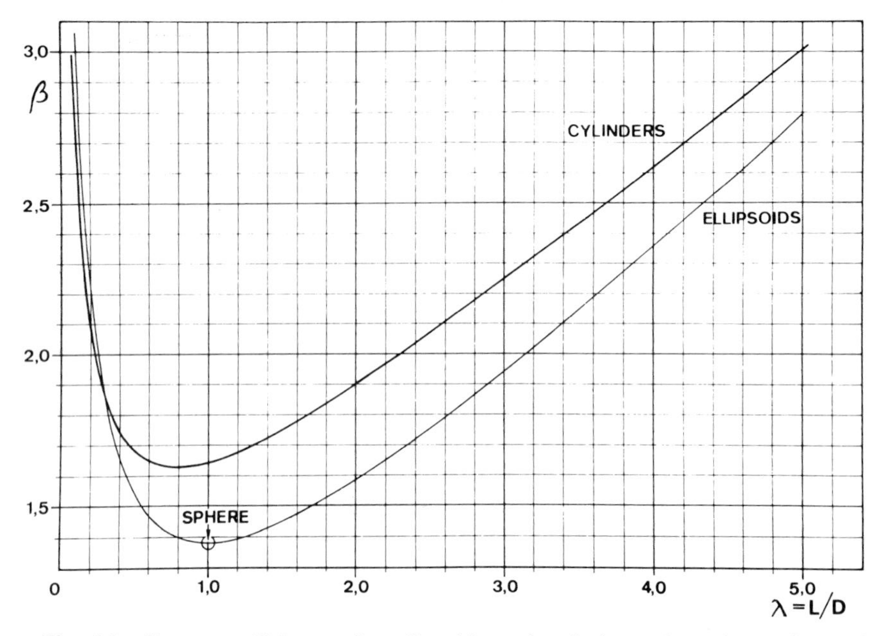

Fig. 6.8 Shape coefficient β for ellipsoids and cylinders of varying ratio λ of length to diameter; $\lambda < 1$ for oblate and $\lambda > 1$ for prolate ellipsoids. (From Weibel, 1969).

For spherical particles

$$\beta_s = \sqrt{\frac{6}{\pi}} \approx 1.38 \qquad (6.24)$$

The graph of Fig. 6.8 gives β values for rotational ellipsoids and circular cylinders expressed as the ratio λ of length to diameter; values for a number of other shapes are given in Table 6.2.

Of these two methods, the first one is to be preferred if the number of particles is the critical parameter; it should, in general, be used in conjunction with a careful analysis of size distribution that simultaneously yields $\overline{D}$ and N_A. While the first method is most easy to use with spherical

TABLE 6.2 SHAPE COEFFICIENTS

Shape	β *(Eq. 6.22, 6.23)*	k *(Eq. 6.33)*
Cube	1.84	2.25
Octahedron	1.86	2.16
Dodekahedron	1.55	1.78
Sphere	1.38	1.50

particles, Table 6.2 also gives a set of coefficients k that allow $\overline{D}$ to be inferred from the mean linear intercept $\overline{L_3}$ (Eq. 6.33), which in turn is related to the surface-to-volume ratio of the particles (Eq. 6.43) described below. If in the context of a more general study a relatively rough estimate is sufficient, then the second method may prove to be quite useful, particularly if V_V is already determined. It should be noticed, however, that this method tends to yield underestimates, because N_A is underestimated due to the inability to recognize and to count very small profiles. If the smallest recognizable profile is less than half the largest profile in diameter, this underestimation will not be larger than ~ 10 to 15%, because these small profiles are relatively rare (Fig. 6.1).

Both these methods presuppose that the sections are infinitely thin, or, in more practical terms, that the ratio of section thickness, T, to mean particle diameter, $\overline{D}$, is small, for example, smaller than 1/10. If this condition is not fulfilled, T has to be taken into account. In electron microscopy of cells, this is certainly the case for ribosomes or pinocytotic vesicles. Abercrombie (1946) and Hennig (1957) have given the formula

$$N_V = N_{AT} \cdot \frac{1}{\overline{D} + T} \qquad (6.25)$$

where N_{AT} is the number of particle "profiles" or projections counted on the unit area of a slice of thickness T. If $T \rightarrow 0$, this formula tends to Eq. (6.21).

In thick slices, it is often not possible to count "cap sections," i.e., particles of which only a small polar cap is included in the slice. If the height, h, of the smallest recognizable cap section can be estimated, this can be accounted for by using the formula of Floderus (1944) or Haug (1967):

$$N_V = N_{AT} \cdot \frac{1}{\overline{D} + T - 2h} \qquad (6.26)$$

The main difficulties in using these formulas are the estimation of the three factors in the denominator and, furthermore, the fact that they are truly applicable only for spherical particles.

2.5. Estimation of Particle Dimensions

2.5.1. *The Shape of Structures*

If the size of discrete particles must be derived from a section analysis, the first and indispensible requirement is to refine their shape as precisely as possible. This is probably the most difficult task in stereology for two reasons: (1) it is extremely difficult to derive quantitative parameters characterizing shape so that much will depend on a sound subjective judgement; (2) the spatial shape of the particle may not be evident at first glance from the shapes of profiles on sections, sometimes not even after a long and careful analysis.

Often it will be necessary to start by postulating a certain shape and to consider the frequency of profile forms that must be expected to result from a random sectioning process (Elias *et al.*, 1954). By comparing the observed distribution of profile forms with that expected from the model, the assumed shape can be either accepted or rejected. For example, if it is assumed that certain structures are very long circular cylinders, then it is expected that 25% of the elliptic section profiles should have ratios of long to short axis greater than 2. If the fraction of these ellipses is smaller, then the structures are probably some kind of ellipsoid; but what is its axial ratio?

Hennig (1967) has derived graphs predicting the frequency distribution of axial ratios of elliptic profiles for rotation ellipsoids of different axial ratio. This type of analysis becomes very problematic, however, if the axial ratio of the ellipsoids varies within the same population. Elias (1972) discusses the general criteria for shape interpretation on sections.

2.5.2 *Mean Diameter and Size Distribution of Spheres*

Spherical shape can be assumed if all profiles on sections are circular. Fig. 6.1 shows that the profile diameter, d, depends on the sphere diameter, D, and on the distance of the section plane from the center of the sphere. If all spheres are of equal size, then the largest profile will correspond to an equatorial section on which D can be directly measured, or Eq. (6.1) can be used to calculate D from the mean profile diameter $\bar{d}$. Here again, it must be taken into consideration that small profiles are lost, so that $\bar{d}$ directly obtained will be somewhat underestimated.

If the population of spheres is made up of several classes of different sizes, each class will contribute a distribution of profiles of the type of Fig. 6.1; on a section, one can observe a compound distribution as shown in Fig. 6.9. How can this be unraveled?

When given a measured distribution of profile diameters, it is evident that the class of largest profiles represents equatorial sections of the largest particle class. Using the distribution pattern of Fig. 6.1, the contribution of this class of spheres to the smaller profile classes can be calculated and subtracted from the histogram. One can then proceed the same way with the second, third, etc., largest class until the profile distribution is "used up."

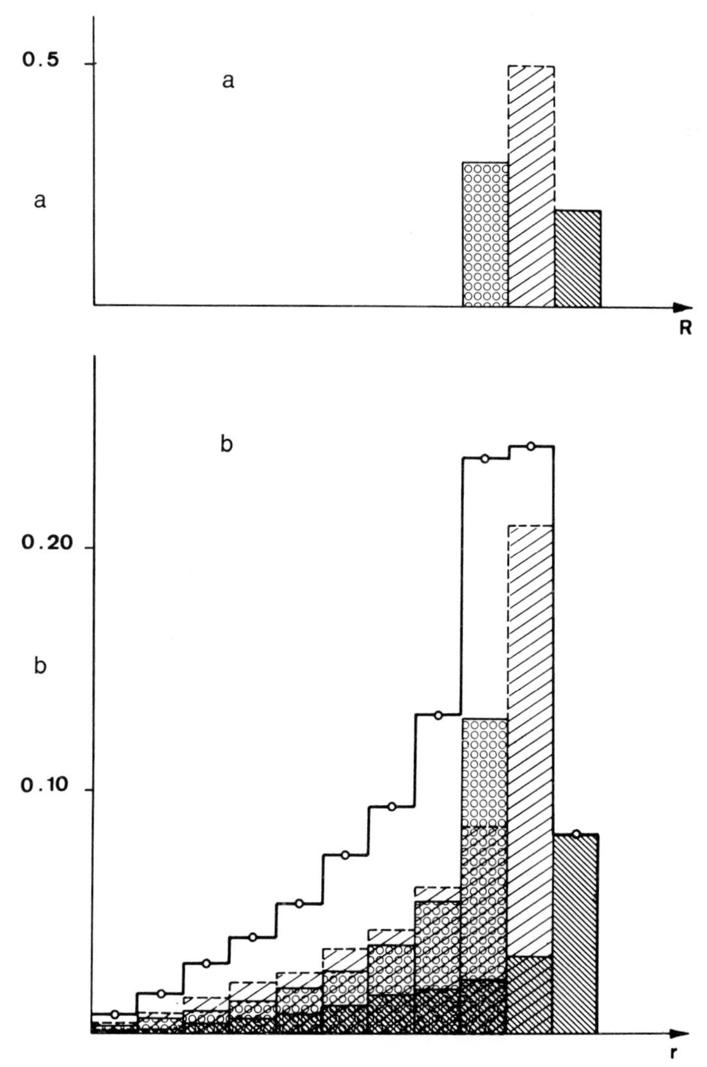

Fig. 6.9 Profile size distribution (b) generated by three size classes of spheres (a). The heavy line with open circles represents the compound distribution of profile radii observed on sections; it is the result of the addition of three partial distributions of the type shown in Fig. 6.1. (From Weibel, 1969).

The first method based on this principle was given by Wicksell (1925), who also generalized it for ellipsoids (Wicksell, 1926). This method is still used in biology (e.g., Baudhuin and Berthet, 1967), and has the advantage that corrections for section thickness and loss of small profiles can be introduced. In materials sciences other similar methods are preferred, mainly those of Schwartz (1934), Scheil (1935), and Saltykov (1958). Bach (1967) has given the problem a thorough mathematical treatment and has proposed solutions of various kinds. A detailed elaboration on these methods, which would be necessary to present them in applicable form, would exceed the space available for this chapter. These methods have been reviewed in detail by Underwood (1968; 1970).

In order to obtain an adequate approximation of the particle size distribution by one of the methods described above, a minimum of 1000 profiles must be measured. Very often, however, it is not necessary to know the actual distribution of particle size; it is sufficient to estimate its parameters. Hennig and Elias (1971) have presented graphs of sample distributions of profiles that can be compared with experimentally obtained distributions.

Giger and Riedwyl (1970) have developed a rather simple semigraphical method that is very useful if it can be assumed that (1) the particles are similar to spheres, (2) their diameters are normally distributed, and (3) the section thickness is negligible. This method shall be presented in detail without elaborating on the theoretical background.

Practical procedure:

1. Determine on independent* random sections the diameter, d, of a large number of profiles (several hundred). Form a histogram (Fig. 6.10) by grouping them into size classes i, the interval between classes being a and the class mean $d_i = i \cdot a$ A$i = 0, 1, 2, \ldots$). The first class comprises all diameters $d_o < a/2$. Choose a so that the histogram is made of no less than 10 and no more than 16 classes. The number of profiles per class is designated by N_i.

2. Very small profiles will usually be missing from the distribution. Complete the distribution by extrapolating linearly towards zero, as shown in Fig. 6.10 by the broken line. There is some uncertainty attached to this procedure, but the possible error thus introduced can be tested secondarily (point 6).

3. Calculate the mean profile diameter $\bar{d}$ from the *completed* distribution and derive a first estimate of the mean particle diameter by

$$\bar{D}_1 = \frac{4}{\pi} \bar{d} \tag{6.27}$$

Draw a vertical line on the histogram at $\bar{D}_1$ (Fig. 6.10).

*The same particle may not be sectioned by more than one section.

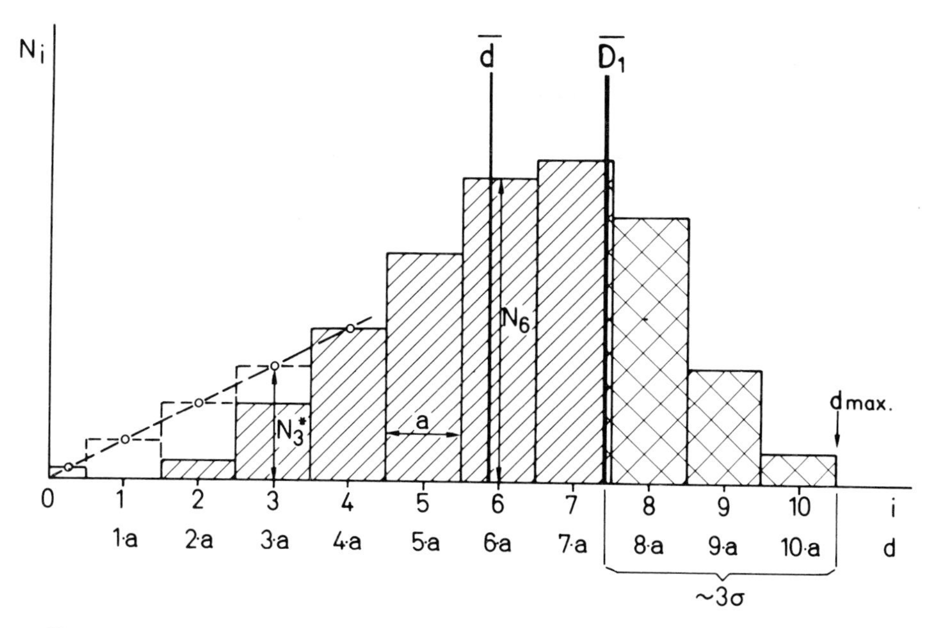

Fig. 6.10 Example of profile size distribution as measured on sections of spherical particles, which is used to estimate $\bar{D}$ and σ by the method of Giger and Riedwyl (1970).

4. The fraction of the area of the histogram to the right of $\bar{D}_1$ is a measure for the standard deviation σ (square root of the variance). This is evident if one considers that all the profiles larger than $\bar{D}_1$ have been derived from particles larger than $\bar{D}_1$, i.e., from (about) the larger half of the particle population. Since the range $\pm$ 3 σ encompasses 99.73% of the normally distributed particle population, and since the maximal profile diameter, $d_{\max}$, should correspond to the largest particle diameter, a first rough estimate of σ is

$$\sigma_1 \approx \frac{d_{\max} - \bar{D}_1}{3} \tag{6.28}$$

5. For more precision, determine the area of the completed histogram A_0 and the area A_1 to the right of $\bar{D}_1$ (cross-hatched in Fig. 6.10) to obtain

$$P = A_1/A_0 \tag{6.29}$$

The graph of Fig. 6.11 allows one to read off two coefficients, $F(P)$ and $f(P)$, which can be used to determine the correct mean particle diameter and standard deviation by

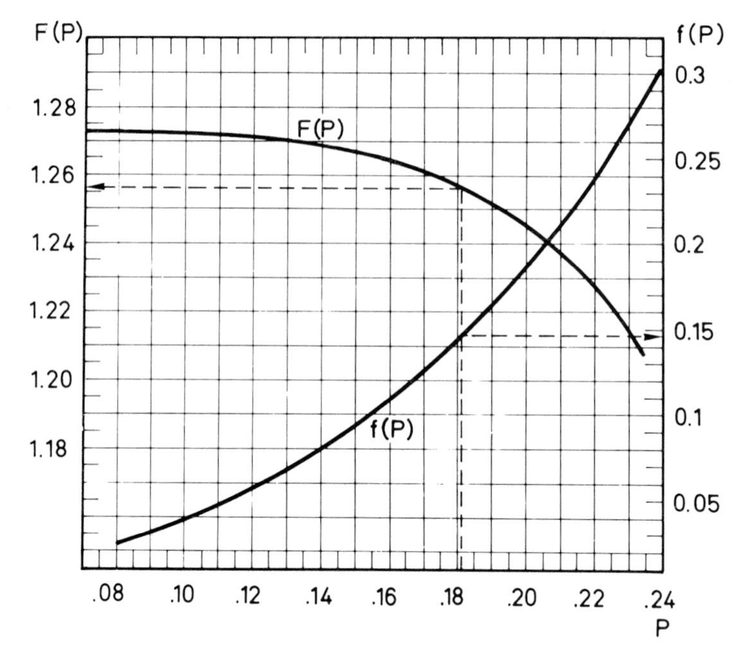

Fig. 6.11 Nomogram for determining coefficients used in Giger-Riedwyl method. (From Giger and Riedwyl, 1970).

$$\overline{D} = F(P) \cdot \bar{d} \tag{6.30}$$

$$\sigma = f(P) \cdot \bar{d} \tag{6.31}$$

It should be noted that $\overline{D} \approx \overline{D}_1$ if $P < 0.1$.

6. One can now test whether the completion of the histogram for missing small profiles had been reasonable. If N is the total number of profiles in the completed histogram, the numbers in the small classes should be

$$N_i^* \approx \frac{i \cdot a^2}{\overline{D^2}} \cdot N \left(i = \frac{1}{2}, 1, 2, \ldots\right) \tag{6.32}$$

If the corrected numbers deviate considerably from these calculated values, the completion should be corrected accordingly and the calculations of parameters should be repeated.

2.5.3. *Size Distribution of Nonspherical Particles*

The methods described above can be used with some degree of approximation whenever the particles are reasonably described by spherical models, i.e., if they are not grossly elongated or flattened. Methods for determining the size distribution of ellipsoids were given by Wicksell (1926), De Hoff (1962), and De Hoff and Bousquet (1970). Meyers (1967) and Saltykov (1967) have considered the case of cubic particles that are complicated mainly because of the sectioning of corners.

2.5.4. *Particle Size and Random Linear Intercepts*

It was stated earlier that V_V can be estimated from the fraction L_L of a test line intercepting the component under investigation. V_V is the sum of all particle volumes divided by the total test volume; L_L is the sum of all chords or intercepts with the particles divided by the total test line length. From this it follows intuitively that a certain relationship must exist between the mean volume of particles and the mean length of chords or linear intercepts formed by random test lines $\overline{L}_3$. In fact, it can be shown that the mean caliper diameter of particles $\overline{D}$ is directly proportional to L_3 by

$$\overline{D} = k \cdot \overline{L}_3 \tag{6.33}$$

where k is a shape coefficient (Table 6.2); for spheres, $k = 3/2$.

Underwood (1970) has presented an extensive Table (p. 90 to 93) listing the relationship between mean linear intercept length and some characteristic particle dimensions for solids of revolution and regular polyhedra.

Of particular interest for biologists is the relationship between the mean linear intercept length and the thickness, D, of plates or disks whose radius is much larger than D. This shape occurs frequently, for example, in the form of cytoplasmic sheets or of cisternae of endoplasmic reticulum. A well developed example is the air-blood barrier in the lung. For such plates it is found that

$$\overline{D} = \frac{1}{2}\overline{L}_3 \tag{6.34}$$

$\overline{D}$ being the arithmetic mean plate thickness.

In biological considerations such plates often form barriers for exchange processes; in this case, the resistance for diffusion across the barrier is not proportional to the arithmetic mean thickness but rather to the harmonic mean thickness D_h (Weibel and Knight, 1964; Weibel, 1970 and 71). It is found that D_h is proportional to the harmonic mean linear intercept L_h:

$$D_h = \frac{2}{3} L_h \tag{6.35}$$

The harmonic mean is the mean of the reciprocal values; with n measurements it is found by:

$$\frac{1}{L_h} = \frac{\sum^{n} \frac{1}{L}}{n} \tag{6.36}$$

It should be noted that

$$\overline{D} \geqq D_h \tag{6.37}$$

the two means being equal if all plates are of identical thickness. This principle can also be used to estimate the thickness of sheets of variable thicknesses.

In practice, there are two ways of determining the mean linear intercept length $\overline{L_3}$. The first is to lay out long test lines on sections, to measure a large number of intercepts and to compute their mean; this is the most direct way. The second way is often more practical. It makes use of two facts: (1) the ends of the intercepts or chords are marked by two intersections, so that the number of intercepts per unit length of test line is

$$N_L = \frac{1}{2} I_L \tag{6.38}$$

where I_L is the number of intersections of profile boundaries per unit length of test line, as defined above; and (2) the density of intercepts per test line length, L_L, is equal to V_V and hence also equal to the fraction of test points falling on profiles:

$$L_L = P_P \tag{6.39}$$

Combining the two equations, one obtains

$$\overline{L_3} = L_L/N_L = 2\, P_P/I_L \tag{6.40}$$

If one uses a coherent test line system, as described below, where each test point is connected with a test line element of length ℓ_T, the total test line length is

$$L_T = P_T \cdot \ell_T \tag{6.41}$$

It is then easily derived that

$$\overline{L_3} = 2 \cdot \ell_T \cdot \frac{P_i}{I_i} \tag{6.42}$$

The mean linear intercept is obtained directly by counting test points falling on profiles and intersections of test lines with profile boundaries. From Eq. (6.42) it is evident that $\overline{L_3}$ is related to the volume-to-surface ratio of the particles (v/s):

$$\bar{L}_3 = 4 \cdot (v/s) \tag{6.43}$$

Since one can determine the mean particle size from the mean linear intercept length, it must be expected that a relationship exists between the size distribution of particles and the distribution of random intercept lengths. In fact, a number of methods for estimating particle size distribution on the basis of intercept distributions have been proposed; these methods have been extensively reviewed by Underwood (1968 and 1970).

2.6. The Connection between Structural Parameters, Test Probes, and Stereological Formulas

The stereological principles developed in the preceding paragraphs were aimed at estimating the basic structural parameters of volume, surface, line, and numerical density, as well as the mean particle diameter on sections. To accomplish this, very specific, quantitatively defined test probes (the unit area of section, A_T, the unit length of (straight) test lines, L_T, and the unit test point set, P_T) were employed. The systematic combination of test probes and parameters leads to the following closed system of formulas:

Test Probe

$$\begin{array}{ccccccc}
\text{Parameter} & V_T & A_T & & & L_T & P_T \\
V & V_V = & A_A & = & & L_L = & P_P \\
S & S_V = & \dfrac{4}{\pi} \cdot B_A & = & 2 \cdot I_L & & \\
M & M_V = & 2 \cdot Q_A & & & & \\
N & N_V = & \left[\dfrac{1}{\bar{D}} \cdot N_A\right] & & & &
\end{array}$$

The symmetry in this system of basic formulas is due to a fundamental law relating the dimensions of the structural parameters of the objects, d_o, and of the probe, d_p, to those of the trace on the probe, d_t. In three-dimensional space the following law holds (Weibel, 1967a):

$$d_t = d_o + d_p - 3 \tag{6.44}$$

Very general mathematical treatments of the foundations of these basic stereological formulas have been presented by Giger (1970) and Miles (1972).

3. STEREOLOGICAL TEST SYSTEMS

3.1. Choice of Optimal Probe

In applying stereological principles, three fundamental test probes are needed:

A_T : test area ($d_p = 2$)

L_T : test line ($d_p = 1$)

P_T : test point set ($d_p = 0$)

Which one is to be chosen for a given task? For reasons of efficiency, a probe that forms a trace of dimension $d_t = 0$ would seem preferable, because this would reduce the measuring process to mere counting. However, one wonders whether this is the *best* procedure because it seems that the counting of points falling on profiles for volume density estimation is much less precise than measuring the profile area with a planimeter. Nevertheless, a careful statistical consideration shows that this is only true with respect to an individual profile, not, however, with respect to an analysis of a large number of profiles.

It was stated earlier that an individual profile is not representative. It is, therefore, more important that many profiles are included in the sample rather than that individual profiles are measured with great accuracy. If the expected error in each method is related to the cost of labor involved in obtaining the measurements, it is found that the point counting procedures are superior to the more laborious methods of area or line measurement. Accordingly, the following presentation of test systems is limited to those permitting a point count analysis.

The probes of choice are:

P_T : Set of test points for analysis of volume density V_V by the formula

$$V_V = P_P = P_i/P_T \qquad (6.45)$$

L_T : Test lines for estimating S_V by the formula

$$S_V = 2 \cdot I_L = 2 \cdot I_i/L_T \qquad (6.46)$$

A_T : Test area for estimating M_V by the formula

$$M_V = 2 \cdot Q_A = 2 \cdot Q_i/A_T \qquad (6.47)$$

and for estimating N_V by any formula involving

$$N_A = N_i/A_T \qquad (6.48)$$

It is evident that optimal test systems should comprise all three probes in such a way that A_T, L_T, and P_T are related quantitatively to each other. Such test systems will be called *coherent test systems.*

3.2. Coherent Square Lattices

3.2.1. *General Properties*

The three probes A_T, L_T, and P_T are realized simultaneously in a simply related fashion with a square lattice of lines of distance d as shown in Fig. 6.12a. If the intersections of the lines are considered as "points" for point counting, it is easily seen that there are as many squares of area d^2, and twice as many line segments of length d, as there are points. This test system is defined as follows:

test points : P_T

test line length : $L_T = P_T \cdot 2d$ (6.49)

test area : $A_T = P_T \cdot d^2$ (6.50)

It should be noted that the broken line of the frame square in Fig. 6.12a is not part of the test lines and produces no test points. To avoid this difficulty, the same test system can be drawn differently, as shown

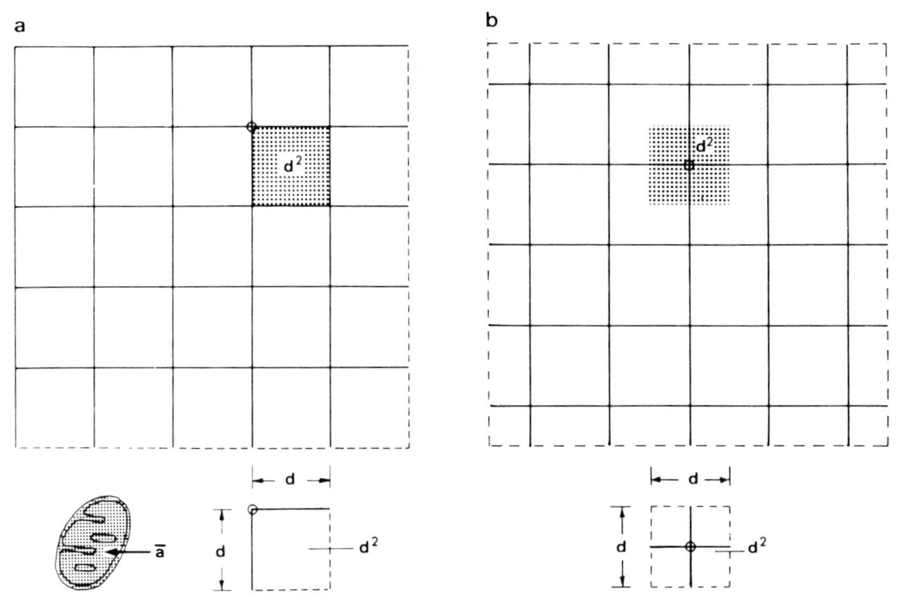

Fig. 6.12 Two identical variants of simple quadratic test system.

in Fig. 6.12b, where the test lines cross in the center of the small unit squares to form one test point per square. The two test systems are quantitatively identical.

3.2.2. *Point Counting Volumetry*

The first pertinent question is how should the test points be spaced. For didactic reasons this technique was introduced by indicating its relation to planimetry. This would imply that a high point density would yield good planimetric results. However, the notion of point counting planimetry should be replaced by one of point sampling of tissue. Each point constitutes a sample, and one asks the question whether the point lies in the component or not. Such a sampling procedure is statistically described by a binomial distribution, and, therefore, it is necessary that each sample, i.e., each point, is independent of the others. Consequently, the points should be spaced in such a manner that no more than one point can fall on the same profile. This is achieved if d is chosen so that

$$d^2 > a_m \qquad (6.51)$$

the area of the largest profile encountered. The inaccuracy with which individual profiles are "measured" with such coarse point lattices is compensated by the fact that the point set covers a very large sample of profiles, thereby securing a representative sample.

One may have to deviate from this general rule in studying rare components where one needs to make sure that no profiles remain undetected by the test system; in this case one may choose d^2 similar to or even smaller than the mean profile area,

$$d^2 \sim \bar{a} \quad \text{or} \quad d^2 < \bar{a} \qquad (6.52)$$

The total number of test points that must be applied to one representative sample unit (see following) depends on V_V and on the relative error, $E(V_V)$, one is willing to accept. This relationship is shown in the graph of Fig. 6.13. The number of test points per representative sample should be approximately

$$P_T = 0.453 \cdot \frac{(1 - V_V)}{V_V \cdot E^2(V_V)} \qquad (6.53)$$

It is easy to see that $\sim$ 400 test points are needed to estimate the volume density of a component where $V_V \sim 0.1$, if one is willing to accept a 10% error; if $E(V_V)$ should be only 5%, then 1600 points are

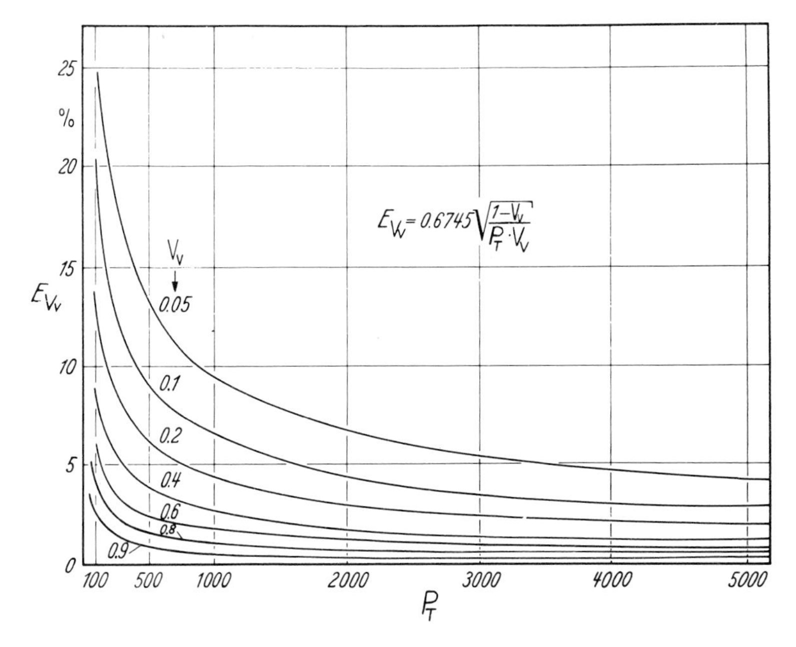

Fig. 6.13 Nomogram relating expected relative error in estimating volume density (E_{V_V}) to test point number (P_T) and volume density (V_V). (From Weibel, 1963).

needed. In practice, one can roughly estimate the order of magnitude of V_V on a few micrographs and then decide on P_T.

It is concluded from the foregoing, that very different point densities and total point numbers are needed for different structures. In liver cells, for example, mitochondria make up about 20% of the cell volume, whereas microbodies contribute only 1 to 2%. Accordingly, the optimal point density is quite different for the two components. This difficulty can be overcome by using *coherent double lattice test systems;* an example is shown in Fig. 6.14. The principle is quite simple. Every *g-th* line is drawn heavier than the others, thereby yielding two superimposed test point sets. If there are P_T intersections of the heavy lines (coarse point lattice), there will be a total of

$$P_T' = g^2 \cdot P_T \tag{6.54}$$

fine line intersections (fine point lattice) in the test system. The advantages of such test systems are evident—the volume density of a sparse organelle, V_{vs}, is estimated by counting those fine points P'_s that fall on profiles of *s*

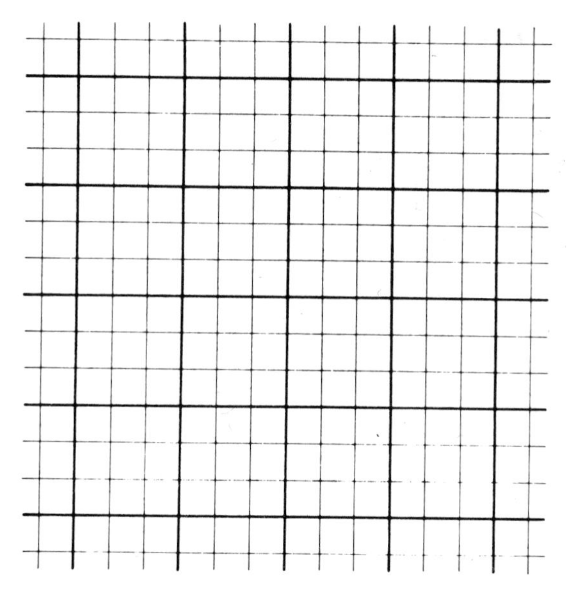

Fig. 6.14 Coherent double lattice test system with lattice ratio $q^2 = 9$. (From Weibel, 1969).

$$V_{Vs} = P_s'/g^2 \cdot P_T \qquad (6.55)$$

whereas the coarse points are used to estimate V_{Vf} of more frequent organelles and

$$V_{Vf} = P_f/P_T \qquad (6.56)$$

The estimates are, nevertheless, coherent in the sense that the test points used differ by a constant factor, g^2. This becomes particularly important if the data are to be expressed not with respect to the total tissue volume, but only with respect to one phase, for instance, the cytoplasm. By making use of some of the considerations in foregoing sections it can be easily shown, that the volume density of the rare organelle in cytoplasm, V_{Vs}^*, is found by

$$V_{Vs}^* = P_s'/g^2 \cdot P_c \qquad (6.57)$$

where P'_s are the *fine* points falling on profies of *s*—one needs only to count these—and P_c is the number of *coarse* points falling on cytoplasm. This procedure is rather efficient.

In practice, one should first decide on the spacing, *d*, needed to estimate the rarest and/or smallest organelles, then the double lattice factor, *g*, is decided on by considering the spacing needed for the other larger components.

3.2.3. *Intersection Counting*

The lines of square lattices may be used to estimate the surface density S_V of components by counting the intersections, I, with profile boundaries where

$$S_V = 2 \cdot I_L = 2 \cdot I/L_T = I/P_T \cdot d \tag{6.58}$$

In analogy to point counting volumetry, the spacing of the test lines depends on the surface density of the component and on the acceptable error. According to Hilliard (1965), the total number of intersections that must be counted to achieve a given accuracy, estimated by the standard deviation of the estimate $\sigma(S_V)$, is

$$I \approx 0.4[S_V/\sigma(S_V)]^2 \tag{6.59}$$

so that the necessary test line length per representative section area is

$$L_T \approx \frac{0.4 \cdot \overline{I_L}}{\sigma^2(I_L)} \tag{6.60}$$

The number of intersections counted with a square lattice may be excessive, because for every point falling inside a profile there will be at least four intersections with the boundary of this profile. From a statistical point of view these intersections are not independent. Part of this difficulty can be overcome by counting only intersections with the horizontal or vertical lines, but then the test line length is only

$$L_T' = P_T \cdot d \tag{6.61}$$

If one wishes to determine the surface density of a component i within the cytoplasm, one must count the point falling on cytoplasm, P_c, along with the intersection points I_i:

$$S_{Vi}{}^* = I_i/P_c \cdot d \tag{6.62}$$

It is evident that intersection counting on double lattice test systems follows the same rules. In using the fine line grid of spacing d for counting intersections, I'_i, and the coarse points for estimating cytoplasmic area, the surface density in cytoplasm is obtained from

$$S_{Vi}{}^* = I_i'/P_c \cdot g^2 \cdot d \tag{6.63}$$

3.2.4. *Counting Transections and Profiles*

In order to estimate the length density, M_V, of lines or curves in space, one must count the number of times these objects are transected by the section plane unit area of section plane. In a square test system the test area is well defined (Eq. 6.50) so that

$$M_{Vi} = 2 \cdot Q_i/A_T = 2 \cdot Q_i/P_T \cdot d^2 \qquad (6.64)$$

or, if expressed in relation to cytoplasmic volume,

$$M_{Vi}^* = 2 \cdot Q_i/P_c \cdot d^2 \qquad (6.65)$$

Similarly, the density of profiles per unit section area can be determined by counting the profile number N_i within the frame of the test system:

$$N_A = N_i/P_T \cdot d^2 \qquad (6.66)$$

In this connection, one must consider carefully an error introduced by the fact that some profiles are intersected by the frame of the test area. If $N_i > 20$, then it suffices to count all profiles intersected by the left and upper margins (as in counting red cells) and to disregard all those intersected by the lower and right edges. If the profiles are large with respect to the field, so that their number is small, then a correction for edge effect can be performed (Giger, 1970). One needs to count separately N^*, the number of profiles completely within the area and, $\widehat{N}$, the number of profiles intersected by the frame. In addition, it is necessary to estimate the fraction of the edge ℓ_t, covered by profiles; this should be equal to P_P. The numerical density of profiles is

$$N_A = \left[N^* + \frac{\widehat{N}}{2} - P_P \right] / A_T \qquad (6.67)$$

In general, it proves to be optimal to choose the test area so that $20 < N < 50$ and, in this case, the edge effect correction ($- P_P$) can be disregarded.

3.3. A Coherent Multipurpose Test System

3.3.1. *General Properties and Design*

It has been noted that square lattices contain an excessive density of lines when compared with the test point density. An alternative coherent test system (Fig. 6.15), which has originally been called the *multipurpose test system* (Weibel *et al.*, 1966), overcomes this difficulty. It is com-

Fig. 6.15 (a) Construction of smallest unit of coherent multipurpose test system composed of lines of length z contained in (nearly) square frame ($P_T = 42$). (b) Larger units with $P_T = 168$ and $P_T = 100$. (From Weibel *et al.*, 1966).

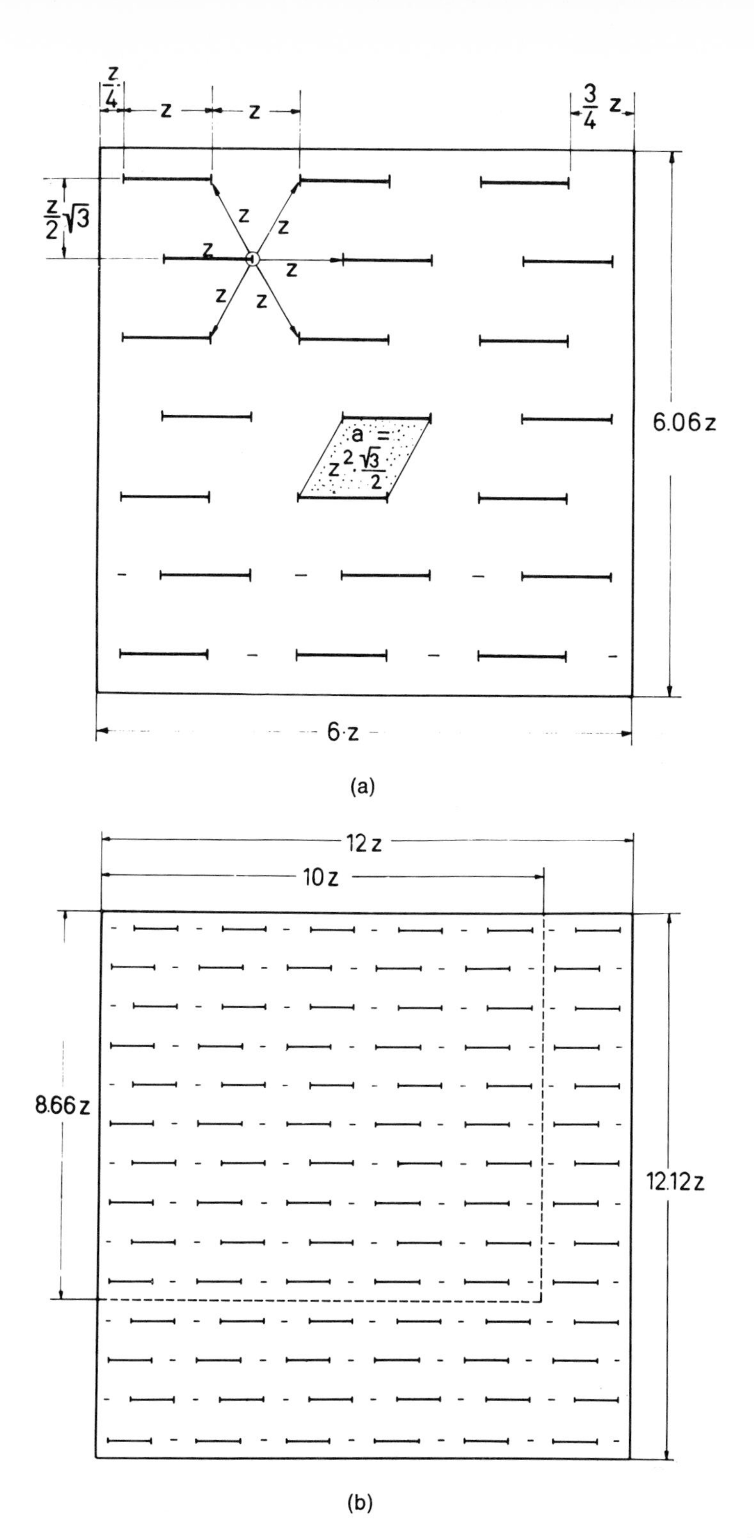

(a)

(b)

posed of discrete short test lines of length z whose end points are arranged in a regular triangular lattice. The lattice unit is not a square, but rather an equilateral rhombus, with angles of 60°, and of area

$$a_T = \frac{\sqrt{3}}{2} z^2 \tag{6.68}$$

It is evident from Fig. 6.15a that there is only one test line segment of length z per every second rhombus, and hence the test line unit per test point is

$$\ell_T = \frac{1}{2} z \tag{6.69}$$

This basic unit allows a number of coherent test systems to be designed. The smallest test areas that are (nearly) square can be obtained by using an array of seven rows of three test lines, as shown in Fig. 6.15a; they contain 42 test points and 21 lines. A larger square test system is obtained by joining four such units to contain 168 points and 84 lines (Fig. 6.15b). It may be convenient to have just 100 points in the test system, as shown in the rectangular area in Fig. 6.15b. This test system has the following characteristics:

test points : P_T

test line length : $$L_T = P_T \cdot \frac{1}{2} \cdot z \tag{6.70}$$

test area : $$A_T = P_T \cdot \frac{\sqrt{3}}{2} \cdot z^2 \tag{6.71}$$

and it is coherent in the sense defined above.

3.3.2. *Practical Use*

For point counting volumetry, the end points of the test lines are used as markers. The same rules as given above for square lattices apply; namely, the spacing of points is optimally chosen so that, in general,

$$a_T > a_m \tag{6.72}$$

the maximal profile area of the structure of interest. Also, the same requirements concerning the total number of test points to be used on a representative sample applies. It is evident that this test system is less flexible with respect to point spacing than square lattices, because double lattice test systems are not easily designed, although it is not impossible.

This test system has its main advantage in surface area estimations. If we count the intersections of a surface trace with the test line segments, I_i, the following is obtained:

$$S_V = 2 \cdot I_L = 4 \cdot I_i / P_T \cdot z \tag{6.73}$$

Since the test system is coherent, it is possible to estimate S_V in the cytoplasm, for example, by introducing cytoplasmic points, P_c, instead of P_T:

$$S_V^* = 4 \cdot I_i / P_c \cdot z \tag{6.74}$$

This test system was originally designed (Weibel and Knight, 1964) to apply the principle of Chalkley *et al.* (1949), which uses short "needles" to estimate the volume-to-surface ratio from a count of intersections (I_i) and point hits (P_i) on the structures:

$$\frac{v}{s} = \frac{z \cdot P_i}{4 \cdot I_i} \tag{6.75}$$

This parameter has been shown to be directly related to the mean intercept length (Eq. 6.43) and thereby to mean particle size. It is especially useful in estimating the arithmetic mean thickness, D, of sheets of irregular thickness (Weibel and Knight, 1964) where

$$\overline{d} = \frac{z \cdot P_i}{2 \cdot I_i} \tag{6.76}$$

The coefficient 4 has been replaced by 2, because the tissue volume is related to both surfaces of the sheet and, consequently, intersections with both surfaces must be counted.

The frame of the test system evidently can be used for counting either transsections of filamentous structures, Q_A, or the numerical density of particle profiles, N_A; exactly the same rules apply as those explained above.

3.4. Other Regular Test Systems

A nearly unlimited number of different test systems may be constructed. They will be coherent if the following conditions are satisfied:

1. The total test area must be subdivided into unit areas of equal size a_T.
2. There must be an equal number of test points, p_T, per unit area, a_T, so that each point is characterized by an area a_p where

$$p_T = a_T / a_p \tag{6.77}$$

Very often $p_T = 1$ and $a_p = a_T$.

3. Each unit area must be related to a unit test line length, ℓ_T, so that

$$\ell_T = c \cdot \sqrt{a_T} \tag{6.78}$$

c being an arbitrary proportionality constant.

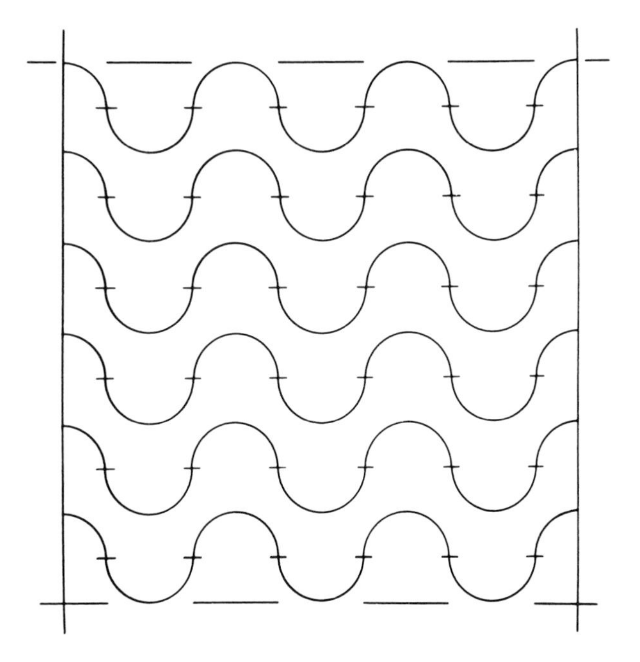

Fig. 6.16 Curvilinear test system of Merz (1968).

A rather interesting test system composed of semicircles (Fig. 6.16) has been developed by Merz (1968). It has the advantage that the inherent anisotropy of oriented, straight test lines is eliminated. It can thus be used for counting surface intersections on anisotropic profiles such as those shown in Fig. 6.5. The test system contains a square lattice of test points, their spacing, d, corresponding to the diameter of the semi-circles, so that the characteristics of this test system are:

test points : P_T

test line length : $L_T = P_T \cdot \frac{\pi}{2} d$ (6.79)

test area: $A_T = P_T \cdot d^2$ (6.80)

3.5. Random Point Lattices

It has been convincingly demonstrated by Hilliard and Cahn (1961) and by Hennig (1967) that regular point lattices are superior to random point systems for statistical reasons. However, the periodic array of the points introduces serious problems if the structures themselves also show some kind of periodic arrangement. In such cases, it may be indicated

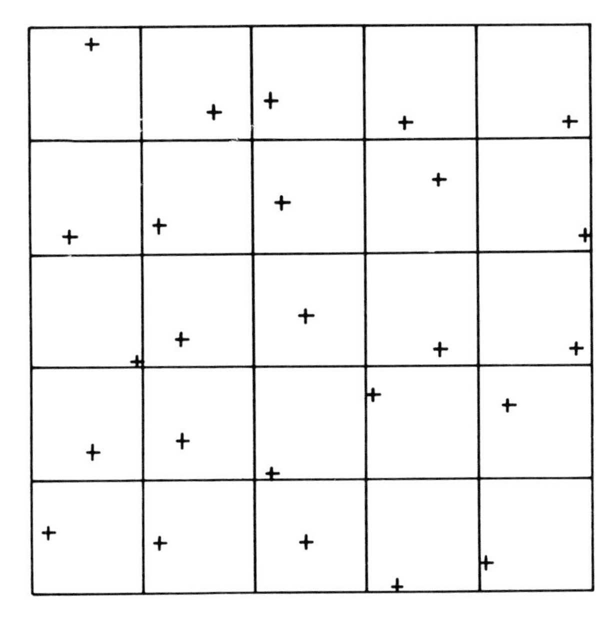

Fig. 6.17 Test system with 25 random points. (From Weibel, 1969).

to use random point lattices. These can be produced by marking points in a coordinate system and locating them by random numbers taken from random number tables.

It will, however, still be advantageous to assure an even distribution of the points over the test field. This is achieved by subdividing the field into unit squares and marking one random point in each, as shown in Fig. 6.17. Note that this test system is still coherent. Using the same approach, a lattice of short test lines may be spread in a random fashion.

3.6. Test Systems for Profile Measurements

As stated above, it will often be adequate for particle sizing to regard particles as "spherical" in spite of considerable deviations from this shape. In this case, all calculations are based on assumed circular cross-sections, although they may be polygonal or distorted. The most practical way for obtaining a "mean diameter" for each profile is to fit a circle of equal area and record its diameter. This is the procedure used in the Zeiss particle size analyzer (Gahm, 1972). If this is not available, a transparent plastic stencil of graded circles, as used for graphic work, or a set of concentric circles drawn on a clear plastic sheet, will be sufficient. Proper choice of interval will allow direct classification into 12 to 15 size classes. Measurement is thus reduced to counting.

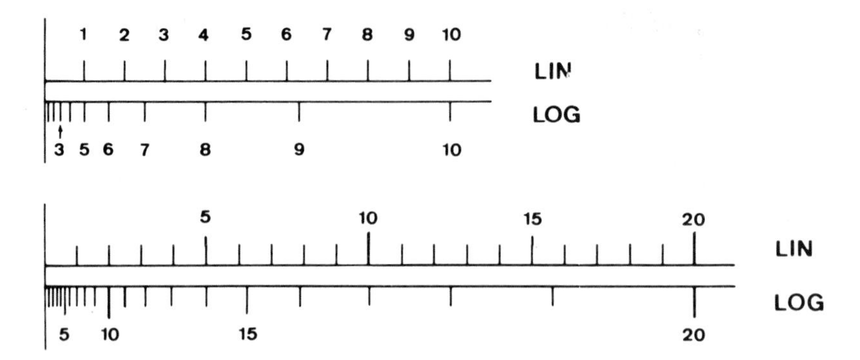

Fig. 6.18 Comparison of linear and logarithmic scales for 10 and 20 classes.

Where measurements of linear intercepts are needed, it is, likewise, more efficient to use special scales that allow direct classification of the data into 10 to 15 size classes by counting. Very often, data obtained with ordinary rulers (e.g., in mm) must, anyway, be divided into classes in order to construct histograms. If arithmetic means of these data are computed, linear scales are appropriate. If, however, the harmonic mean is relevant, as in applying Eq. 35, it is better to use a logarithmic scale. The reason for this is that small values must be measured with more precision than large ones because their reciprocal value weighs more. Figure 6.18 compares linear to logarithmic scales.

4. SAMPLING AND DESIGN OF STEREOLOGICAL ANALYSIS

4.1. The Statistical Nature of Stereology

Four fundamental premises set forth in the preceding paragraphs have already indicated that statistical considerations play a central role in stereology, and they are:

a. Stereological methods must deal with *large numbers* of related structures.
b. They produce *average* values of parameters characterizing the structures.
c. The procedure of sectioning allows only a restricted *sample* of the structures to be analyzed.
d. The principles developed were all based on considerations of geometric *probability*.

If this is borne in mind, it becomes evident that the successful application of stereological methods critically depends on sound procedures of sampling, and that the validity of the estimates thus obtained must be subjected to a statistical evaluation.

In practice, sampling usually proceeds through several stages:

1. Organs are diced into small tissue cubes.
2. A small number of these cubes are cut into sections (thin slices), *one* of which is used for analysis.
3. Electron micrographs are recorded from each section and usually comprise only a fraction of the section area.
4. Point counting procedures are applied which sacrifice a large part of the information contained in the micrographs, because all those components not hit by test probes are disregarded.

Each of these stages must be attended with great care, because it will contribute to the over-all accuracy of the estimates. It is, however, difficult to give specific rules because experience shows that each tissue has its own particular problems. This presentation is, therefore, restricted to a few general recommendations.

4.1.1. *Randomly Dispersed Structures*

From a stereological point of view, structures are considered randomly dispersed if the individual units are not specifically oriented with respect to the plane of section. This is the case, for example, in the liver and in most glandular tissues, although their functional units appear highly ordered. In such cases, no special precautions are required either for sampling tissue blocks or for obtaining sections or micrographs.

On sections, fields are best sampled in regular steps by recording the micrographs in predetermined corners of the meshes of the supporting copper grid (Fig. 6.19). At higher magnifications with relatively small sections, one may use as many as four corners. Locating the micrograph with respect to the supporting grid ensures that the sample is unbiased, i.e., that no structures are preferentially included in the micrograph, even subconsciously.

4.1.2. *Layered Structures*

Epithelial sheets often show a layering of specific structures, which is a characteristic and functionally important property of the tissue. In general, a section perpendicular to the layers or to the surface will cut through all levels and will faithfully represent their relative depths.

The sampling procedures will depend on the problem to be solved. For surveys it may be best to record micrographs in strips covering the

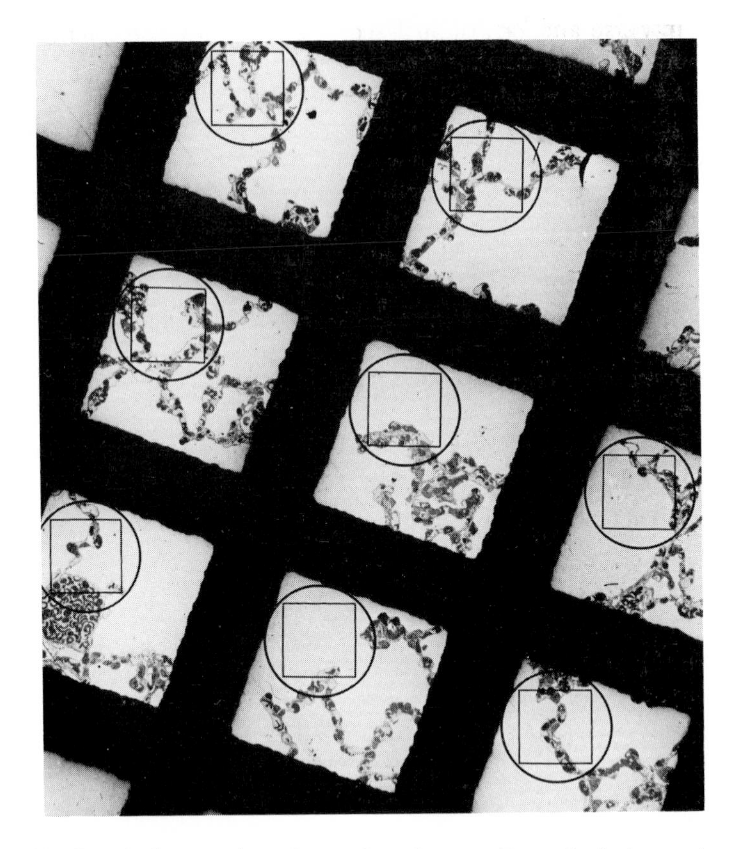

Fig. 6.19 Practical procedure for systematic sampling of electron microscope fields. (From Weibel *et al.*, 1966).

entire depth and overlapping the upper and lower boundary of the tissue. Alternatively, one may sample from different, well-specified depths; this procedure is described in detail by Schroeder and Münzel-Pedrazzoli (1970) for studies of the gingival mucous membrane.

It should be noted that for point counting volumetry the angle of sectioning is not critical, whereas in estimating surface density, preferential orientations (anisotropy) of membranes must be considered and, when necessary, special procedures applied (Hilliard, 1967; Sitte, 1967).

4.1.3. *Fasciculated Structures*

In striated muscle, tendons, nerves, or kidney medulla, long cylindrical units are arranged in parallel bundles. A complete analysis must be based on both transverse and longitudinal sections. On both, the micrographs can be recorded by the sampling method described previously.

In this case, it also applies that point counting volumetry can be performed on any section, but surface estimations critically depend on the angle of sectioning. Two special procedures may overcome this difficulty:
1. Intersection counts can be performed on *transverse sections* using any one of the line lattices presented. Remember, that by doing this, one actually determines the density of surface traces on the section area

$$B_A = \frac{\pi}{2} I_L \qquad (6.81)$$

If the structures resemble long cylinders of any configuration on cross-section, then one can easily show that $B_A = S_V$ so that

$$S_V = \frac{\pi}{2} I_L \qquad (6.82)$$

Likewise, the number of long, parallel filaments or tubules crossing the transverse area is equal to their density in the unit volume:

$$M_V = Q_A \qquad (6.83)$$

Hilliard (1967) has proposed the use of a special test figure of nearly elliptical shape for surface density estimation on longitudinal sections of anisotropic tissues.

The sarcoplasmic reticulum of muscle fibers is an oriented membrane system, which shows the additional complexity of a periodic array. Furthermore, its study on longitudinal sections is aggravated because the section thickness is greater than the tubule diameter. Therefore, a special procedure, based on oblique sections, has been developed for estimating surface density of the sarcoplasmic reticulum (Weibel, 1972).

4.2. Systematic vs. Random Sampling

It has been repeatedly shown that a well-conceived procedure of systematic or stratified sampling yields a smaller error than simple random sampling if: (a) the systematic lattice is randomly applied to the material, and (b) the material has no inherent periodicity that could interfere with that of the sampling lattice (Cochran, 1953; Hilliard and Cahn, 1961; Hennig, 1967; Ebbesson and Tang, 1967), because the independence of the samples is better assured in systematic sampling. In simple random sampling, chance may cluster the samples in one particular area, leaving others unsampled. This is avoided in systematic sampling because the separation of the samples by their even distribution over the material is predetermined in the sampling lattice. Ebbesson and Tang (1967) have shown that the standard error in determining the number of nucleoli in pancreas decreases more rapidly with increasing sample size in systematic rather than in simple random sampling.

In stratified sampling, the material is first divided into regions or strata from which one or more samples (e.g., tissue blocks) are picked by simple random sampling. This method is intermediate between simple random and systematic sampling and the decrease in relative error with regard to sample size is also intermediate between the two (Ebbesson and Tang, 1967). There is a useful procedure for obtaining the primary sample of tissue blocks: a few sample slices are taken from different regions of the organ, these are diced and processed in individual vials, and, from these stratified pools of blocks, a few are randomly picked before embedding or sectioning. This is the routine procedure in the author's laboratory.

4.3. Micrographs and Representative Sample Size

The magnification of electron micrographs on which a stereological analysis is to be performed must be high enough to allow unambiguous localization of test points within organelles or easy counting of intersections between membrane traces and test lines. A trial with micrographs of different magnification is necessary to decide on an optimal magnification.

On the other hand, the magnification should be as low as possible to ensure that as large a section area as possible is included in the sample micrograph. Very often, however, magnification is too high for an individual micrograph to be representative of the material, since it may comprise only part of its components. In this case, a representative sample is obtained by combining a certain number of individual micrographs.

An unambiguous definition of "representative sample size" is very difficult to propose, because it depends on many properties of the material that only partly can be quantified, and on the level of precision required. The minimum requirement for representativity of a sample is that *none of the components may be missing.* In general, such a requirement must be stated in terms of probability, for instance, in 95% of the samples. Often, the data obtained from more than one micrograph must be pooled to form a representative sample.

In practice, it is advisable to have at least one representative sample from each section, and at least five such samples for each object, preferably on five or more sections. Furthermore, the total number of test points required for obtaining a certain precision (Fig. 6.13) must be applied to each representative sample unit and must, therefore, be distributed over several sections.

4.4. Statistical Evaluation

It must be realized that stereological data are merely estimates, obtained from minute samples of tissue. How well this estimate corresponds to the true value may be assessed by determining the standard error of the estimate, S.E., which is the standard deviation of the data, S.D., divided by the square root of the number of estimations. For this analysis, the sample unit is defined as corresponding to one representative sample. Designating the parameter with x, and having studied x on n representative samples one obtains:

$$\bar{x} = \frac{\Sigma x}{n} \qquad (6.84)$$

$$\text{S.D.} = \sqrt{\frac{\Sigma x^2 - x\Sigma x}{n-1}} \qquad (6.85)$$

$$\text{S.E.} = \sqrt{\frac{\Sigma x^2 - x\Sigma x}{(n-1)n}} \qquad (6.86)$$

For further detail on statistical analysis of the data, the reader is referred to standard textbooks of statistics, such as Snedecor (1956), Cochran (1953), and Dixon and Massey (1957).

5. SOURCES OF SYSTEMATIC ERROR

5.1. Effect of Section Thickness

Basic stereological principles are based on the assumption that the measurements are obtained on sections of zero thickness, whereas, in reality, they are "slices" having a certain thickness, T. The micrographs, therefore, represent projections of the entire content of these slices onto the photographic plate or screen. Consequently, opaque structures may be over-represented, whereas translucent structures may be "covered" by opaque or dense components; the resulting overestimation of opaque components, called *Holmes effect*, depends on the ratio of T to the diameter, D, of the structures under consideration. Using the example of spheres, Fig. 6.20 shows that this effect is insignificant if $T \ll D$, but that

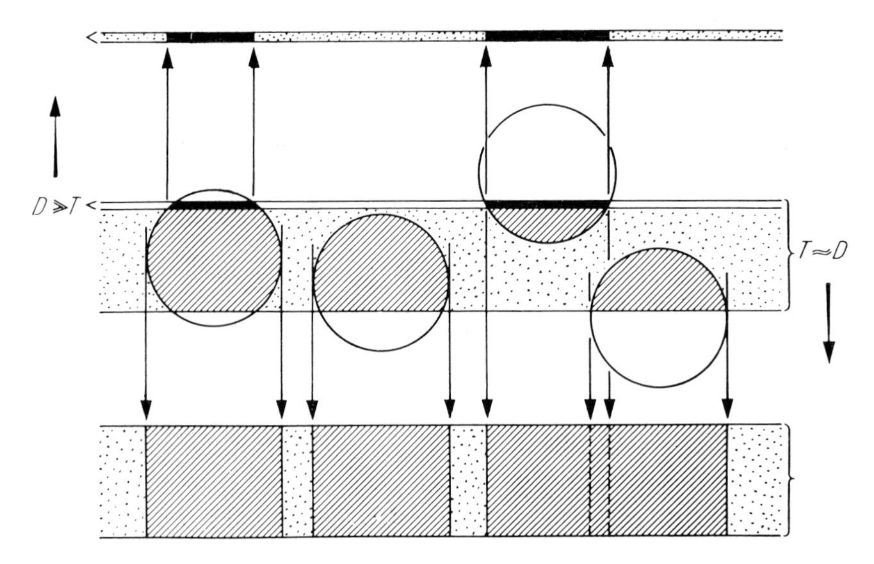

Fig. 6.20 Effect of section thickness, *T*, and particle diameter, *D*, on content of projection image, the Holmes effect. (From Weibel and Elias, 1967).

it becomes important if $T \approx D$. If the volume density of an opaque, granular component is estimated from the areal density A_{Ao}' of its projected profiles by determining P_{Po}' with a point lattice, then the correct value of V_{Vo} is obtained as follows (Holmes, 1927; Hennig, 1957):

$$V_{Vo} = \frac{P_{Po}'}{1 + \frac{3T}{2D}} \qquad (6.87)$$

The volume density of the translucent phase, V_{Vt}, must be corrected in the other direction, i.e., what has been removed from P_{Po}' must be added to P_{Pt}', the point density on the translucent phase.

Hennig (1969) has studied the overestimation of opaque platelike structures of thickness *D*, and found the correction to be

$$V_{Vo} = \frac{P_{Po}'}{1 + \frac{T}{2D}} \qquad (6.88)$$

Cahn and Nutting (1959) approached this problem somewhat differently. They realized that the degree of overestimation somehow depends on the amount of surface of the structures because the error occurs on the edges; they introduced a correction that uses a surface density estimation. If the intersections with test lines, I_{Lo}', are counted in addition to P_{Po}' the correction formula is as follows:

$$V_{Vo} = P_{Po}' - \frac{1}{2} I_{Lo}' \cdot T \qquad (6.89)$$

When should such corrections be applied? The answer to this question once again depends on how large an error one is willing to accept. Very often one is only interested in comparing measurements between two or more experimental groups. If one can expect that this error has affected all groups to the same extent (same section thickness, D not appreciably altered), then one can usually forego a correction. But, if absolute values are needed, it is advisable to introduce a correction if the ratio $T/D > 1/10$. If the ratio is smaller, a correction is probably not necessary, mainly because the loss of small profiles due to contrast deficiency may have compensated this overestimation.

5.2. Tissue Preparation and Sectioning

The validity of stereological data critically depends on the faithful preservation of cell and tissue dimensions. This necessitates the greatest care in tissue preparation. It is essential that the osmolarity of all aqueous solutions is measured and adjusted (cf. Volume 1, Hayat, 1970), that dehydration is started with 70% ethyl-alcohol, avoiding lower grades (Weibel and Knight, 1964), and that the entire procedure is strictly standardized. Once stereological work is started, it is best not to change the methods of tissue preparation, because any change may preclude future comparisons.

Another consideration for stereological work is section compression, which may amount to as much as 20%. It does not affect estimates of volume density if one assumes that all components are equally compressed. However, it does introduce errors in surface density estimations.

Because of the Holmes effect described above, it is important that section thickness is maintained as constant as possible. Measurements of section thickness are important in advanced stereological work; Gillis and Wibo (1971) have constructed a special interference microscope for such measurements. Hayat (1970) has discussed in depth the compression and thickness of sections.

6. DATA COLLECTION AND HANDLING

Stereological analysis starts with micrographs obtained by conventional electron microscopy and leads to quantitative results through a sequence of operations that can be summarized as follows:

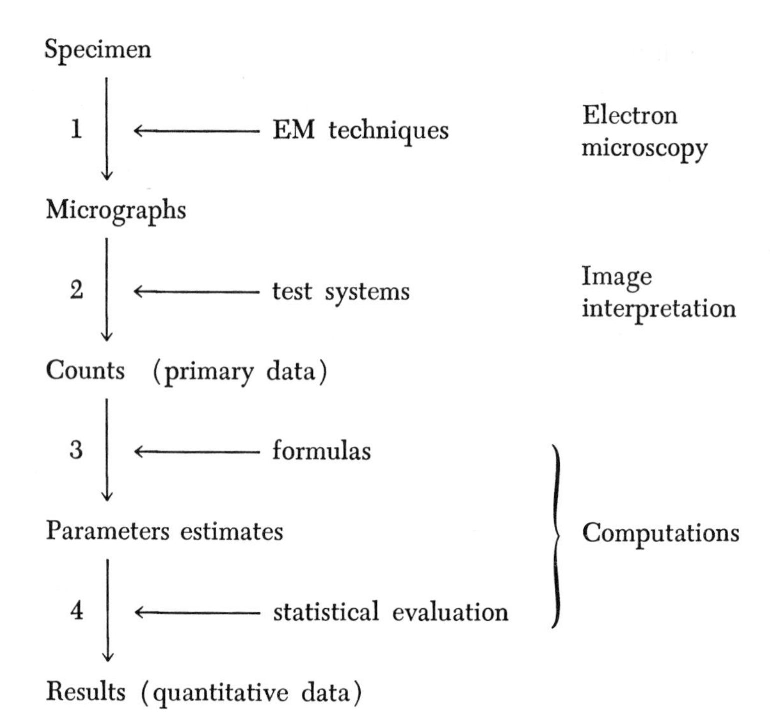

In this section, the practical application of steps 2 to 4, i.e., of image interpretation and computations, shall be discussed. It is evident that there are various approaches. Perfectly good stereology can be performed with improvised means, with nothing more than a pencil and pad, but it may also involve sophisticated techniques of automatic image analysis and computer evaluation.

6.1. Simple Improvised Techniques

The beginner may want to perform a stereological analysis of micrographs routinely produced and printed on photographic paper. For this purpose, he may produce a test system by drawing a square lattice, as shown in Fig. 6.12, on a transparent plastic film using India ink, or prepare one photographically. These films can easily be superimposed on the micrographs, using paper clips to keep them in place. One may also photographically print a lattice onto the micrographs themselves (Loud, 1968).

The simplest means for recording the point counts is to follow the test points systematically, line by line, and record on a tick chart their occurrence in the tissue components. Alternatively, one may employ simple mechanical counters as they are customarily used in hematology; this is much easier because the investigator can learn to operate the counter keys blindly, thereby devoting his full visual attention to image interpretation. While stereological parameters can be calculated by hand, statistical evaluation should be carried out with a desk calculator or computer.

6.2. Advanced Technology

Once larger experimental series are to be subjected to a thorough stereological analysis, improvised techniques will be inefficient. First of all, large numbers of micrographs must be printed—a costly and time-consuming affair—and the simple counting procedures will be tiresome. The stereological laboratory unit shown in Fig. 6.21 is very efficient,

Fig. 6.21 A laboratory unit for stereological work in electron microscopy consisting of a projection unit with interchangeable screens, a counter unit with keyboard, and an automatically tabulating typewriter; card punch not shown.

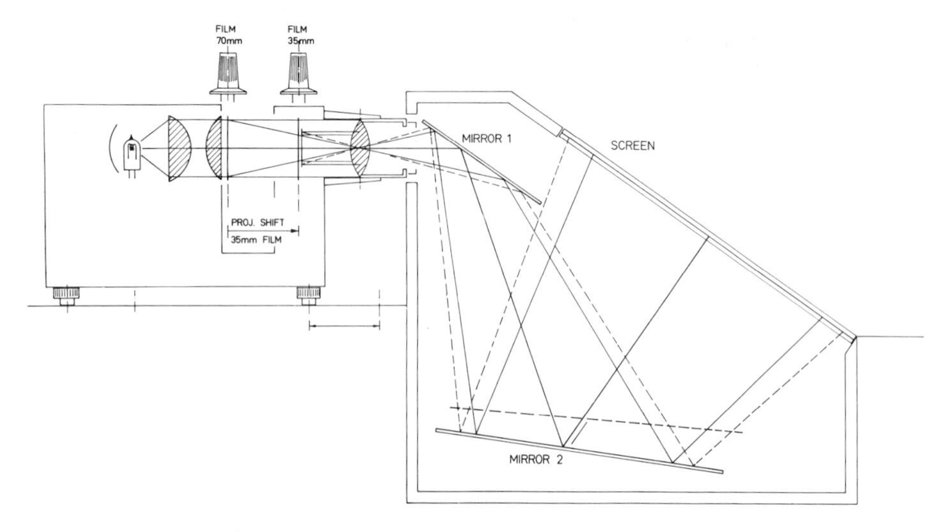

Fig. 6.22 Diagram of projection unit of Fig. 6.21.

mainly because every step that does not require skillful intervention by the investigator is automated (Weibel, 1967b).

The following describes the procedures used in the author's laboratory. Instead of printing the micrographs on paper, they are recorded on 35 mm film; the negative films are then contact printed onto film in a long light box and the positives are projected by a projection unit onto a ground glass screen, 30 × 30 cm, containing a test system. The diagram of the projection unit is shown in Fig. 6.22. The screens are made of two optical ground glass plates between which a photographic transparency of the test system is sandwiched. The screens are exchangeable to allow greater flexibility.

The data-compiling unit consists of ten counters and two totalizers, which are activated from a keyboard (Weibel, 1967c). The read-out facility transfers the data to a printed sheet, punch cards, or tape. Starting with data on punch cards or tape, all the computations can be performed automatically by means of the special program POCOSTER, which contains stereological formulas and provides for statistical evaluation and group comparisons (Gnägi *et al.*, 1970). Similar programs have been developed for use with programmable desk calculators; although these are more limited, they are often more convenient for day-to-day checking of results and for computations with a small volume of data.

This semiautomated system has proven to be very useful over the past years, mainly because of its very great flexibility. It has been successfully employed in the experimental study of very different materials, such as the normal and diseased lung (Weibel *et al.*, 1966; Weibel, 1970/71; Burri and Weibel, 1971; Kistler *et al.*, 1967; Kapanci *et al.*, 1969), the liver (Weibel *et al.*, 1969; Stäubli *et al.*, 1969; Hess *et al.*, 1972; Bolender and Weibel, 1972), the transformation of immunoblasts to plasma cells (Simar and Weibel, 1972), and the structural transformation of protozoans (trypanosoma) (Hecker *et al.*, 1972).

6.3. Automatic Image Analysis

Currently available automatic image analysis devices are still of limited use in biological electron microscopy, because they discriminate structures essentially by contrast, i.e., by levels of gray. If the preparation allows this type of discrimination, then the instruments based on television scanning are excellent means for the efficient screening of large series of micrographs. They operate with a linear scan and, consequently, can easily perform volume and surface density estimations. The advanced devices also include facilities for various kinds of profile size distribution analysis, for profile counting, and for topological study of sections. In some, observer intervention is possible thereby reducing image analysis to a semiautomatic procedure while retaining the facility for high-speed measurement of profiles (Fisher, 1972).

These instruments are currently in a phase of rapid development and it is difficult to forecast the utility they may eventually attain in electron microscopy. The novice in stereology should know that these devices can improve efficiency in the stereological study of *simple* preparations, but, presently, they have not been successfully applied to complex biological electron micrographs.

7. SPECIAL APPLICATIONS OF STEREOLOGICAL METHODS IN CYTOLOGY

The space available does not permit one to outline in detail the practical aspects of applying stereological methods. Therefore, the discussion is confined to a few general remarks, referring the reader to specific papers, and leaving a good part of the specific experimental planning to his own imagination!

7.1. Subcellular Organelles: Morphometric Characterization of Liver Cells as an Example

In order to illustrate the practical application of stereological methods to the study of special cell types, the main points of an experimental protocol are given below. This study was designed to quantitatively correlate structural and biochemical changes in liver cells during and after phenobarbital treatment (Weibel *et al.*, 1969, Stäubli *et al.*, 1969; Bolender and Weibel, 1972).

The biochemical studies were performed on homogenates of whole tissue, relating the data to a unit of liver weight (1 gm). Correspondingly, stereological analysis was performed on random tissue samples, relating the data to the unit volume (1 ml) of tissue.

Because of the wide range of structural dimensions (lobules are ~ 1 mm in diameter while tubules of the endoplasmic reticulum are ~ 300 Å wide), a method of sampling at four different levels of magnification was adopted:

I. Paraffin sections were used to estimate the volume density of lobular tissue, $V_{V}\ell$, in the light microscope at 200 × magnification, using a 100 point, simple square lattice test system on the screen of the WILD M 501 automatic sampling stage microscope (Weibel, 1970).

II. From a pool of osmium-fixed and Epon-embedded tissue cubes, a random sample of five blocks per animal was chosen and sections were cut at 1 μm. After staining with toluidin-blue, these sections were evaluated with the light microscope (1000 × magnification) to determine the number and size distribution of nuclei, from which the number of "mononuclear" cells could be calculated. If the electron microscope can be operated at this low magnification, this step can advantageously be performed on electron micrographs (Bolender and Weibel, 1972).

III. Ultrathin sections (600 to 900 Å) cut from the same blocks were mounted on 200 mesh copper grids. Six to ten micrographs, taken from each section, following the systematic sampling procedure of Fig. 6.19, were recorded at a primary magnification of ~ 2500 ×; the final magnification on the screen of the projection unit was ~ 22,500 ×. Only fields containing lobular tissue were recorded. A 9 : 1 double lattice test system containing 99 coarse points (Fig. 6.19) allowed the estimation of the volume densities in hepatocyte cytoplasm or lobular tissue of nuclei, and of larger cytoplasmic organelles (mitochondria, microbodies, lysosomes), as well as extracellular components; the fine mesh was used only for the rare microbodies, lysosomes, and bile capillaries.

IV. A second sample of six to ten micrographs was recorded at a higher magnification of ~ 10,000 × on the film, so that the final magnification on the projection screen was ~ 90,000 ×. The rules for positioning the specimen on the screen of the microscope were the same as in stage III; however, fields that contained less than ~ 50% cytoplasm were discarded, since, at this level, all measurements were related to cytoplasmic volume. The stereological analysis was mainly concerned with

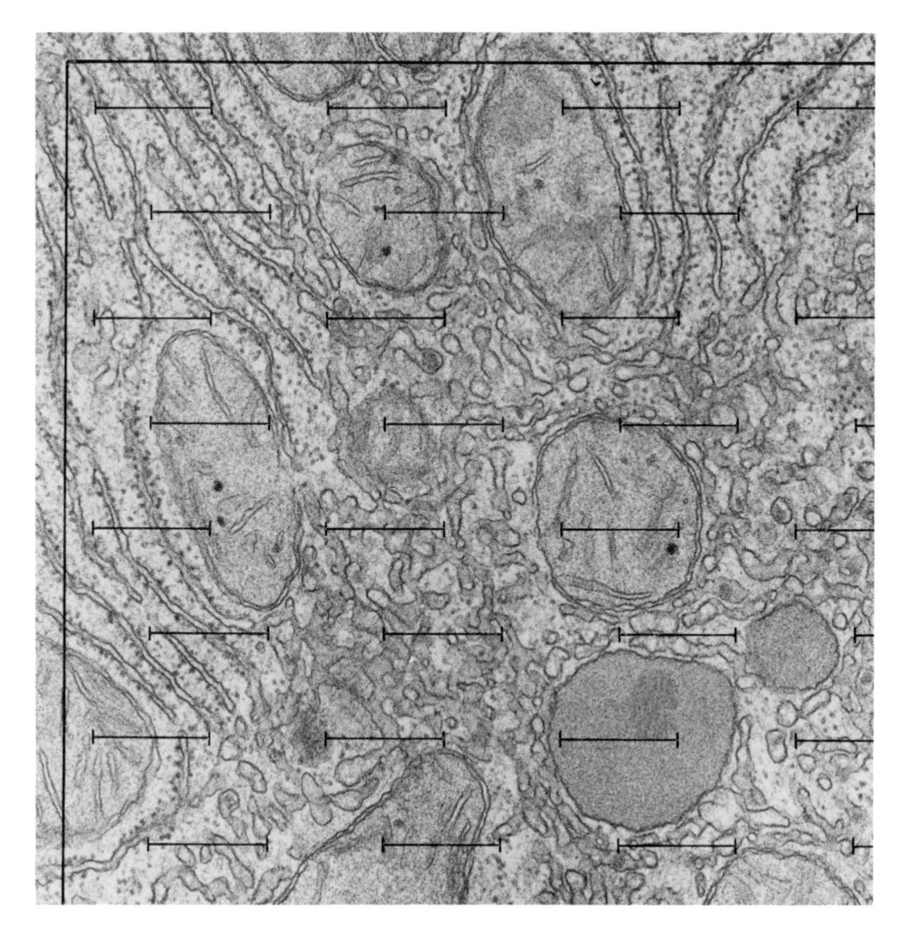

Fig. 6.23 An electron micrograph with part of a multipurpose test system showing intersections with membranes of ER and mitochondria.

the estimation of volume and membrane surface densities of the endoplasmic reticulum and mitochondria. Ribosomal counts were also performed at this stage, and, for this purpose, a coherent multipurpose test system was used (Fig. 6.23). Since the densities were calculated with respect to cytoplasmic volume, the total number of test points and the total length of the test line were determined by the number of test points falling on cytoplasmic constituents, as discussed in relation to Eq. (74).

The procedure of using four sampling levels is efficient and yields representative samples for all components, while, at the same time, providing adequate resolution. In the following, the calculation procedure is illustrated using the example of volume and surface densities of the endoplasmic reticulum; four sampling levels are described.

Level IV: Counted: test points on ER (P_{er}), and on total cytoplasm (P_c^{IV}), which includes points on ER; intersections of ER membranes with test lines (I_{er}).

Calculated: densities in cytoplasm

$$V_{Ver}{}^{c} = P_{er}/P_c^{IV} \tag{6.90}$$

$$S_{Ver}{}^{c} = 4 \cdot I_{er}/P_c^{IV} \cdot z \tag{6.91}$$

Level III: Counted: points on cytoplasm (P_c^{III}), on nuclei (P_n^{III}), and on total lobular tissue (P_l^{III}).

Calculated: cytoplasmic density in lobular tissue

$$V_{Vc}{}^{l} = P_c^{III}/P_l^{III} \tag{6.92}$$

$$V_{Vh}{}^{l} = (P_c + P_n)^{III}/P_l^{III} \tag{6.93}$$

Level II: Counted: profiles of hepatocyte nuclei (N_n), points on nuclei (P_n) and total cytoplasm (P_c^{II}).

Calculated numerical density and mean volume of "mononuclear" hepatocytes

$$N_{Vh}{}^{l} = \frac{1}{1.38} \cdot \left(\frac{N_n}{A_T}\right)^{3/2} \left(\frac{P_T}{P_n}\right)^{1/2} \tag{6.94}$$

$$= \frac{1}{1.38 \cdot P_T \cdot d^3} \cdot \frac{N_n^{3/2}}{P_n^{1/2}} \tag{6.95}$$

$$\bar{v}_h = \frac{V_{Vh}}{N_{Vh}} = \frac{P_n + P_c^{II}}{P_T \cdot N_{Vh}} \tag{6.96}$$

Level I: Counted: test points on lobular tissue (P_l^I).

Calculated: volume density of lobules in liver

$$V_{Vl} = P_l^I/P_T^I \tag{6.97}$$

From these data, one can now calculate the following parameters characterizing the endoplasmic reticulum:

densities in total liver

$$V_{Ver} = V_{Ver}{}^{c} \times V_{Vc}{}^{l} \times V_{Vl} \tag{6.98}$$

$$S_{Ver} = S_{Ver}{}^{c} \times V_{Vc}{}^{l} \times V_{Vl} \tag{6.99}$$

and volume and surface per average "mononuclear" hepatocyte

$$V_{er}{}^{(h)} = V_{V_{er}}{}^{c} \times \frac{V_{V_c}{}^{l}}{V_{V_h}{}^{l}} \times \bar{v}_h \qquad (6.100)$$

$$S_{er}{}^{(h)} = S_{V_{er}}{}^{c} \times \frac{V_{V_c}{}^{l}}{V_{V_h}{}^{l}} \times \bar{v}_h \qquad (6.101)$$

If one knows the total liver volume, V_L, the absolute volume and surface of ER can be obtained:

$$V_{er} = V_{V_{er}} \times V_L \qquad (6.102)$$

$$S_{er} = S_{V_{er}} \times V_L \qquad (6.103)$$

By dividing these absolute values by the body weight one may calculate "specific" dimensions related to the unit body weight.

Experimental studies should consist of an adequate number of animals (or specimens) in order to account for individual variability. In these studies, a minimum of five animals per group were used, which allowed means and standard errors to be estimated and groups to be compared by suitable statistical tests (t-test, rank test, etc.). Further details on procedures and calculations may be found in the original papers (Weibel *et al.*, 1969; Stäubli *et al.*, 1969; Hess *et al.*, 1972; Bolender and Weibel, 1972).

7.2. Other Cell Types

Essentially, the same methodology has been successfully applied to a number of very different cell types. In the lung, the composition of cells and tissues under normal and pathological conditions has been extensively studied (Weibel and Knight, 1964; Weibel, 1963, 1970/71; Kistler *et al.*, 1967; Kapanci *et al.*, 1969; Burri and Weibel, 1971; Eggermann and Kapanci, 1971; Gil and Weibel, 1972). Endothelial cells were studied with respect to a specific organelle (Fuchs and Weibel, 1966; Burri and Weibel, 1968; Steinsiepe and Weibel, 1970). Recently, the development of endoplasmic reticulum in immunoblasts was followed morphometrically, whereby a number of special stereological problems had to be solved (Simar and Weibel, 1972). A morphometric study of the transformation of protozoans (Trypanosoma) also required particular solutions (Hecker *et al.*, 1972). The sarcoplasmic reticulum was investigated with stereological methods by Pager (1971) and, more recently, by Weibel (1972) who explored some of the difficulties encountered in working with such a highly oriented and periodic tissue. Sitte (1967) discussed the problems of the stereological study of kidney, and Haug (1967 and 1972) dealt with the stereological analysis of nervous tissue.

7.3. Subcellular Fractions

Baudhuin and Berthet (1967) have developed a method for the morphometric evaluation of subcellular fractions collected by a special filtration technique. Investigating mitochondrial and lysosomal fractions, they used the method of Wicksell (1925) to derive particle size distributions (Baudhuin, 1968). This approach has recently been extended to microsomal fractions by Wibo *et al.* (1971) (also see Deter in this volume).

7.4. Cytochemistry

It is evident that much could be gained in cytochemistry if the distribution of reaction products were quantitatively estimated. This could be easily achieved with stereological point counting procedures by estimating the relative volume densities of reaction product in various compartments. Unfortunately, many cytochemical methods are not sufficiently quantitative to allow this.

7.5. Immunocytochemistry

In studies involving the use of electron dense, tracer-tagged antibodies, the surface area of cells or organelles binding these antibodies could be quantitated by using a simple method. If counts are made of the intersections occurring between random test lines and surface traces of membranes marked, I_a, or unmarked, I_o, with antibody, then the fraction of marked surface is

$$S_{Sa} = \frac{I_a}{I_a + I_o} \qquad (6.104)$$

Any test system of linear probes may be used for this purpose.

7.6. Autoradiography

The need for better quantification in autoradiography has long been recognized and while various procedures have been proposed (Whur *et al.*, 1969; Williams, 1969), they are not very satisfactory. Two aspects must be considered:

1. The radioactive sources incorporated into the tissue are "point probes." If they were randomly distributed, a differential count of their location would be equivalent to a stereological point count; it would measure the relative volumes of the compartments. In autoradiography, however, one usually works from the hypothesis that a certain labeled compound is preferentially concentrated in a given subcellular com-

partment, i.e., that there are more "point sources" in this compartment than one would expect from a random distribution.

It is evident that this preferential concentration could be quantified by combining a "source count" with a stereological random point count. For this purpose, one must determine on an autoradiograph the number of radiation sources, R_i, in a component i, and the total number of radiation sources, R_T. Simultaneously, one superimposes a test system with test points, P_T, and determines the number of these points, P_i, falling on profiles of i. One then obtains the volume density by

$$V_{Vi} = P_i/P_T \qquad (6.105)$$

and the relative frequency of radiation sources in i by

$$R_{Ri} = R_i/R_T \qquad (6.106)$$

If the radiation sources are randomly distributed, one will find $R_{Ri} = V_{Vi}$. It follows that any preferential accumulation of the radiation source in i can be tested by calculating the radiation source density

$$R_{V_i} = \frac{R_i}{V_i}$$

which can be obtained by

$$R_{V_i} = \frac{R_{Ri}}{V_{V_i}} = \frac{R_i \cdot P_T}{P_i \cdot R_T} \qquad (6.107)$$

The more R_{Vi} deviates from 1 the higher will be the concentration of the labeled compound in the component i.

2. One of the major analytical problems associated with autoradiography is the fact that the location of the developed silver grain indicates the location of the radiation source in terms of a certain probability (Caro, 1962; Salpeter and Bachmann, 1964; Bachmann *et al.*, 1968; Whur *et al.*, 1969). Although opinions differ regarding resolution (Salpeter and Bachmann, 1972), there is general agreement that the grains are distributed around the radiation source in a probability pattern as shown in Fig. 6.24a. In analogy to the work of Bachmann *et al.* (1968) one may, therefore, draw three circles around the grain in such a way that they delimit the fields containing the radiation source with a 30, 60, and 90% probability. In terms of the probability of containing the source, the three shaded areas of Fig. 6.24b are equal, each representing 30%.

In order to estimate the probability with which the radiation source lies within one of the tissue components, one should estimate the composition of each of these areas and weigh it according to the decreasing probability. The arguments regarding the relationship between test points and test area put forth in earlier sections make it apparent that one may

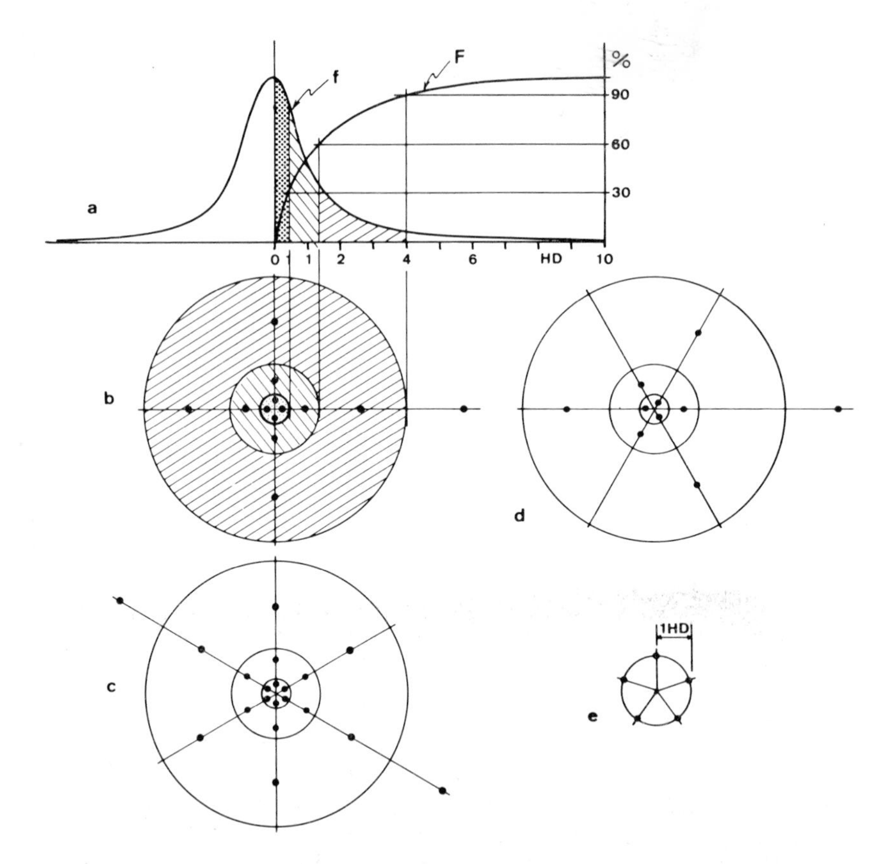

Fig. 6.24 (a) Distribution of grain density (f) around the radiation source and probality (F) that the radiation source lies within a circle around the developed grain, expressed as half-distances (HD); drawn according to Bachmann *et al.* (1968). (b-d) Test point systems for estimating the probability of location of the radiation source. (e) Simplified test system of five points on a circle having a half-distance radius.

obtain this weighted estimate directly by using the same number of test points in each of these areas. In Figure 6.24b-d three examples of test point sets that fulfill this condition are shown. There are either 3, 4, or 6 points per field, and, in addition, 1 or 2 points marked beyond the 90% circle to account for the 10% probability that the radiation source lies outside this field.

By centering such a test point system around a silver grain on an autoradiograph, one can determine the relative probabilities that the radiation source lies within a certain component by counting the number of test points, G_i, falling onto that component. This is a random point count, however, one biased with respect to the silver grains, or with respect to the location of the radiation source. If the "grain point set" contains g test points ($g = 10$ in Fig. 6.24d) then the following relationships hold:

$$G_T = R_T \cdot g \tag{6.108}$$

$$G_i = R_i \cdot g \tag{6.109}$$

so that Eqs. (6.106) and (6.107) become

$$R_{Ri} = G_i / G_T \tag{6.110}$$

$$R_{V_i} = \frac{G_i \cdot P_T}{P_i \cdot G_T} \tag{6.111}$$

It is evident that R_T is the total number of grains on the autoradiographs and G_T the total number of points of the "grain point sets" counted.

The "grain point sets" discussed so far may estimate the probability of source location around the individual grain with too much precision. After all, one is dealing with a statistical problem that reaches a certain significance only if a large enough number of grains have been evaluated. If the autoradiographs contain enough grains, it may be advantageous to simplify the procedure of estimation. In a recent study (Stäubli and Weibel, 1972), a grain point set having only five points (Fig. 6.24e) was placed on a circle at about the "half-distance" around the grain; half-distance being defined as the circle containing the source with 50% probability. In practical application of this method, it should be noted that the section thickness is very critical because the number of radiation sources is directly proportional to section thickness.

8. CONCLUDING REMARKS

Many biologists shy away from using stereological methods because of the apparent need for advanced mathematical knowledge. This chapter was, for the most part, restricted to the use of elementary arithmetics. In fact, the only mathematics needed for applying stereology is an elementary knowledge of statistics.

References

Abercrombie, M. (1946). Estimation of nuclear population from microtomic sections. *Anat. Rec. 94*, 239.

Bach, G. (1967). Kugelgrössenverteilung und Schnittkreisverteilung; ihre wechselseitigen Beziehungen und Verfahren zur Bestimmung der einen aus der anderen. *In:* "Quantitative Methods in Morphology" (Weibel, E. R. and Elias, H., ed.), p. 23. Springer, Berlin-Heidelberg, New York.

Bachmann, L., Salpeter, M. M., und Salpeter, E. E. (1968). Das Auflösungsvermögen elektronenmikroskopischer Autoradiographien. *Histochemie 15,* 234.

Baudhuin, P. (1968). L'analyse morphologique quantitative de fractions subcellulaires. Thèse présentée en vue de l'obtention du grade d'Agrégé de l'Enseignement Supérieur, Université Catholique de Louvain.

———, and Berthet, J. (1967). Electron microscopic examination of subcellular fractions. II. Quantitative analysis of the mitochondrial population isolated from rat liver. *J. Cell Biol. 35,* 631.

Bolender, R. P., and Weibel, E. R. (1973). A morphometric study of the removal of phenobarbital induced membranes from hepatocytes after cessation of treatment. *J. Cell Biol.* (in press).

Burri, P., and Weibel, E. R. (1968). Beeinflussung einer spezifischen cytoplasmatischen Organelle von Endothelzellen durch Adrenalin. Z. Zellforsch. *88,* 426.

———, and Weibel, E. R. (1971). Morphometric estimation of pulmonary diffusion capacity. II. Effect of PO_2 on the growing lung. Adaptation of the growing rat lung to hypoxia and hyperoxia. *Respir. Physiol. 11,* 247.

Cahn, J. W., and Nutting, J. (1959). Transmission quantitative metallography. *Trans. Am. Inst. Mining Met. Engrs. 215,* 526.

Caro, L. B. (1962). High resolution autoradiography. II. The problem of resolution. *J. Cell Biol. 15,* 189.

Chalkley, H. W. (1943). Methods for the quantitative morphologic analysis of tissues. *J. Nat. Cancer Inst. 4,* 47.

———, Cornfield, J., and Park, H. (1949). A method for estimating volume-surface ratios. *Science 110,* 295.

Cochran, W. G. (1953). "Sampling Techniques." John Wiley, New York.

De Hoff, R. T. (1962). The determination of the size distribution of ellipsoidal particles from measurements made on random plane sections. *Trans. Am. Inst. Mining Met. Engrs. 224,* 474.

———, and Rhines, F. N. (1961). Determination of the number of particles per unit volume from measurements made on random plane sections: the general cylinder and the ellipsoid. *Tr. AIME 221,* 975.

———, and Rhines, F. N. (1968). "Quantitative Microscopy." McGraw-Hill Book Company, New York.

———, and Bousquet, P. (1970). Estimation of the size distribution of triaxial ellipsoidal particles from the distribution of linear intercepts. *J. Microscopy 92,* 119.

Delesse, M. A. (1847). Procédé mécanique pour determiner la composition des roches. *C. R. Acad. Sci.* (Paris) *25,* 544.

Dixon, W. J., and Massey, F. J. (1957). Introduction to Statistical Analysis. McGraw-Hill Book Company, New York.

Ebbesson, S. O. E., and Tang, D. B. (1967). A Comparison of Sampling Procedures in a Structured Cell Population. In: "Stereology," (Elias, H., ed.), p. 131. Springer, New York.

Eggermann, J., and Kapanci, Y. (1971). Experimental pulmonary calcinosis in the rat: ultrastructural and morphometric studies. *Lab. Invest. 24,* 469.

Elias, H. (1972). Identification of structure by the common-sense approach. *J. Microscopy 95* H, 59.

———, Sokol, A., and Lazarowitz, A. (1954). Contributions to the geometry of sectioning. II. Circular cylinders. *Z. wiss. Mikr. 62,* 20.

Fisher, C. (1972). Current capabilities and limitations of available stereological techniques. IV. Automatic image analysis for the stereologist. *J. Microscopy 95* H, 385.

Floderus, S. (1944). Untersuchungen über den Bau der menschlichen Hypophyse mit besonderer Berüchsichtigung der quantitativen mikromorphologischen Verhältnisse. *Acta path. microbiol. scand.* Suppl. 53.

Fuchs, A., and Weibel, E. R. (1966). Morphometrische Untersuchung der Verteilung einer spezifischen cytoplasmatischen Organelle von Endothelzellen in der Ratte. *Z. Zellforsch. 73,* 1.

Gahm, J. (1972). Current capabilities and limitations of available stereological techniques. I. Stereological measurements with lines, circles and structural standards. *J. Microscopy 95,* 368.

Giger, H. (1970). Grundgleichungen der Stereologie I. *Metrika 16,* 43.

———, und Riedwyl, H. (1970). Bestimmung der Grössenverteilung von Kugeln aus Schnittkreisradien. *Biometr. Zschr. 12,* 156.

Gil, J., and Weibel, E. R. (1972). Morphological study of pressure-volume hysteresis in rat lungs fixed by vascular perfusion. *Respir. Physiol. 15,* 190.

Gillis, J. M., and Wibo, M. (1971). Accurate measurement of the thickness of ultrathin sections by interference microscopy. *J. Cell Biol. 49,* 947.

Glagoleff, A. A. (1933). On the geometrical methods of quantitative mineralogic analysis of rocks. *Tr. Inst. Econ. Min. and Metal,* Moscow, 59.

Gnägi, H. R., Burri, P. H., and Weibel, E. R. (1970). A multipurpose computer program for automatic analysis of stereological data obtained on electron micrographs. *7ème Congrès Internat. de Microscopie Electronique,* Grenoble, p. 443.

Haug, H. (1967). Probleme und Methoden der Strukturzählung im Schnittpräparat. In: "Quantitative Methoden in der Morphologie" (Weibel, E. R. und Elias, H., ed.), p. 58: Springer, Berlin-Heidelberg-New York.

———. (1972). Stereological methods in the analysis of neuronal parameters in the central nervous system. *J. Microscopy 95,* 165.

Hayat, M. A. (1970). "Principles and Techniques of Electron Microscopy: Biological Applications." Vol. 1. Van Nostrand Reinhold Company, New York and London.

Hecker, H., Burri, P. H., Steiger, R., and Geigy, R. (1972). Morphometric data on the ultrastructure of the pleomorphic bloodforms of *Trypanosoma brucei,* Plimmer and Bradfort. *Acta Tropica 29,* 182.

Hennig, A. (1957). Zur Geometrie von Schnitten. *Z. wiss. Mikrosk. 63,* 362.

———. (1967). Fehlerbetrachtungen zur Volumenbestimmung aus der Integration ebener Schnitte. In: "Quantitative Methods in Morphology" (Weibel, E. R. and Elias, H., ed.), p. 99. Springer, Berlin-Heidelberg-New York.

———. (1969). Fehler der Volumenermittlung aus der Flächenrelation in dicken Schnitten. *Mikroskopie 25,* 154.

———, and Elias, H. (1971). A rapid method for the visual determination of size distribution of spheres from the size distribution of their sections. *J. Microscopy 93,* 101.

Hess, F. A., Weibel, E. R., Gnägi, H. R., and Preisig, R. (1972). Morphometry of dog liver. I. Normal base-line data. II. Evaluation of the validity of needle biopsies. (in preparation).

Hilliard, J. E. (1965). In: "Recrystallization, Grain Growth and Textures," p. 267, Am. Soc. Metals, Metals Park, Ohio.

———. (1967). Determination of Structural Anisotropy. In: "Stereology" (Elias, H., ed.), p. 219, Springer, New York.

———, and Cahn, J. W. (1961). An evaluation of procedure in quantitative metallography for volume-fraction analysis. *Trans. Am. Inst. Min. Engrs. 221,* 344.

Holmes, A. H. (1927). "Petrographic Methods and Calculations." Murby, London.

Kapanci, Y., Weibel, E. R., Kaplan, H. P., and Robinson, F. R. (1969). Pathogenesis and reversibility of the pulmonary lesions of oxygen toxicity in monkeys. II. Ultrastructural and morphometric studies. *Lab. Invest. 20,* 101.

Kistler, G. S., Caldwell, P. R. B., and Weibel, E. R. (1967). Development of fine structural damage to alveolar and capillary lining cells in oxygen-poisoned rat lungs. *J. Cell Biol. 32,* 605.

Loud, A. V. (1968). A quantitative stereological description of the ultrastructure of normal rat liver parenchymal cells. *J. Cell Biol. 37,* 27.

Merz, W. A. (1968) Streckenmessung an gerichteten Strukturen im Mikroskop und ihre Anwendung zur Bestimmung von Oberflächen-Volumen-Relationen im Knochengewebe. *Mikroskopie 22,* 132.

Miles, R. E. (1972). Multi-dimensional perspectives on stereology. *J. Microscopy 95,* 181.

Myers, E. J. (1967). Size Distribution of Cubic Particles. In: "Stereology" (Elias, H., ed.), p. 187, Springer, New York.

Pager, J. (1971). Étude morphométrique du système tubulaire transverse du myocarde ventriculaire de rat. *J. Cell Biol. 50,* 233.

Rosiwal, A. (1898). Ueber geometrische Gesteinsanalysen. Ein einfacher Weg zur ziffermässigen Feststellung des Quantitätsverhältnisses der Mineralbestandteile gemengter Gesteine. *Verh. K. K. Geol. Reichsamt, Wien,* p. 143.

Salpeter, M. M., and Bachmann, L. (1964). Autoradiography with the electron microscope. A procedure for improving resolution, sensitivity and contrast. *J. Cell Biol. 22,* 469.

———, and Bachmann, L. (1972). Radioautography. In: "Principles and Techniques of Electron Microscopy: Biological Applications." Vol. 2 (Hayat, M. A., ed.). Van Nostrand Reinhold Company, New York and London.

Saltikov, S. A. (1958). "Stereometric Metallography," Second Edition, 446 p., Moscow.

———. (1967). A Stereological Method for Measuring the Specific Surface Area of Metallic Powders. In: "Stereology," (Elias, H., ed.), p. 63, Springer, New York.

Scheil, E. (1935). Statistische Gefügeuntersuchungen. *Zschr. Metallk. 27,* 199.

Schroeder, H. E., and Münzel-Pedrazzoli, S. (1970). Application of stereologic methods to stratified gingival epithelia. *J. Microscopy 92,* 179.

Schwartz, H. A. (1934). The metallographic determination of the size distribution of temper carbon nodules. *Metals and Alloys 5,* 139.

Simar, L., and Weibel, E. R. (1972). Morphometric study of endoplasmic reticulum during transformation of immunoblasts to plasma cells. (in preparation).

Sitte, H. (1967). Morphometrische Untersuchungen an Zellen. In: "Quantitative Methods in Morphology" (Weibel, E. R. and Elias, H., eds.), p. 167. Springer, New York.

Snedecor, G. W. (1956). Statistical methods applied to experiments in agriculture and biology. The Iowa State University Press, Ames, Iowa.

Stäubli, W., Hess, R., and Weibel, E. R. (1969). Correlated morphometric and biochemical studies on the liver cell. II. Effects of phenobarbital on rat hepatocytes. *J. Cell Biol. 42,* 92.

———, and Weibel, E. R. (1972). Combined autoradiographic and morphometric study of peroxisome proliferation. (in preparation).

Steinsiepe, K. F., und Weibel, E. R. (1970). Elektronenmikroskopische Untersuchungen an spezifischen Organellen von Endothelzellen des Frosches (Rana temporaria). Z. *Zellforsch. 108,* 105.

Tomkeieff, S. I. (1945). Linear intercepts, areas and volumes. *Nature 155,* 24.

Underwood, E. E. (1968). Particle-Size Distribution. In: "Quantitative Microscopy" (De Hoff, R. T. and Rhines, F. N., ed.), p. 149. McGraw-Hill Book Company, London.

———. (1970). "Quantitative Stereology." Addison-Wesley, Reading Mass.

Weibel, E. R. (1963). "Morphometry of the Human Lung." Springer, Berlin-Göttingen-Heidelberg, and Academic Press, New York.

———. (1966). Postnatal growth of the lung and pulmonary gas-exchange capacity. *Ciba Fdn. Symp. Lung Developm.* 131.

———. (1967a). Structure in Space and Its Appearance on Sections. In: "Stereology" (Elias, H., ed.), p. 15. Springer, New York.

———. (1967b). A semi-Automatic System for Stereologic Work in Light and Electron Microscopy. In: "Stereology," (Elias, H., ed.), p. 275. Springer, New York.

———. (1967c). Automatic Sampling Stage Microscope and Data Print-Out Unit. In: "Stereology" (Elias, H., ed.), p. 330. Springer, New York.

———. (1969). Stereological principles for morphometry in electron microscopic cytology. *Int. Rev. Cytol. 26,* 235.

———. (1970). An automatic sampling stage microscope for stereology. *J. Microscopy 91,* 1.

———. (1970/71). Morphometric estimation of pulmonary diffusion capacity. I. Model and method. *Respir. Physiol. 11,* 54.

———. (1972). A stereological method for estimating volume and surface of sacroplasmic reticulum. *J. Microscopy 95,* 229.

———, and Gomez, D. M. (1962). A principle for counting tissue structures on random sections. *J. Appl. Physiol. 17,* 343.

———, and Knight, B. W. (1964). A morphometric study on the thickness of the pulmonary air-blood barrier. *J. Cell Biol. 21,* 367.

———, Kistler, G. S., and Scherle, W. R. (1966). Practical stereological methods for morphometric cytology. *J. Cell Biol. 30,* 23.

———, Stäubli, W., Gnägi, H. R., and Hess, F. A. (1969). Correlated morphometric and biochemical studies on the liver cell. I. Morphometric model, stereologic methods and normal morphometric data for rat liver. *J. Cell Biol. 42,* 68.

Whur, P., Herscovics, A., and Leblond, C. P. (1969). Radioautographic visualization of the incorporation of galactose-^{3}H and mannose-^{3}H by rat thyroids in vitro in relation to the stages of thyroglobulin synthesis. *J. Cell Biol. 43,* 289.

Wibo, M., Amar-Costesec, A., Berthet, J., and Beaufay, H. (1971). Electron microscope examination of subcellular fractions. III. Quantitative analysis of the microsomal fraction isolated from rat liver. *J. Cell Biol. 51,* 52.

Wicksell, S. D. (1925). On the size distribution of sections of a mixture of spheres. *Biometrica 17,* 84.

———. (1926). On the size distribution of sections of a mixture of ellipsoids. *Biometrica 18,* 151.

Williams, M. A. (1969). The assessment of electron microscopic autoradiographs. *Advs. in Optical and Electron Microscopy 3,* 219. Academic Press, London.

7 Critical Point-Drying Method

M. A. Hayat
Department of Biology, Newark State College, Union, N.J.

B. R. Zirkin
Department of Biology, Illinois Institute of Technology, Chicago, Ill.

Before a substance can be observed with the electron microscope, all volatile materials must be removed. In the case of biological materials, water must be removed. When using conventional procedures for the preparation of biological specimens for electron microscopy, the specimens are usually fixed, embedded, and sectioned. The advantages and disadvantages of chemical fixation of biological specimens for transmission electron microscopy have been discussed in detail (Hayat, 1970). For basic procedures for the preparation of a wide variety of biological specimens, the reader is referred to Hayat (1972).

For studies of whole cells or whole-mounts of certain cell structures, techniques are available to circumvent the conventional sectioning techniques. The critical step in these procedures is in drying the specimens. If biological specimens are allowed to air-dry, gross distortion of cells usually occurs due to the action of surface tension forces as the receding surface of the evaporating water passes the specimen. By comparing Fig. 7.1 (air dried) with Fig. 7.2 (critical point dried), the collapse of the spore wall is apparent in the former. In order to avoid serious distortion, the specimens must be brought to the dry state without the disrupting effect of surface tension forces. This can be accomplished by using freeze-drying and critical point-drying methods. Freeze-drying methods have been discussed in depth by Rebhun (1972). The critical point-drying method, introduced for transmission electron microscopy by Anderson (1951), will be discussed in this chapter.

When a given liquid in equilibrium with its vapor is heated in a confined space, a temperature is reached above which it is impossible to liquify the vapor regardless of the applied pressure. This temperature is the "critical temperature." As the pressure is increased during the heating, the vapor density increases until at the "critical pressure" the density of the vapor and liquid phases are equal. The conjunction of critical temperature and pressure, the "critical point," results in the surface tension of the liquid becoming zero, and there is no distinction between vapor and liquid phases. The gas can then be released. Condensation will not occur during the release as long as the temperature is kept above the critical temperature. Thus, in the method described by Anderson (1951), specimens immersed in liquid CO_2 are brought through the critical point by increasing temperature and pressure. The CO_2 gas is slowly bled off above the critical temperature for CO_2. The specimens are dried without disruption by high surface forces resulting from a receding interface.

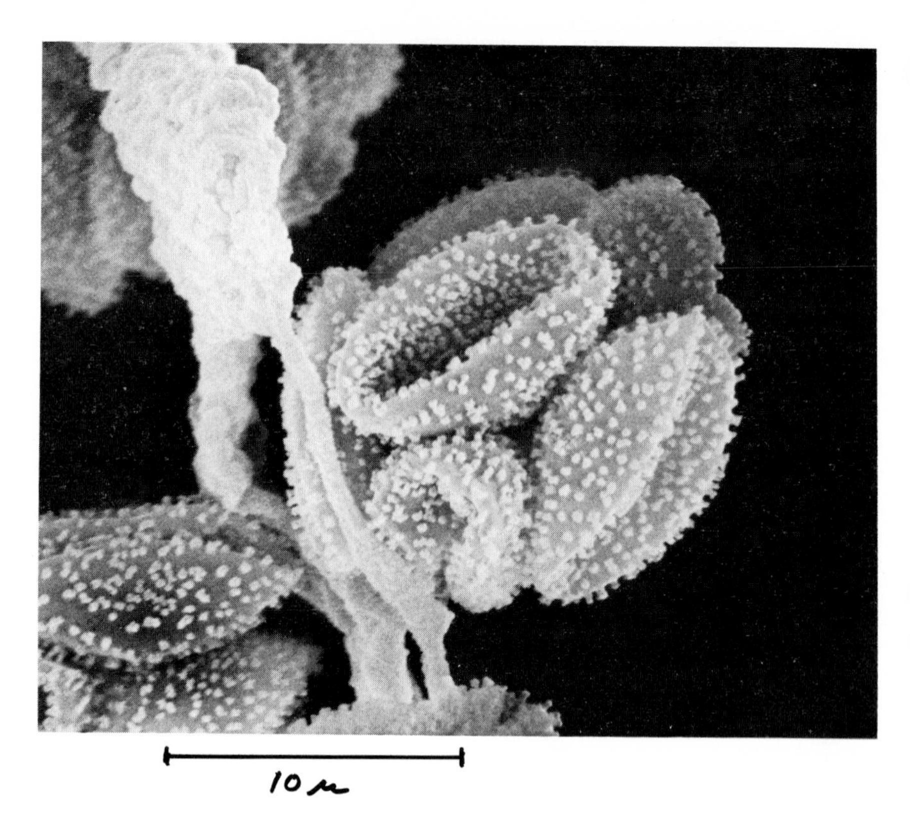

Fig. 7.1 The spores of myxomycete, *Badhamia utricularis* in the sporangium attached to a portion of the capillitium—air dried. Note the collapse of the spore wall. 2,793 X. *Courtesy of A. L. Cohen.*

The critical temperature of water is 374°C, far too high to be of practical use in the critical point-drying method. In Anderson's technique, the water in biological material is replaced by a liquid of suitable critical temperature and pressure. The specmen is dehydrated in a series of ethyl alcohols of increasing concentrations. The alcohol is then replaced by amyl acetate which, in turn, is replaced by liquid carbon dioxide. The critical temperature and pressure of carbon dioxide are 36.5°C and 1080 lbs/in^2, respectively. Amyl acetate is used because it is miscible both with ethyl alcohol and liquid carbon dioxide. The specimen, immersed in liquid carbon dioxide under pressure, is then passed through the critical temperature of the carbon dioxide. At or above this temperature, the liquid is converted to gas. After release of the gas, the specimen, now dry, can be viewed with the electron microscope.

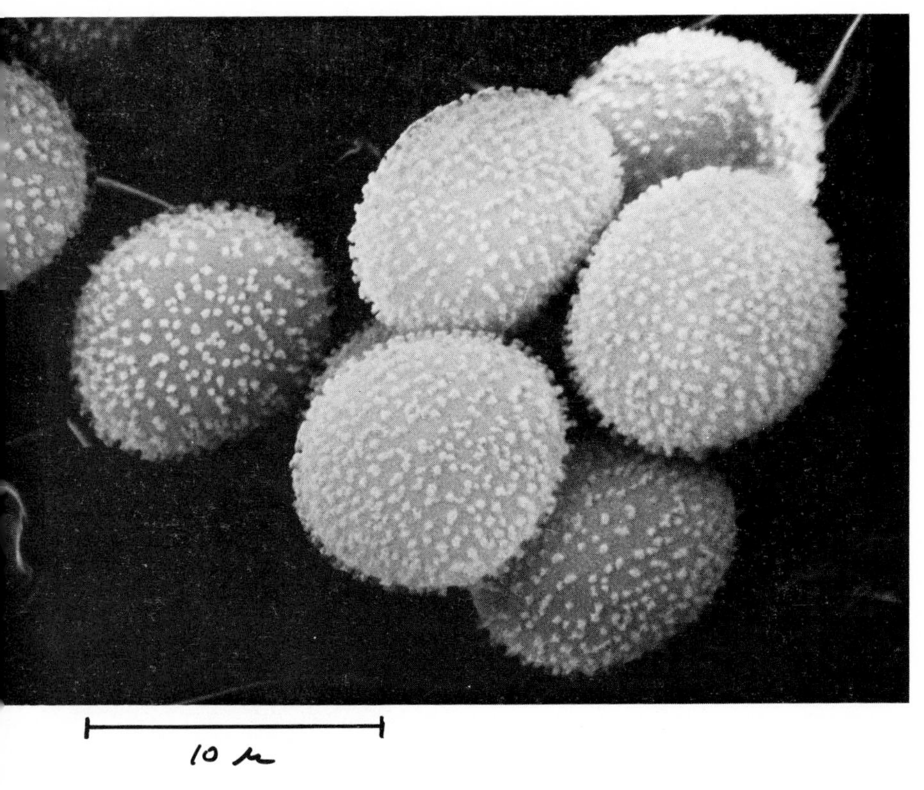

Fig. 7.2 The spores of myxomycete, *Badhamia utricularis*—critical point dried (Freon 13 as transitional fluid). 5,700 X. *Courtesy of A. L. Cohen.*

Reagents other than liquid carbon dioxide also have been used in this method. Koller and Bernhard (1964) used liquid nitrous oxide; since nitrous oxide is miscible with water, the use of amyl acetate is unnecessary in this procedure. Fluorocarbon fluids such as the Freons have relatively low critical pressures and temperatures and also have been used in critical point-drying. Cohen *et al.* (1968) investigated several Freons and found Freon 13 ($CClF_3$) suitable, with a critical pressure of 561 lbs/in^2 and a critical temperature of 28.9°C. Freons are also miscible with absolute ethanol, are nearly odorless, nontoxic, noninflammable and inert. In the Freon method, the amyl alcohol step is eliminated, and the method is faster than the CO_2 method.

The critical point-drying technique has been particularly useful in studies of intact chromosomes. The technique was first used for chromosome studies by Ris (1961). The isolation of chromosomes on the surface of a Langmuir trough, followed by critical point-drying, was introduced by Gall (1963) as a means of observing recognizable whole chromosomes with the electron microscope.

By the Gall method, cells are transferred to the distilled water surface of a Langmuir trough. The surface tension forces at the air-water interface break open cells and spread the contents in a thin layer over the surface. The spread structures are then picked up by touching grids to the surface, dehydrated, immersed in amyl acetate, and dried by Anderson's critical point-drying technique. The step-by-step procedure for the Langmuir trough-critical point technique for isolation of chromosomes, using liquid carbon dioxide and a critical point-drying apparatus (Fig. 7.3), is given below.

a

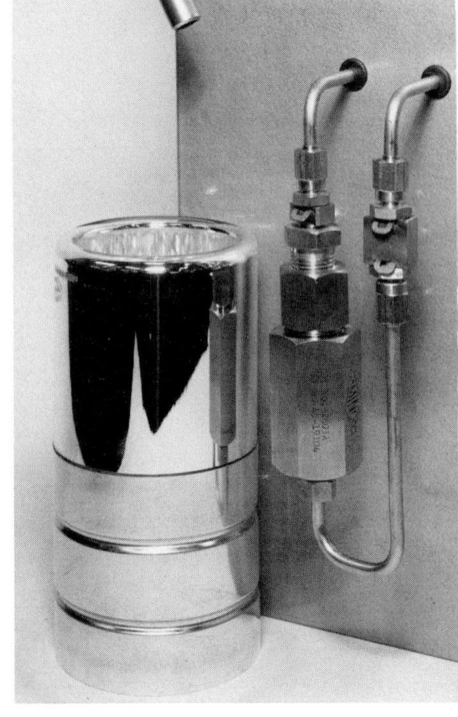

b

Fig. 7.3 Critical point drying apparatus. *Courtesy of American Instrument Company.*

Fig. 7.3a Front view showing valves B and C; specimen bomb is immersed in water.

Fig. 7.3b Specimen bomb.

c

Fig. 7.3c Back of the apparatus showing liquid CO_2 tank.

1. Fill the trough (Fig. 7.4a) with distilled water until a raised meniscus is formed.
2. Clean the distilled water surface by moving barriers (A and B) across the surface repeatedly. To determine whether the surface has been cleaned adequately, place one barrier (B) and the foil indicator (C) at one end of the trough as shown, and advance the second barrier (A) across the surface, toward the foil indicator. If the foil moves before the barrier is ~ 1/16 in away, the surface is not sufficiently clean.
3. Certain cells, such as nucleated red blood cells, can be transferred directly to the surface of the trough with a glass rod. After cleaning the water surface, and with barriers and foil positions as shown in Fig. 7.4a, place a drop of blood on the end of a clean glass rod and pass the rod through the surface of the trough. This results in the formation of a film on the surface.

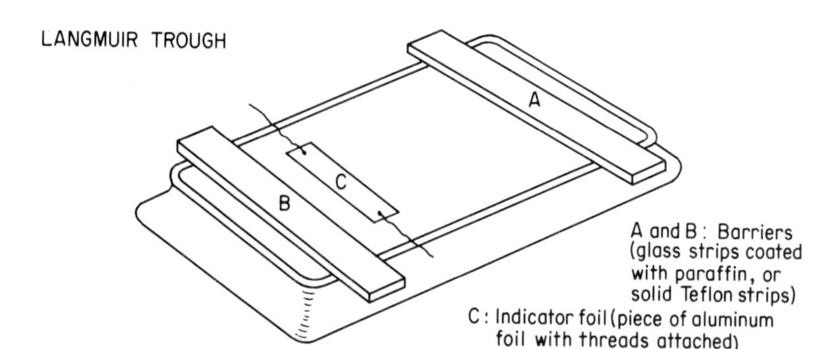

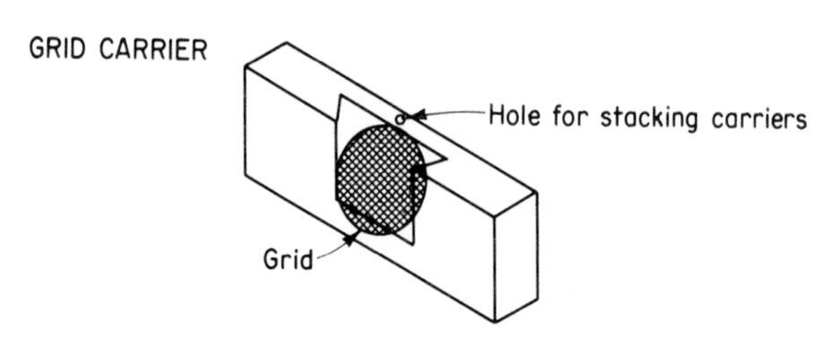

Fig. 7.4 Langmuir trough.

Many cells cannot be spread on the trough in this way without pretreatment of some kind. In the case of milkweed bug testes, for example, spreading can be achieved by grinding the tissue between the ends of two glass slides and then passing the slides through the surface of the trough (Wolfe and Hewitt, 1965). Plant root tips must be prefixed prior to spreading on a trough (Wolfe, 1968); most animal cells, on the other hand, will not spread if prefixed. Some sperm have been spread after pretreatment in thioglycolate (Lung, 1968), EDTA (Solari, 1965), proteases (Walker, 1971), and urea (Walker, 1971); other sperms have been spread without pretreatment (Zirkin, 1971).

4. After transferring material to the surface of the trough, compress the film by slowly moving one barrier (A) toward the second (B) until wrinkles are just visible on the surface in reflected light. If the surface of the trough has become occupied by material, the indicator foil should move as the barrier is moved.
5. Pick up the film by touching Formvar-carbon grids to different portions of the surface. Place the grids directly in staining solution (15 min in 1% or 2% aqueous uranyl acetate will give good results). After staining, wash the grids in distilled water.
6. Transfer the grids, still wet, to a series of ethyl alcohols of increas-

ing concentrations for dehydration. These transfers can be carried out conveniently in grid carriers, an example of which is shown in Fig. 7.4b. Load the grids into the carrier under water or 30% ethanol, then place the carrier into 70% and 95% ethanol (about 2 min each) and 3 changes of 100% ethanol (10 min each). After dehydration, transfer the grids to amyl acetate (3 changes, 10 min each).

7. A diagram of a critical point apparatus is shown in Fig. 7.5. Open valve A (tank valve), bleed out some carbon dioxide by opening valve B, and then close valves B and C.
8. Blot excess amyl acetate from the grid carrier. Load the carrier in the pressure vessel of the apparatus.
9. Manipulate the valves so that the specimen is flooded with liquid carbon dioxide. Specifically: (a) Open valve B slowly until the pressure stabilizes (read on the meter). (b) Open valve B entirely. (c) Open valve C slowly to permit a continuous flow of carbon dioxide.
10. With valve C cracked, raise a bucket of cold water around the pressure vessel to maintain the temperature well below the critical temperature (31°C) of the carbon dioxide. Wait about 20 min for the

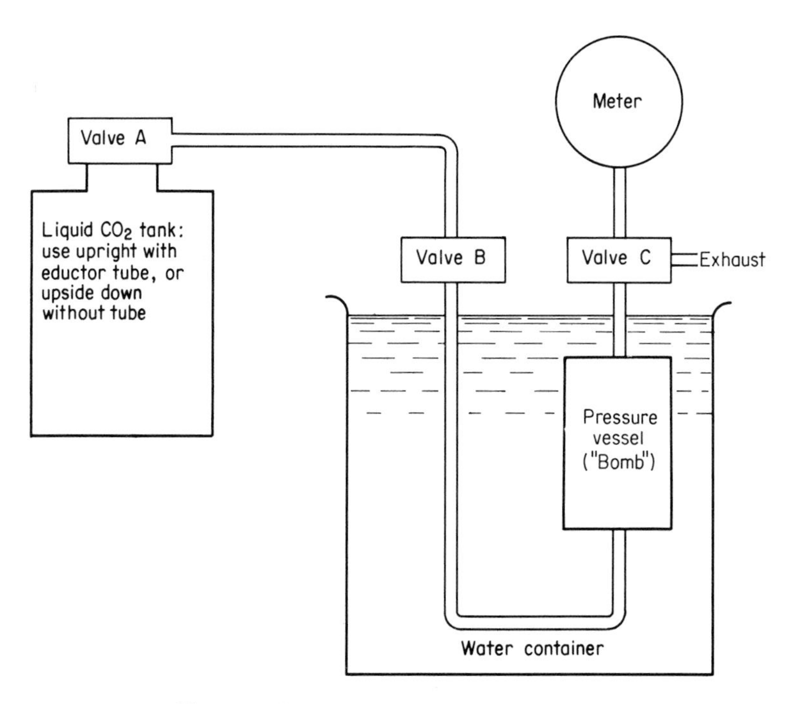

Fig. 7.5 Critical point drying apparatus.

liquid carbon dioxide to replace the amyl acetate. Valve C must be opened more widely from time to time to clear ice that forms and blocks the vapor bleed.

11. After the 20 min, shut valve C and then valve B. Immerse the pressure vessel in water at 45°C. As the temperature of the carbon dioxide increases, the pressure will increase. This step increases the temperature and pressure to the critical point.
12. After the pressure stabilizes, open valve C slowly in order to bleed off gaseous carbon dioxide. Keep the warm water around the apparatus while doing this.
13. When the pressure reading is zero, remove the warm water bucket, dry the pressure vessel carefully, and remove the specimens.

The Langmuir trough-critical point method has been used extensively for studies of metaphase chromosomes and interphase chromatin (for reviews see Ris, 1969; Ris and Kubai, 1970; Wolfe, 1969). A modified Langmuir method was recently employed to study quantitatively a whole metaphase plate of human chromosomes (Golomb and Bahr, 1971) (Fig. 7.6). In the best preparations, the morphology of chromosomes prepared by this technique is very similar to the appearance of chromosomes under the light microscope. This has been shown particularly clearly by Wolfe and Hewitt (1966) who were able to identify the stages of meiosis, including various stages of meiotic prophase, in milkweed bug testes prepared by this technique. Portions of a somatic cell nucleus from the testes of *Rana pipiens* and a metaphase chromosome from bovine kidney cells, demonstrating the typical fibrous appearance of chromatin prepared by the Langmuir trough-critical point technique, are shown in Figs. 7.7 and 7.8 respectively.

In addition to studies of chromosomes and interphase chromosomes, the Langmuir trough-critical point technique has been valuable in studying chromatin from mature sperm nuclei. In sections, sperm nuclear elements often are so closely packed that they are extremely difficult or impossible to resolve. This difficulty can be circumvented by spreading unfixed sperm on a trough and then drying by the critical point technique. Sperm nuclei of many organisms show considerable degrees of spreading (Fig. 7.9), allowing nuclear structures to be easily resolved (Fig. 7.10). In addition, organelles such as ciliary basal bodies (Argetsinger, 1964), microtubules (Gall, 1966; Wolfe, 1967), nuclear and cell membranes (DuPraw, 1965; Wolfe, 1967), and cortical structures of ciliate protozoa (Grim, 1967) have been observed by this technique.

The critical point technique has already been of substantial use in

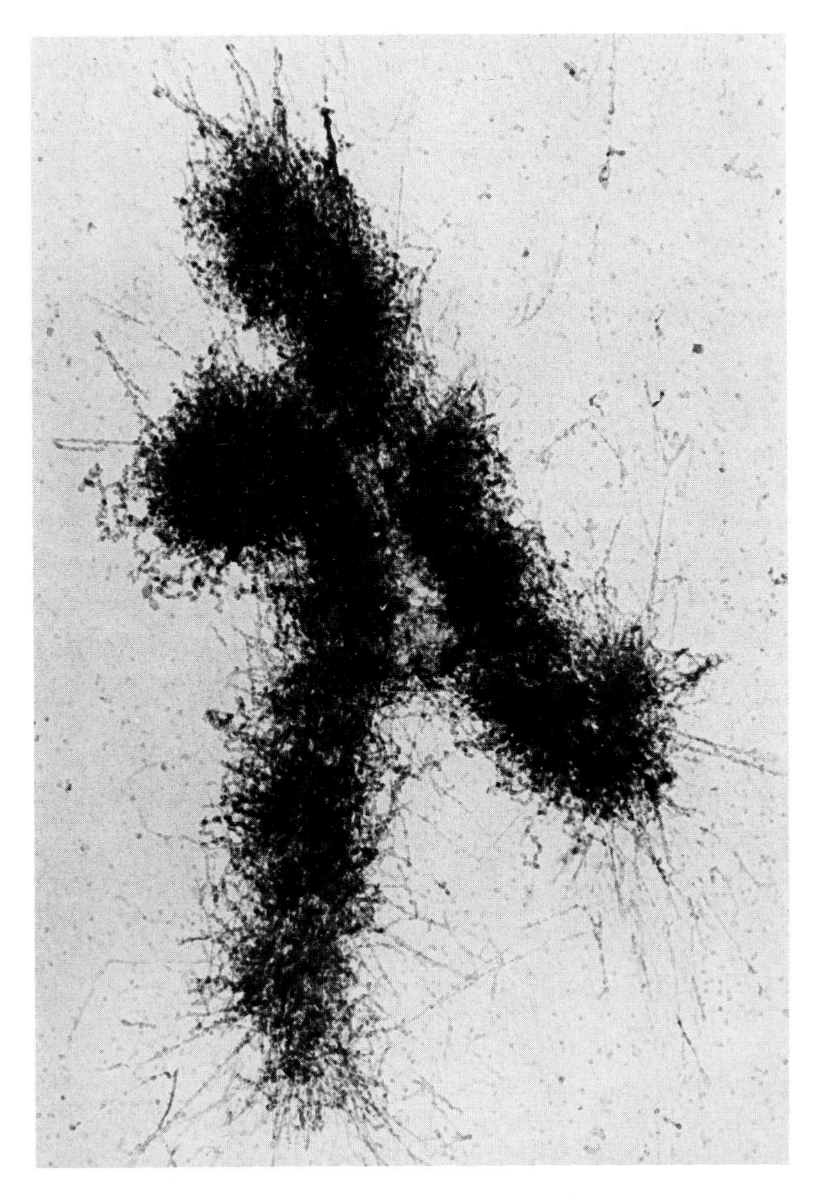

Fig. 7.6 Human chromosomes prepared by surface spreading and critical-point dried in natural contrast. Note the fibrous nature of the chromatids; bumpy, twisting loops of chromatin fibers constitute the entire chromosome. 24,000 X. *Courtesy of G. F. Bahr.*

Fig. 7.7 Somatic cell nucleus in the frog testis, showing typical chromatin fibers. 94,000 X. Micrograph by B. R. Zirkin.

the preparation of tissues for transmission electron microscopy. In addition, inert dehyrdation of organelles, followed by critical point-drying, has been used recently for the electron microscopy of whole-mounted mitochondria (Fig. 7.11) and granules in rat mast cells (Pihl and Bahr, 1970). The technique has also been used to prepare materials for observation under the scanning electron microscope. Turnbill and Philpott (1970) have utilized this technique to prepare tissue culture cells for scanning electron microscopy. They were successful in preparing even

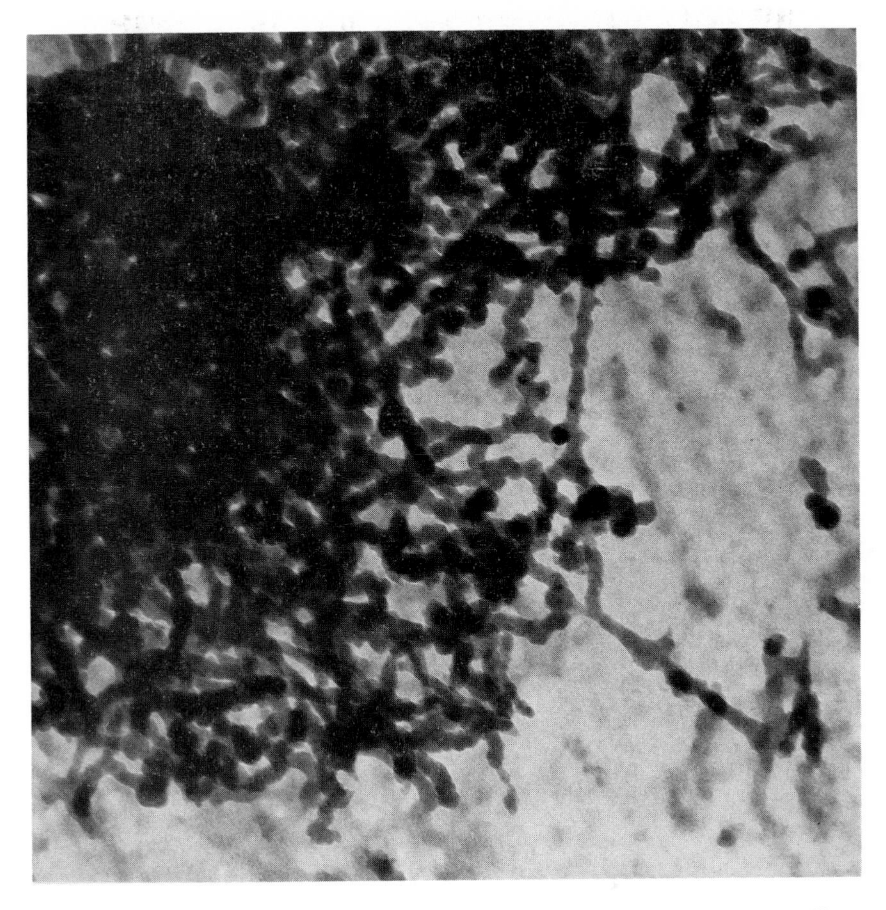

Fig. 7.8 A portion of an isolated bovine kidney chromosome showing fibers similar in appearance and dimensions to those in interphase nuclei. 80,000 X.
Figs. 7.8-7.10 From Zirkin, B. (1971). *J. Ultrastruct. Res.* 36, 237. Used by permission.

gross specimens such as complete mouse heart and starfish tube feet. The duration of preparation was ~ 1 hr, and the spatial arrangement in the specimens was little affected.*

The advantages of critical point-drying and the availability of commercial drying apparatus will almost certainly result in wider application of this method in the next few years. Critical point-drying apparatus can be obtained commercially from the following sources:

*For a detailed discussion on the application of this technique to scanning electron microscopy, the reader is referred to Cohen (1973).

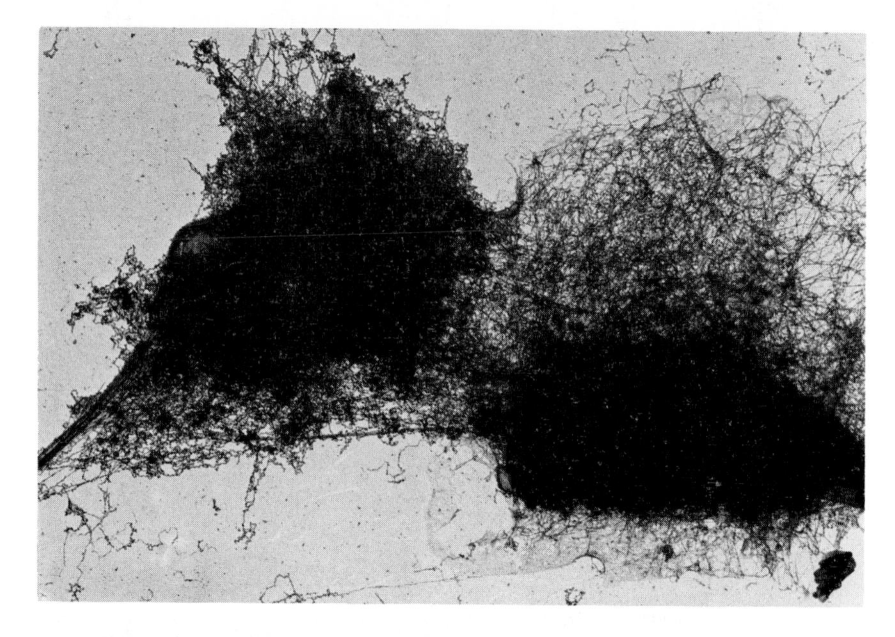

Fig. 7.9 Two isolated goldfish sperm nuclei. 10,400 X.

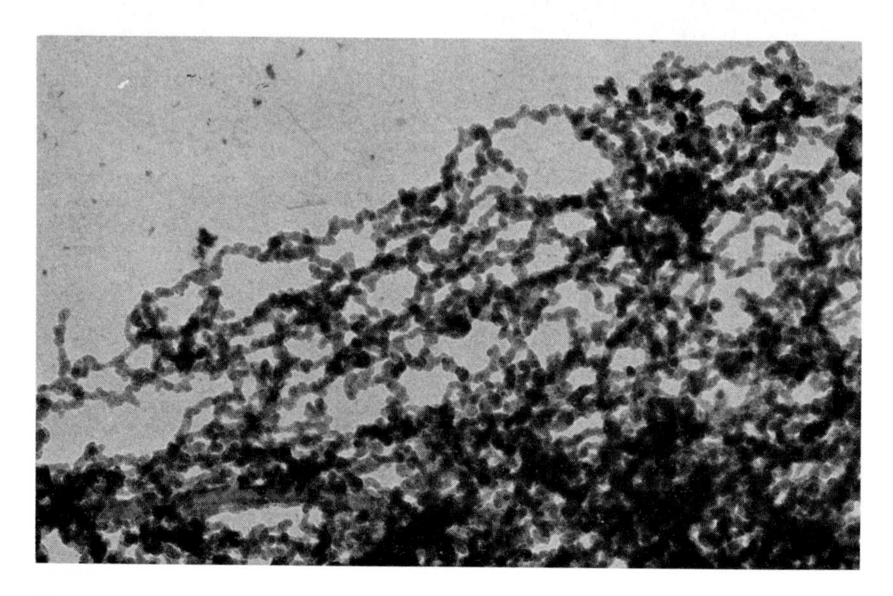

Fig. 7.10 A portion of an isolated goldfish sperm nucleus at higher magnification showing the fibrous structure typical of histone-containing sperm nuclei. 80,000 X.

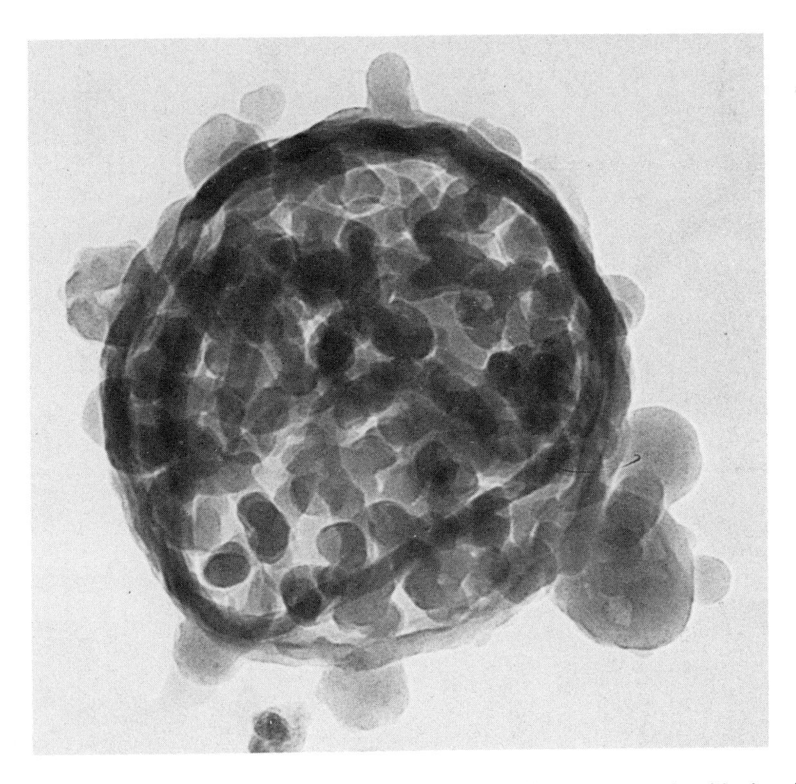

Fig. 7.11 Rat liver mitochondrion isolated in 0.25 M sucrose and critical point dried out of Cellosolve after 10 sec glutaraldehyde fixation. The many "end profiles" reflect the three-dimensional arrangement of the strands. 48,000 X. From Pihl, E., and Bahr, G. F. (1970). *Exptl. Cell Res.* 63, 391. Used by permission.

American Instrument Company, Silver Spring, Maryland.
Bomar Company, Tacoma, Washington.
Denton Vacuum Company, Inc., Cherry Hill, New Jersey.
Polysciences, Inc., Warrington, Pennsylvania.

References

Anderson, T. F. (1951). Techniques for the preservation of three-dimensional structure in preparing specimens for the electron microscope. *Trans. N.Y. Acad. Sci. 13,* 130.

Argetsinger, J. (1964). The isolation of ciliary basal bodies (kinetosomes) from *Tetrahymena pyriformis. J. Cell Biol. 24,* 154.

Cohen, A. L. (1973). Scanning Electron Microscopy: Principles and Methods

(Hayat, M. A. ed.). Van Nostrand Reinhold Company, New York and London.

Cohen, A. L., Marlow, D. P., and Garner, G. E. (1968). A rapid critical point method using fluorocarbons ("Freons") as intermediate and transitional fluids. *J. Microscopie* 7, 331.

DuPraw, E. J. (1965). The organization of nuclei and chromosomes in honeybee embryonic cells. *Proc. Nat. Acad. Sci.* 53, 161.

Gall, J. G. (1963). Chromosome fibers from an interphase nucleus. *Science* 139, 120.

———. (1966). Chromosome fibers studied by a spreading technique. *Chromosoma* 20, 221.

Golomb, H. M., and Bahr, G. F. (1971). Analysis of an isolated metaphase plate by quantitative electron microscopy. *Expt. Cell Res.* 68, 65.

Grim, N. (1967). Ultrastructure of pellicular and ciliary structures of *Euplotes eurostomis*. *J. Protozool.* 14, 625.

Hayat, M. A. (1970). "Principles and Techniques of Electron Microscopy: Biological Applications," Vol. 1. Van Nostrand Reinhold Company, New York and London.

———. (1972). "Basic Electron Microscopy Techniques." Van Nostrand Reinhold Company, New York and London.

Koller, T., and Bernhard, W. (1964). Séchage de tissus au protoxyde d'azote (N_2O) et coupe ultrafine sans matiere d'inclusion. *J. Microscopie* 3, 589.

Lung, B. (1968). Whole-mount electron microscopy of chromatin and membranes in bull and human sperm heads. *J. Ultrastruct. Res.* 22, 485.

Pihl, E., and Bahr, G. F. (1970). A new approach to the study of cell organelles with the electron microscope. *Exp. Cell Res.* 59, 379.

Rebhun, L. I. (1972). Freeze-Substitution and Freeze-Drying. *In:* "Principles and Techniques of Electron Microscopy: Biological Applications," Vol. 2 (Hayat, M. A. ed.). Van Nostrand Reinhold Company, New York and London.

Ris, H. (1961). Ultrastructure and molecular organization of genetic systems. *Can. J. Genet. Cytol.* 3, 95.

———. (1969). The Molecular Organization of Chromosomes. *In*: "Handbook of Molecular Cytology "(Lima-de-faria, A. ed.). North-Holland Pub. Company, Amsterdam and London.

———, and Kubai, D. F. (1970). Chromosome structure. *Ann. Rev. Genet.* 4, 263.

Solari, A. J. (1965). Structure of the chromatin in sea urchin sperm. *Proc. Nat. Acad. Sci.* 53, 503.

Turnbill, C., and Philpott, D. E. (1970). The critical point-drying method applied to scanning electron microscopy of L-929 cells. *Proc. 28th. Ann. Meet. Electron Micros. Soc. Amer.*, p. 278. Claitor's Publishing Division, Baton Rouge.

Walker, N. H. (1971). Studies on the arrangement of nucleoprotein in elongate sperm heads. *Chromosoma* 34, 340.

Wolfe, S. L. (1967). Microtubular elements in cell membranes isolated from nucleated erythrocytes. *J. Ultrastruct.* Res. *17,* 588.

———. (1968). The effect of prefixation on the diameter of chromatin fibers isolated by the Langmuir trough-critical point method. *J. Cell Biol. 37,* 610.

———. (1969). Molecular Organization of Chromosomes. *In*: "The Biological Basis Of Medicine "(Bittar, E. E. ed.). Academic Press, London.

———, and Hewitt, G. M. (1966). The strandedness of meiotic chromosomes from *Oncopeltus. J. Cell Biol. 31,* 31.

Zirkin, B. R. (1971). The fine structure of nuclei in mature sperm. I. Application of the Langmuir trough-critical point method to histone-containing sperm nuclei. *J. Ultrastruct. Res. 36,* 237.

Author Index

Subject Index